HISTORICAL STUDIES

IN THE PHYSICAL SCIENCES

2

Einstein and Bohr

Historical Studies
in the
Physical Sciences

Russell McCormmach, *Editor*

Second Annual Volume 1970

UNIVERSITY OF PENNSYLVANIA PRESS • PHILADELPHIA

NOTICE TO CONTRIBUTORS

Historical Studies in the Physical Sciences, an annual publication issued by the University of Pennsylvania Press, is devoted to articles on the history of the physical sciences from the eighteenth century to the present. The modern period has been selected since it holds especially challenging and timely problems, problems that so far have been little explored. An effort is made to bring together articles that expose new directions and methods of research in the history of the modern physical sciences. Consideration is given to the professional communities of physical scientists, to the internal developments and interrelationships of the physical sciences, to the relations of the physical to the biological and social sciences, and to the institutional settings and the cultural and social contexts of the physical sciences. Historiographic articles, essay reviews, and survey articles on the current state of scholarship are welcome in addition to the more customary types of articles.

All manuscripts should be accompanied by an additional carbon- or photocopy. Manuscripts should be typewritten and double-spaced on 8½″ x 11″ bond paper; wide margins should be allowed. No limit has been set on the length of manuscripts. Articles may include illustrations; these may be either glossy prints or directly reproducible line drawings. Articles may be submitted in foreign languages; if accepted, they will be published in English translation. Footnotes are to be double-spaced, numbered sequentially, and collected at the end of the manuscript. Contributors are referred to the *MLA Style Sheet* for detailed instructions on documentation and other stylistic matters. (*Historical Studies* departs from the MLA rules in setting book and journal volume numbers in italicized Arabic rather than Roman numerals.) All correspondence concerning editorial matters should be addressed to Russell McCormmach, Department of History and Philosophy of Science, University of Pennsylvania, Philadelphia, Penna. 19104.

One hundred and fifty free reprints accompany each article.

Historical Studies in the Physical Sciences incorporates *Chymia,* the history of chemistry annual.

The publishers are grateful to the Ehrenfest family for permission to reproduce the Einstein-Bohr photographs and to Professor Martin J. Klein for supplying the negatives. The restoration of the negatives and production of the prints were done by William R. Whipple.

Contents

Editor's Foreword   ix

MARTIN J. KLEIN
The First Phase of the Bohr-Einstein Dialogue   1

RUSSELL MC CORMMACH
Einstein, Lorentz, and the Electron Theory   41

STANLEY GOLDBERG
In Defense of Ether: The British Response to Einstein's
Special Theory of Relativity, 1905–1911   89

ROMUALDAS SVIEDRYS
The Rise of Physical Science at Victorian Cambridge
[With Commentary by Arnold Thackray and Reply by
Romualdas Sviedrys]   127

PAUL FORMAN
Alfred Landé and the Anomalous Zeeman Effect,
1919–1921   153

YEHUDA ELKANA
Helmholtz' "Kraft": An Illustration of Concepts in Flux   263

ELIZABETH WOLFE GARBER
Clausius and Maxwell's Kinetic Theory of Gases   299

EDWARD E. DAUB
Entropy and Dissipation   321

Note on Contributors   355

Editor's Foreword

This second volume in the series, *Historical Studies in the Physical Sciences,* contains eight articles on the development of physics from the middle of the nineteenth century. Since three of these directly concern Einstein, I am taking this occasion to discuss some promising directions for historical work suggested by Einstein's career and context. In the discussion I have mentioned several recent writings on Einstein, but the selection is meant to be illustrative, not comprehensive; e.g., I have not commented on Einstein biographies, such as the fine one by Boris Kuznetsov. The article literature is a useful indicator of the problems historians of science currently find interesting in connection with Einstein; the selection of problems exposes the limitations of contemporary writing in the history of modern science, and these are what I want primarily to draw attention to.

Whereas ten years ago historians of science were not writing on Einstein, since that time at least fifteen good historical studies on his work and career have appeared. Despite this growth in the literature, Einstein still draws only a fraction of the attention of his Scientific Revolution counterparts, Newton and Galileo. There are some frustrations in approaching Einstein historically, but these are not insuperable. To begin with, there is no edition of Einstein's publications in any Western European language, though there is a Readex microprint edition; in any case, most of his papers are published in journals that are easily obtainable. Selections of his scientific correspondence have just begun to be published, notably his exchanges with Max Born and Arnold Sommerfeld; the large collection of Einstein letters at Princeton is, however, accessible to scholars. Newton's private papers have had to wait for over two hundred years for their proper handling, so that the delay over Einstein's seems hardly exceptional. Einstein's later physics is mathematically demanding, but then so is much of seventeenth century science. The fact is that historians of science are only now beginning to confront really modern topics. If this second volume of *Historical Studies* is any indication, Einstein and twentieth century physics are certain to receive a greater share of historical interest in the immediate future. This would seem an appropriate time to discuss some of the central problems that students of Einstein and very recent physics may wish to take up.

Within the canon of Einstein scholarship the major interest has centered on the introduction of special relativity and light quanta in the years around 1905 (contributions in this volume suggest that the trend may be changing). This instance of the familiar clustering effect in the writing of the history of science is understandable, but hardly comforting. The interest in historical continuity has been incomparably weaker than the interest in precipitous change; and the historiography of science is deficient as a result. To make the point I need only call attention to the absence of studies concerned with entire scientific careers. Einstein's general theory of relativity and his gravitational and unified field theories have not attracted the attention of historians, yet these theories occupied him for the better part of his career, and he regarded them as his most characteristic and original achievement. Our understanding of Einstein's career would be enormously enhanced if the structure and development of his later work were charted as thoroughly as his earlier has been.

It is generally acknowledged that personality factors pervade the structure of a scientist's work, yet slight effort has been made to assess these factors. I want to suggest a possible approach to this assessment, one based on certain tensions within Einstein; I intend this only as an example of a type of historical approach that might be undertaken. Early in life Einstein revealed a passionately religious longing, yet he had an equally passionate critical and anti-authoritarian spirit that denied him the certainties of organized faiths. Later he became an ardent Zionist, but never a follower of the Jewish religion. He supported democracy, and at the same time he believed that the majority of men are sheep, incapable of forming or expressing an independent opinion, and saved from tyranny only by their inner contradictions. He had a passionate feeling for humanity, yet he conceded that in his personal life he belonged to no one, not even his family. He disliked the idea of a full-time job in pure research, yet for most of his career he was paid to do nothing but pure research. He devoted his life to comprehending the rationality of nature, yet he believed that one of the most beautiful experiences man can have is of the mysterious. He was unpolitical and preferred scholarly isolation above all, yet he played a critical role in inaugurating the modern era of the engaged scientist with his letters to Roosevelt on atomic weaponry. He fully recognized and accepted the parts of his nature that were in some degree unreconciled. The enumeration of his contrary, or seemingly contrary, qualities can be extended, and in particular to his scientific thought. Einstein was far more able to hold contraries in the same field of vision than most of his contemporaries in physics. Keats isolated this capability as the essential quality in the poet's make-up; and without suggesting that there is a constant creative type throughout history, it does seem that this capability is often a common denominator of the personalities of greatly original modern thinkers in diverse fields, in science as well as in poetry, art, and statesmanship. Martin Klein has highlighted Einstein's early unique insight into the pervasive, seemingly contradictory wave-particle duality of nature and the immensely productive use he made of it. The psychology of thinkers who can contain and exploit unreconciled concepts that others find only disconcerting may help clarify the sources of Einstein's originality. Instead of asking how physics was precipitously changed by Einstein's unique genius, we might first ask how it was possible that a certain personality type was able to seize upon the tensions within physics at the turn of the century and productively transform them into a new synthesis.

It is a commonplace that the rise of the professionalized scientific disciplines in the latter nineteenth century markedly changed patterns of recruitment and research. But there has been little attempt to relate professionalism to the quality of individual scientific thought and experience. Paul Forman's study of Alfred Landé and the anomalous Zeeman effect is the first to center on the influence of professional exigencies on the texture of scientific work. While there exists no comparable study of Einstein, he has left us his subjective impressions, and these are highly revealing. He had generally sharp things to say about the effects of professionalism; the overspecialization and narrowly competitive spirit that professionalism encouraged were repugnant to him. More to the point, the professional conditions of physics did not seem essential or useful to his work. He carried out the researches that brought him to the notice of the world physics community in his off hours as a Swiss patent clerk. He did not see another theoretical physicist until he was thirty, long after he had laid the groundwork for much of twentieth century research. For the whole of his precocious years in the patent office, he was removed from direct contact with professional physics circles. Even his professional initiation at the elite technical school, the Zürich Polytechnic, did not seem very purposeful to him. True, he worked in the physical laboratory and sat through systematic lectures on theoretical physics and thus had the advantage of the most recent innovations in physics training in Europe. But he found the theoretical professor pettifogging and behind times, and the main benefit of his Zürich years was the leisure they gave him to read Maxwell and other authors who were not taught. Klein has perceptively commented on the professional doubts of Einstein, especially his disgust with the education he received; the cramming he had to do for examinations so repelled him that he had little taste for physics for a year after.

Einstein's extraordinary independence was due partly to the recent emergence of theoretical physics as a distinct discipline. As a theoretical physicist he was outwardly unencumbered by the need for laboratory and staff, a condition that harmonized with the inner freedom he sought through a vocation in physics. When Einstein graduated in 1900 there were specialist chairs in theoretical physics, but not many of them. It was only after a decade of prolific originality that he was offered his first professorship in theoretical physics. After that he rose quickly; and in 1913 he was called to the center of German and world physics, Berlin, to become a salaried member of the Berlin Academy and soon after to head the new Kaiser Wilhelm Institute for Physical Research. The position combined high status and unlimited research freedom, and most young physicists would have thought it ideal. But Einstein felt strong reservations, as he did about his later Princeton Institute position. Einstein's arrangements were a kind of apotheosis of the research ethos the scientific community had nurtured from the early nineteenth century. His reservations stemmed from his belief that the highest form of productivity in science could not be exchanged for security or be elicited by the promise of professional advancement. Einstein was disdainful of the overproduction of immature research prompted by the competition for prestige and by a faulty reward system.

The self-images of scientists and the images society holds of scientists are closely related to the professional experience. These images bear on such central issues as intellectual motivation and career selection and satisfaction. We need to understand why people with certain personalities in particular cultural and economic settings turn to a particular science and find emotional gratification in advancing its field of knowledge. While there are as yet no studies that depict how physics might look in comparison with other vocations to a gifted, perceptive student around 1900, we have Einstein's recollection of why he entered physics then. Physics appealed to him and to others he admired because it dealt with a reality that lay outside man. It beckoned to him as a liberation from the merely personal and from all intellectual subservience and social custom. The objective reality of physics was an other-wordly one, and it elicited from him the rapt devotion that religion had in his youth. The background of his career decision was the clear distinction that had been argued out in German intellectual circles in the nineteenth century between the inner world of consciousness and the outer world of physical reality. Einstein believed that the deepest motivation of the dedicated physicist was akin to that of the poet, the artist, and the philosopher. Each fashioned a cosmos intelligible and satisfying to himself and made it the core of his emotional being; and each found an escape from the everyday world's dreariness and crudity this way. Einstein recognized that the vast majority of physicists were different than he was. They were impelled by professional ambition and by the sport of exercizing their mental muscles, or by purely practical ends. But for circumstance they might have been engineers or businessmen. Einstein observed that men like this had built much of the temple of science, but by no means all of it. If the rank and file were all there were, there would be no temple. Its existence depended on a few finely tempered, but odd and solitary men, men whose scientific commitment lay outside the competitive inducements of the profession. They were an elite which included science's natural leaders. The authority of the natural leaders was not rooted in official position, but in capacity and in idealistic dedication to the search for truth, in innate superiority that others recognized without feeling diminished by it. Einstein found in physics the measure of social identity and community he desired. The ground of his satisfaction was the kinship he felt with physics' natural leaders, above all with the learned and gentle Dutch theoretical physicist H. A. Lorentz. Although he did not say it in these words, it is clear that Einstein was drawn to natural science because of its internationalist credo, a credo that exactly suited his independent temperament. At the same time that he began preparing for a scientific career, he repudiated the statism and militarism of Wilhelmian Germany by giving up his natural citizenship. It was entirely in keeping with his early joint commitment that he later became physics' staunchest internationalist in both science and politics throughout the bitterly divisive interwar years.

The physics profession made ample room for Einstein, and in Berlin he participated in its normal self-regulating activities. But his ties with the powerful academic establishment were always weak. He belonged to the Berlin University faculty, but he did not have a regular appointment and had no say in faculty deliberations, and as a rule his advice was not solicited on appointment decisions in the German university system. In both the Berlin Academy and in the Princeton Institute he seems to have had little if any influence. He sought no institutional authority and in general had few wants. He had next to no students and formed no research school. He might therefore seem to fall largely outside the concerns of those who study the social activity of science. For some time there has been a movement away from the preoccupation with great men in the history of science. The history of the modern physical community will be written for the most part as the history of large numbers of capable workers who subscribed to the values of their profession. But the odd and solitary member like Einstein offers valuable insights into the social and institutional mechanisms of the profession precisely because he has a kind of natural authority and commands unique adjustments. He reacts on the values of the profession, too, in his role as an inspirational model for new recruits entering the field; and the scientific ideas that underlie his natural authority serve to shape research strategies for the discipline and to define intraprofessional alignments.

For all of his extraordinary gifts Einstein was very much a product of his time. This should be unnecessary to remark on, except that there is a tendency to concentrate on what is unique in intellectual leaders and ignore what is common. For one thing Einstein worked on problems that the physics discipline recognized as central problems. For another he had an iconoclasm that was rooted in the collective outlook of his generation of German physicists no less than in his individual personality. That numbers of Einstein's physics contemporaries were a self-proclaimed revolutionary generation has largely been overlooked, due to the fact that Einstein's innovations proved more fruitful and outlived the others'. Einstein in fact came along somewhat late, after the more influential of his colleagues had largely agreed that a radical transformation had occurred in the outlook of physics. It was widely accepted that the two-century old mechanical view of nature had been overturned, and that it and mechanics had now to yield to the electron theory and the electromagnetic view of nature. For a time Einstein was regarded by his contemporaries as unresponsive to change, for he seemed to want to retain mechanical concepts in the foundations of physics. Einstein differed from most in regarding electromagnetism no less than mechanics as an unacceptable basis for all of physics, and consequently he did not participate in the widespread effort to reduce mechanics to electromagnetism; rather he worked productively with both sciences while seeking a plane of insight that would resolve their mutual contradictions. To understand the orientation of Einstein and his contemporaries, we need to look at the whole activity of physics around 1900; a study with this objective would be most helpful at this time. There are studies of specific problems of physics bearing on the immediate origins of the quantum and relativity theories; and Gerald Holton, Klein, and others have written on the relations of Einstein with other individual physicists, such as Lorentz, Bohr, Henri Poincaré, Max Planck, Ernst Mach, and August Föppl. But there is no single study of this period that treats all of the branches of physics and their connections, the leading research directives, and physical world views. A valuable start is Tetu Hirosige's survey of the state of electrodynamics prior to Einstein's special theory of relativity, together with his and Sigeko Nisio's article-counts for the various fields of physics in the same period.

Einstein's fame is linked with transformations in total world views. His formative years coincided with a time of much explicit discussion of world views in physics, and he dedicated his career to consolidating a relativistic world view of his own and trying to convert the physical community to it. The historical literature on the Scientific Revolution devotes a good deal of comment to the establishment of the mechanical world view. But little attention has been paid to the world view controversies that accompanied the dissolution of the mechanical consensus two hundred years later; these were in part a concerted effort to end the anomy in physical research that resulted from the dissolution. Even in a modern, highly technical field like physics, it is necessary to look beyond the terms in which the scientific debate is carried out to understand the function and satisfaction of world views. The reasons favoring immaterial, electromagnetic concepts over the inert, material imagery of mechanics must be sought in the intellectual and esthetic currents of the time as well as in specific configurations of scientific problems. Holton and Michael Polanyi have written persuasively on the strong esthetic determinants in Einstein's work on relativity. The determinants can be recognized in the work of Einstein's contemporaries, too; the particular forms they take depend jointly on personality type and contemporary cultural conditions. The debate over world views needs to be related to the cultural crisis that German intellectuals increasingly responded to from about 1890. The philosophical and scientific expressions of positivism and materialism were widely attacked, as were all forms of specialized learning. There were calls for idealist world views to counter the technological, materialist civilization that was rapidly emerging in Germany. The German natural scientists responded to their environment, sometimes consciously and sometimes not; their world-view antagonisms reflected their participation in the general crisis mentality of their class. The interplay between science and the larger culture operates on the level of the deepest intellectual motivations, and its study is one of the more difficult and pressing in the history of recent science. The historical comparison of scientific and ordinary languages is one way to isolate the interplay.

Our understanding of Einstein or of any other individual scientist will remain insecure until we have a history of the physics discipline in the nineteenth and twentieth centuries. The discipline mediates between the conditions of the non-scientific world and the terms of reference of the practicing physicist. Although the history of the discipline has not yet been written, some of the matters it must take up are evident enough. Einstein's career again supplies concrete illustrations pointing to the need for enlarging the historiography of science. Einstein was born at the right time; his critical, nonconformist outlook exactly suited the state of physics. Contrary to the myth of Victorian complacency, late nineteenth century physics was intensely critical; the discipline sanctioned a probing of the foundations of physics by making it an essential activity of the theorist. Einstein could question the truth claims of the received physical theories and at the same time embrace the activity of theoretical physics. The critical mood of nineteenth century physics was capped by the epistemological writings of Mach and Poincaré; Einstein's powerful critiques of the foundations of classical physics were rooted in the same tradition. He started by examining the foundations of a special branch of physics, thermodynamics, and then went on to examine the foundations of physics as a whole. His purpose in his 1905 relativity and light-quantum studies was to expose the mutual incompatibility of mechanics and electromagnetism and to propose concepts for a common foundation for both specialties. Klein and Holton have pointed to the identity of the critical insights behind both Einstein's quantum and relativity proposals. What is now needed is a deeper understanding of the rise of specialization in nineteenth century scholarship and its bearing on the critical tradition in physics. Physics had become internally specialized to a high degree before Einstein began his career; its individual branches had been intensively cultivated within their own terms of reference, a consequence of the intensive research ethos of the middle nineteenth century. Einstein's lifework was shaped by his critical perception of the resulting disharmony in the parts of physics and by the imperative he felt to fashion a theory for the totality of physical reality. This imperative was in part a consequence of the stage the physics discipline had reached. The physical theorist of the generation before Einstein had become increasingly a specialist of the whole. The discipline encouraged him to see his critical responsibility as spanning the entire state of physical theory; Einstein was highly responsive to this understanding of the theorist's work.

Einstein's career points up the need for exploring the interrelations of specialist disciplines. Unlike Newton, Einstein was not at the same time an original mathematician and an original physical theorist. Again and again he looked to mathematicians for help in formulating physical concepts. His collaboration with mathematicians is a subject in itself, as is the whole interaction of physicists and mathematicians in the construction of relativistic field theories. In general it is clear that histories of physics and mathematics would be enriched if more notice were taken of the relations between the two disciplines. Hirosige has made a valuable start in his analysis of the role of the mathematician Hermann Minkowski in winning acceptance for Einstein's special theory of relativity. Another interdisciplinary problem is to trace the responses of experimental and observational scientists to Einstein's theories. The genesis and reception of Einstein's 1905 work on quanta and relativity have been related to the experimental context by Holton, Klein, Loyd Swenson, Stanley Goldberg, Kenneth Schaffner, and others. Einstein's later theories offer an equally rich source for studies of the historically operative criteria that persuade physicists of the correctness and fecundity of a new idea. Einstein's special theory of relativity moved to the center of physics, while his general theory of relativity remained an admired but remote creation that was hard to test or improve upon or use.

A study of the diffusion of general relativity would illuminate the individual and collective elements in theory conversion. Goldberg's example of tracing the responses of physicists to special relativity to national educational structures ought to be applied to the case of general relativity, too. An important study might be made of the minority of physicists who shared in some degree Einstein's belief in the correctness of seeking a causal, continuum-based theory at a time when most physicists were busy framing a mechanics of quantal discontinuities. The starting point for the study might be V. V. Raman and Paul Forman's suggestive analysis of the origins of Schrödinger's wave mechanics; the authors have used the techniques of the sociology of knowledge, identifying group characteristics of physicists sympathetic to Einstein.

I want finally to point to a number of topics where fruitful lines of communication might be established between historians of science and historians of other specialties. Einstein belongs not only to the history of physics, but to the history of the Zionist movement and to the history of European Jewry and its cultural efflorescence at the end of the nineteenth century. And he belongs to the history of twentieth century anti-Semitism both inside and outside science. As the most famous Jewish scientist and as an internationalist and social-ist as well, he was anathema to the propagandists for racist science. This matter should interest historians of science, and other histor-ians, too. Some work has recently been done on this, notably by Siegfried Grundmann and Armin Hermann.

For the last thirty-five years of his life, Einstein provided the world with its most compelling image of the scientist. The unkempt hair and bohemian dress became the outward signature of scientific genius—thus the public's disorientation when Oppenheimer eclipsed Einstein at the end, when the scientist as magus was replaced by the scientist as government advisor, as technocratic administrator, as the superbly professional personality whose scientific contribution the public had no glimpse of. Science supplies society with images and metaphors for a variety of uses, and society found relativity assimilable in protean forms. Science also supplies society with sym-bols of man's omnipotence or, alternatively, of his impotence. The only mathematical equation that probably every literate man recog-nizes is Einstein's relation between mass and energy, the equation the media habitually superpose on a mushroom cloud. How the lay public puts this together with the picture of an elderly German in a shapeless sweater and sandals studying a blackboard covered with formulas would make a worthwhile topic in modern social history. The findings would be useful to historians of science, too, in their traditionally narrower pursuits; for society's impressionistic ideas about science condition the appeal of scientific careers and the sup-port for scientific research.

Einstein was an intellectual in the larger sense, not just a scientific professional. He saw himself as one of a world intellectual elite that transcended the divisions between literary, scientific, and artistic activity. The common characteristic of the members was their ideal search for truth, the same as that which distinguished the natural leaders of science. The motive of the search counted more for Ein-stein than the knowledge that resulted from it; the search was a moral imperative. The elite was the guardian of Western culture and its democratic ideals, and it was obligated to make public protest when its heritage was threatened. Einstein urged the elite to organize and use its influence in the cause of world freedom and peace. He did not expect the same courage and independence from the larger class of intellectuals who were not among the elite. Brigitte Schröder has shown how intensely conservative and nationalistic European scholars and scientists were during and after the first world war and how Einstein stood apart from the rest. From his unpopular declaration of pacifism in Berlin to his denunciation of witchhunts in America, he was scornful of the majority of his colleagues who refused to speak out on matters of conscience, even at the risk of losing livelihood and reputation. He did not seek political authority within the physics profession, but moral influence in the world at large. His implicit rejection of the current two-cultures distinction, his belief in the essential community of interest of an interdisciplinary intellectual elite, his advocacy of the social responsibility of scientific and literary intellectuals alike, and his political pronouncements and appeals to intellectuals would make a valuable chapter in the social history of twentieth century intellectuals.

I welcome contributions to this periodical on themes related to the ones I have discussed. By this I do not mean contributions on Einstein, though of course they are welcome too. My remarks on Einstein's individuality posed against his context are easily generalizable: The interplay of a particular personality with a particular discipline at a particular moment in time defines a recurrent problem in the history of science, one that is seldom explored with sensitivity. A closely related and generally overlooked set of problems revolves around the historical motivations, self-images, and personalities of scientists. The contextual problems posed by the scientific disciplines and professions and their mutual interaction with the larger culture and society have hardly begun to be studied. It would seem a step forward if historians of science embraced a broader selection of subject matter and approaches; in this way they might make more frequent contact with general historians to the advantage of both.

The First Phase of the Bohr-Einstein Dialogue

BY MARTIN J. KLEIN[*]

1. Niels Bohr and Albert Einstein discussed and disagreed about the paradoxes of the quantum theory for a third of a century. The extraordinary personal and intellectual qualities of the two men and the unprecedented difficulty and depth of the issues they debated make these discussions unique in the history of physics. In this paper I want to analyze the first occasion on which Bohr and Einstein differed over a question of fundamental principle. This occurred during the years 1923 to 1925, just before the new quantum mechanics began to appear. The principles at stake were nothing less than the validity of the laws of conservation of energy and momentum and the existence of the wave-particle duality for radiation.

Bohr wrote an account of his long dialogue with Einstein on the occasion of Einstein's seventieth birthday.[1] This is a precious document for anyone interested in the history of twentieth-century physics, but in reading it we must keep in mind that it appeared in 1949 and that it was written by one of the participants in the dialogue. Bohr naturally told the story as he saw it at the time of writing, and his account shows the insight gained by his decades of rich experience with quantum physics. The principal theme of Bohr's essay was the series of critical attacks that Einstein directed over the years against quantum mechanics, and the successive defeat of each of these at-

---

* Department of the History of Science and Medicine, Yale University, New Haven, Conn. 06520.

1. N. Bohr, "Discussion with Einstein on Epistemological Problems in Atomic Physics," in *Albert Einstein: Philosopher Scientist,* ed. P. A. Schilpp (Evanston, Illinois, 1949), p. 199. Reprinted in N. Bohr, *Atomic Physics and Human Knowledge* (New York, 1958), p. 32.

1

tacks by Bohr and his collaborators, with each exchange leading to a new and deepened understanding of the fundamentals of the new physics. Einstein was never reconciled to the severe restriction of physical theory to a probabilistic goal which seemed to be an essential feature of quantum mechanics. It is hardly surprising then that he appears in Bohr's essay as the more conservative of the two, concerned about "the lack of firmly laid down principles for the explanation of nature, in which all could agree," while Bohr was the one who thought that "we could hardly trust in any accustomed principles, however broad, apart from the demand of avoiding logical inconsistencies."[2]

Perhaps it had also seemed that way in 1924. For one of the most startling suggestions made during the twenties, that decade of startling suggestions about how the laws of physics should be altered, was that "we abandon any attempt at a causal connection between the transitions in different atoms, and especially a direct application of the principles of conservation of energy and momentum, so characteristic for the classical theories." This extreme measure, abandoning causality and the conservation laws in atomic physics, was proposed by Niels Bohr in a paper written early in 1924 in collaboration with H. A. Kramers and J. C. Slater.[3] It is this paper that will be the focus for what I have to say.

In Bohr's description of his dialogue with Einstein, the Bohr, Kramers, Slater paper is mentioned only very briefly. Bohr and Einstein never met to discuss this work, nor did they correspond about it, so far as I know. Nevertheless, the Bohr, Kramers, Slater proposal is an essential part of the story of the Bohr-Einstein relationship, and this is perhaps its greatest historical interest. Bohr and his collaborators were struggling with the paradox that radiation behaves under some conditions like an electromagnetic wave and under others like a particle of energy. It was, of course, Einstein who had proposed the idea of energy particles or light quanta in 1905, and this idea had just acquired a new respectability almost twenty years later as a result of Arthur Compton's work on the modified wavelength of the X rays scattered by free electrons. If we

2. *Ibid.*, p. 228.
3. N. Bohr, H. A. Kramers, and J. C. Slater, "The Quantum Theory of Radiation," *Phil. Mag., 47* (1924), 785. Passage quoted is from p. 791. (A German version of the same paper appeared in *Z. Phys., 24* [1924], 69.)

ask *why* Bohr and his co-workers were willing to give up the validity of the conservation laws, except as statistical averages, I think the answer is clear: it was to save physics from an alternative they considered even less acceptable—the admission of light quanta. The Bohr, Kramers, Slater work was an attempt to preserve certain essential features of the wave theory of radiation in the face of an apparent need for Einstein's light quanta, and it was for this conservative reason that they were willing to take the radical step of abandoning the laws of energy and momentum. Their attempt failed, but an analysis of their proposal in the context of its time may add to our understanding of the ideas of both Bohr and Einstein and the interaction between them.[4]

Both men were fully aware of the complexity of the problems they were struggling with, and neither was ever satisfied with easy answers. This account of the first phase of their long dialogue may suggest something of "the years of anxious searching in the dark, with their intense longing, their alternations of confidence and exhaustion" that must precede "the final emergence into the light," when "the happy achievement seems almost a matter of course and any intelligent student can grasp it without too much trouble."[5]

2.   In 1905 Albert Einstein suggested, in all seriousness, that light be considered as composed of a collection of independent particles of energy.[6] He made this suggestion, despite all the successes of the wave theory of light throughout the nineteenth century, because he was convinced that this new hypothesis of light quanta offered a more fruitful approach to the understanding of processes involving the emission and absorption of light. The classical theory had failed to account for the existence of an equilibrium distribution of

---

4. The Bohr, Kramers, Slater paper has been discussed in some detail by Jammer, Meyer-Abich, and van der Waerden. (a) M. Jammer, *The Conceptual Development of Quantum Mechanics* (New York, 1966), pp. 181–188, 345–350. (b) K. M. Meyer-Abich, *Korrespondenz, Individualität und Komplementarität* (Wiesbaden, 1965), pp. 102–133. (c) B. L. van der Waerden, ed., *Sources of Quantum Mechanics* (Amsterdam, 1967), pp. 11–15. The paper itself is also reprinted by van der Waerden, p. 159.

5. A. Einstein, *The World as I See It* (New York, 1934), p. 108.

6. A. Einstein, "Über einen die Erzeugung und Verwandlung des Lichtes betreffenden heuristischen Gesichtspunkt," *Ann. Phys.*, 17 (1905), 132. English translation by A. B. Arons and M. P. Peppard in *Amer. Jour. Phys., 33* (1965), 367.

energy in blackbody radiation, and Einstein immediately recognized this as a difficulty in the very foundations of physics.[7] He was led to the hypothesis of light quanta by his study of the entropy of radiation in the high-frequency range, where its spectrum was adequately described by Wien's law. When Einstein analyzed the form of this entropy with the help of Boltzmann's relationship between entropy and probability, he concluded that, when Wien's distribution holds, radiation acts as though it consists of quanta of energy, with the energy quantum proportional to the frequency of the radiation.

This new "heuristic viewpoint" on the nature of light justified itself at once; Einstein used it to account for Stokes's law of fluorescence and the known qualitative, and very puzzling, properties of the photoelectric effect. He also proposed the exact relationship between the frequency of the incident light in the photoelectric effect and the voltage that would stop all photoelectrons produced, a linear equation with a universally constant slope.

One indication of the immediate reaction to Einstein's suggestion can be found by a careful reading of Philipp Lenard's Nobel Lecture, delivered a year later, in May 1906.[8] Lenard received the prize for his work on electrons, including his important experiments on the photoelectric effect. Einstein and his work are simply not mentioned by Lenard, although he asserted that he had "tried hard to put into their historical perspective all the publications which in my opinion have made basic contributions to knowledge," and he actually cited two other papers from the same volume of the *Annalen der Physik* that contained Einstein's work.

By 1909 Einstein's "ceaseless" preoccupation with the "incredibly important and difficult"[9] question of the constitution of radiation had led him to a deeper insight into the theoretical situation. He was now convinced that "the next phase of the development of theoretical physics will bring us a theory of light that can be interpreted as a

7. For further discussion see M. J. Klein, (a) "Einstein's First Paper on Quanta," *The Natural Philosopher,* ed. D. Gershenson and D. Greenberg (New York, 1963), 2, 57; (b) "Thermodynamics in Einstein's Thought," *Science, 157* (1967), 509.

8. P. E. A. von Lenard, "On Cathode Rays," *Nobel Lectures. Physics. 1901–1921* (Amsterdam, 1967), p. 105.

9. A. Einstein to J. J. Laub, 1909. Quoted in C. Seelig, *Albert Einstein, A Documentary Biography,* trans. M. Savill (London, 1956), p. 87.

kind of fusion of the wave and emission [particle] theories."[10] Einstein had reasons for holding this opinion, and he explained them at a meeting in Salzburg in September of that year. He had analyzed the implications of Planck's law for the blackbody radiation spectrum, using an approach that was peculiarly his own: the study of fluctuations.[11]

The fluctuations ΔE about the average energy E of the blackbody radiation having frequencies between ν and $\nu + d\nu$, and contained in the subvolume V of the enclosure, could be calculated from a basic result of statistical mechanics:

$$\overline{(\Delta E)^2} = kT^2 \left(\frac{\partial E}{\partial T}\right)_V , \qquad (1)$$

where k is Boltzmann's constant, and T is the temperature of the walls of the enclosure. Einstein derived this result independently in 1904 and had already used it to great advantage in exploring the significance of fluctuation phenomena. Since the energy E could be expressed in the form

$$E = \rho(\nu, T)Vd\nu , \qquad (2)$$

where $\rho(\nu, T)$ is the spectral density of the radiation, the fluctuations were determined by the form of the spectral distribution. When Planck's distribution law,

$$\rho(\nu, T) = \frac{8\pi\nu^2}{c^3} \frac{h\nu}{\exp(h\nu/kT) - 1} , \qquad (3)$$

was substituted, Einstein obtained the result,

$$\overline{(\Delta E)^2} = (Vd\nu)\{h\nu\rho + (c^3/8\pi\nu^2)\rho^2\} . \qquad (4)$$

In these equations h is Planck's constant and c is the velocity of light.

While the existence of these energy fluctuations was to be expected, regardless of the wave or particle nature of the radiation, their particular form led Einstein to his new conclusions. The wave theory should lead only to the second term, as one could easily check

10. A. Einstein, "Über die Entwicklung unserer Anschauungen über das Wesen und die Konstitution der Strahlung," *Phys. Z., 10* (1909), 817. See also his earlier paper, "Zum gegenwärtigen Stand des Strahlungsproblems," *Phys. Z., 10* (1909), 185.

11. For further discussion see M. J. Klein, "Einstein and the Wave-Particle Duality," *The Natural Philosopher, 3* (1964), 1.

by using the wave limit of the spectral distribution, that is, the low-frequency classical form first given by Lord Rayleigh. The first term has a natural interpretation as the fluctuation to be expected if the radiation consisted of a collection of independent particles ($E/h\nu$ in number). This term would dominate in the high-frequency or Wien limit of the distribution law.

Einstein concluded that there were two independent causes producing the fluctuations, and that an adequate theory of radiation would have to provide both wave and particle mechanisms. He confirmed this view by a completely independent argument: he calculated the momentum fluctuations of an object suspended in a cavity at a given temperature, an argument closely following his theory of Brownian motion. These fluctuations also had two terms of the same structure, identifiable as wave and particle contributions, again demonstrating that interfering waves alone could not meet the needs of thermodynamic equilibrium. Einstein was certain that the existence of light quanta was a *necessary* result of the fluctuation properties of blackbody radiation, and not just an assumption *sufficient* for deriving Planck's law. Therefore, he thought, "a profound change in our views of the nature and constitution of light is indispensable." He was also confident that such a new fundamental theory, incorporating quanta and interference phenomena, could be constructed.

Einstein's convictions were not shared. Even those who were generally sympathetic to his other work, such men as Max Planck and H. A. Lorentz, had only sharply critical things to say about his light quanta.[12] It was simply not possible to visualize a quantum theory that could account for interference and diffraction phenomena, and physicists were not prepared to sacrifice the electromagnetic-wave theory, which adequately explained these phenomena, on the basis of something so unsubstantial as fluctuation arguments. In 1916 Robert A. Millikan,[13] in the paper reporting his complete experimental confirmation of Einstein's photoelectric equation, could re-

12. See, for example, Planck's remarks after Einstein's paper at Salzburg, *Phys. Z.*, *10* (1909), 825. See also M. Planck, "Zur Theorie der Wärmestrahlung," *Ann. Phys.*, *31* (1910), 758, and H. A. Lorentz, "Die Hypothese der Lichtquanten," *Phys. Z.*, *11* (1910), 349.

13. R. A. Millikan, "A Direct Photoelectric Determination of Planck's 'h'," *Phys. Rev.*, *7* (1916), 355.

mark on "the astonishing situation that these facts were correctly and exactly predicted . . . by a form of quantum theory which has now been pretty generally abandoned." Millikan was wrong only in suggesting that the idea of light quanta, which he referred to as Einstein's "bold, not to say reckless, hypothesis," had *ever* been accepted.

During that same year, 1916, Einstein published another paper on radiation which showed that he, at least, had not given up his ideas on the quantum structure of radiation.[14] He offered a new derivation of Planck's distribution law, a derivation he described as being "astonishingly simple and general"; he thought it might even properly be called "*the* derivation" of Planck's law.[15] This "purely quantal" treatment was based on statistical assumptions about the processes of emission and absorption, and the basic quantum hypothesis that atomic systems have a discrete set of possible stationary states. The proof turned on the requirement that absorption and emission of radiation suffice to keep a gas of atoms in thermodynamic equilibrium.

The basic idea can be given in a few lines. Suppose that m and n are two atomic states of energies ε_m and ε_n, where $\varepsilon_m > \varepsilon_n$. Guided by the analogy to the behavior of a classical oscillator in the electromagnetic field, Einstein assumed that the probability dW_a that, during the time dt, an atom in state n absorbs energy $(\varepsilon_m - \varepsilon_n)$ from the field, whose spectral radiation density is ρ, and makes a transition to state m, is given by the equation

$$dW_a = B_{mn}\rho dt . \tag{5}$$

Similarly, the probability that, during time dt, an atom in the upper state m emits energy $(\varepsilon_m - \varepsilon_n)$ and drops to state n was assumed to have the form

$$dW_e = (B_{nm}\rho + A_{nm})dt , \tag{6}$$

where the two terms refer respectively to processes stimulated by the radiation field and processes occurring spontaneously.

These radiation processes must preserve thermodynamic equilib-

---

14. (a) A. Einstein, "Zur Quantentheorie der Strahlung," *Phys. Gesellschaft, Zürich, Mitteilungen, 16* (1916), 47. Also in *Phys. Z., 18* (1917), 121. (b) English translation in B. L. van der Waerden, ed., *Sources of Quantum Mechanics,* p. 63.

15. A. Einstein to M. Besso, 11 August 1916.

rium among the atoms, so that we must have equal rates of absorption and emission,

$$B_{mn}\rho g_n \exp\left(-\epsilon_n/kT\right) = \{B_{nm}\rho + A_{nm}\}g_m \exp\left(-\epsilon_m/kT\right), \qquad (7)$$

where g_m and g_n are the statistical weights of the two states. Solving for ρ, assuming that ρ becomes infinite when T does, and using Wien's displacement law, Einstein obtained both the Planck distribution for ρ and Bohr's relationship

$$\epsilon_m - \epsilon_n = h\nu, \qquad (8)$$

where ν is the frequency of the radiation associated with this transition.

Einstein's further analysis brought out a new aspect of the radiation problem. To make his theory fully consistent, he had to make explicit use of the completely directional character of energy quanta. Each emitted quantum, for example, must carry away a momentum $h\nu/c$ in a definite direction. The direction would be that of the external radiation if the emission were stimulated, but it would be a direction determined only by chance if the emission were spontaneous. In either case the emission process would be fully directional, and spherical waves would simply not exist. Einstein considered the directional character of quanta to be the main new result of his work. It strengthened his conviction that "a proper quantum theory of radiation" would have to be constructed, even though he had not yet come "any closer to making the connection with the wave theory." But Einstein's argument for this result, a variation of an old favorite of his, based on the Brownian motion of a molecule in the fluctuating radiation field, did not carry the same weight with other theorists that it did with him.

3.   Einstein tried hard to find a crucial experiment that would distinguish sharply and directly between the wave and particle theories of radiation. He thought he had found one when he presented a paper to the Academy at Berlin in December 1921, "On an Experiment Concerning the Elementary Process of Light Emission."[16] The question at issue was whether or not the light emitted in one elementary process by a moving atom is monochromatic. On the wave theory the

16. A. Einstein, "Über ein den Elementarprozess der Lichtemission betreffendes Experiment," *Berliner Berichte* (1921), p. 882.

frequency ν emitted by an atom moving with velocity v would vary with the angle θ between the velocity and the direction of observation,

$$\nu = \nu_0 \{1 + (v/c) \cos \theta\} , \qquad (9)$$

where ν_0 is the frequency emitted by an atom at rest. This is the first–order Doppler effect. On the quantum theory of light emission, how-ever, a consideration of Bohr's fundamental equation, (8), made Einstein "inclined to ascribe one uniform frequency to every ele-mentary act of emission, including emission from a moving atom."[17] Einstein proposed an experiment that should decide between these two alternatives—a frequency varying with direction, as required by the wave theory, and a single fixed frequency, "as suggested though not required by the quantum theory."

The moving light source would consist of the excited atoms in a beam K of canal rays (positive ions). With the help of a lens L_1 and a screen S with a small opening, one could select light coming from a small portion of the beam. A second lens L_2, placed so that the open-ing in the screen was at its focus, would then make this light into a beam of parallel rays. More precisely, it would make the surfaces of constant phase into planes. These planes would, however, be inclined to one another, in fanlike fashion, if the frequency and wavelength of the emitted light varied with the angle of emission. The light at the top of the beam to the right of L_2 in the figure would be of lower frequency than the light at the bottom. If one were now to allow this beam of light to pass through a dispersive medium, the planes of constant phase would rotate as they traveled, since the lower-fre-quency light at the top of the beam would travel at a different speed than the higher-frequency light at the bottom of the beam in a dis-persive medium. This rotation would manifest itself, according to

17. At the end of his short paper Einstein remarked that although the fre-quency of the single elementary emission process was independent of direction, this was not inconsistent with the existence of the Doppler effect. I must confess that I simply do not understand this remark. It is true that the Doppler effect had not yet been shown to be valid on the basis of a quantum theory, though Erwin Schrödinger would soon give such a derivation. (E. Schrödinger, "Dopplerprinzip und Bohrsche Frequenzbedingung," *Phys. Z., 23* [1922], 301. See also E. Fermi, "The Quantum Theory of Radiation," *Revs. Modern Phys., 4* [1932], 419.) Since equation (9) would be valid for both wave and quantum theories, the effect Ein-stein predicted would not have distinguished between them, if it had existed. It seems as though Einstein missed this point, and, so far as I know, no one else ever remarked on it.

9

Einstein, as a deviation of the light beam from its original direction. He calculated the deviation to be expected and found that it should be possible to obtain an easily measurable deviation of a few degrees under reasonable experimental conditions. No such deviation would be expected if the frequency were independent of the angle of emission, as "suggested" by the quantum theory of emission. If the deviation were not observed, the wave theory would suffer a direct contradiction.

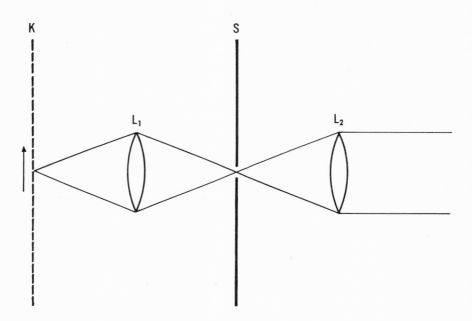

Only a few weeks later Einstein reported to his friend Paul Ehrenfest that the experiment had been tried by Hans Geiger and Walther Bothe in Berlin, and that the outcome was negative: the deviation he had predicted on the basis of the wave theory was not observed.[18] Ehrenfest found this a startling result. "If your light experiment really turns out anticlassically," he commented on a postcard, "—I mean after both theoretical and experimental criticism—then, you know, you will have become really *uncanny* to me. . . . I mean that quite seriously." If the result were genuine, then, Ehrenfest thought,

18. A. Einstein to P. Ehrenfest, 11 January 1922.

Einstein would have discovered "something completely colossal."[19]

Just because this anticlassical result was so disturbing, Ehrenfest could not put it aside. He wrote to Einstein about it again two days later to report a subtlety in the propagation of waves through a dispersive medium which Einstein had missed, and which put his theoretical result in doubt.[20] Einstein had argued as if the wave train were infinitely long, discussing the propagation from the behavior of the phase velocity of the waves. In fact, however, it was a finite wave group or wave packet that was involved in the experimental arrangement, which meant that the group velocity had to be considered. (Ehrenfest was particularly sensitive to the importance of wave packets and group velocity. Some years earlier he had caught an error in the work of the great authority on wave motion, Lord Rayleigh, an error which arose from overlooking just such a point.[21]) Ehrenfest referred Einstein to an old paper by J. Willard Gibbs[22] in which Gibbs had discussed theoretically the measurement of the velocity of light in a dispersive medium by means of Foucault's rotating mirror experiment. Gibbs had shown that "while the individual wave rotates, the wave-normal of the group remains unchanged, or, in other words, that if we fix our attention on a point moving with the group . . . the successive wave-planes, as they pass through that point, have all the same orientation."

Ehrenfest saw that Gibbs's analysis also applied to Einstein's experiment. The increasing inclination of the planes of constant phase would indeed occur for individual waves as they propagated through the dispersive medium, but these waves would cease to exist physically as soon as they propagated out of the moving group. Inside the moving wave group there would always be just the same range of inclinations of these planes as there was when the group entered the dispersive medium.

Ehrenfest knew his friend well enough to preface his discussion by

19. P. Ehrenfest to A. Einstein, 17 January 1922.
20. P. Ehrenfest to A. Einstein, 19 January 1922.
21. P. Ehrenfest, "Misst der Aberrationswinkel im Fall einer Dispersion des Äthers die Wellengeschwindigkeit?" *Ann. Phys., 33* (1910), 1571. See also M. J. Klein, *Paul Ehrenfest. The Making of a Theoretical Physicist* (Amsterdam, 1970), Chapter 7.
22. J. W. Gibbs, "On the Velocity of Light as Determined by Foucault's Revolving Mirror," *Nature, 33* (1886), 582. Reprinted in *The Scientific Papers of J. Willard Gibbs* (New York, 1906; reprinted 1961), *2*, 253.

saying: "Of course you are such a devil of a fellow that naturally you will finally turn out to be right in the end." He also concluded with the remark: "Don't be annoyed with me if I am wrong; and don't be annoyed with me if I am right."

Einstein answered that everyone was now attacking him for his recently announced result.[23] He and Max von Laue had had "a regular duel" over it at the Berlin colloquium. Nevertheless, Einstein remained convinced that he was right. He had carried out a new and more rigorous calculation based on the wave theory, a calculation too long to include in his letter, and he considered his proof to be "certain," or at least "what a theoretical physicist calls certain." He was repentant over past blunders and curious to see what Ehrenfest would have to say about the new proof. Einstein was particularly fond of this proof over which he had really taken a lot of trouble.

Ehrenfest "wished the new proof well" but refused to believe in the deviation predicted by the wave theory.[24] He, too, was convinced of his position, and saw no way of getting around his wave group argument. Since "the experiment operates *essentially* with wave groups, no *individual* wave ever travels from the left end of the tube [containing the dispersive medium, carbon disulfide] to the right end; the waves entering at the left all die out during their wandering through the desert of carbon disulfide—those that do arrive at the right hand end are *completely different individual waves* which were born in the desert. And they come forth at the right end with the same inclinations as those with which their deceased parents entered at the left."

A week later Einstein wrote: "You were absolutely right."[25] He had discovered an error in his new calculations, and when it was corrected the predicted deviation disappeared. But the whole problem was a deceptive one, so that it was probably worth publishing the detailed calculation to clarify it. The opportunity was provided by the next meeting of the Prussian Academy on 2 February 1922.[26] The calculation Einstein presented was quite different in form from the Ehrenfest-Gibbs discussion, but the central point was the use of a

23. A. Einstein to P. Ehrenfest, received 22 January 1922.
24. P. Ehrenfest to A. Einstein, 26 January 1922.
25. A. Einstein to P. Ehrenfest, 30 January 1922.
26. A. Einstein, "Zur Theorie der Lichtfortpflanzung in dispergierenden Medien," *Berliner Berichte* (1922), p. 18.

wave group rather than an infinite train of waves. Einstein had to conclude that the negative result obtained by Geiger and Bothe did not allow one to infer anything about the wave or quantum nature of light emission.

4. It was, in any case, difficult for physicists to escape completely from Einstein's light quanta by the early 1920's, since the quanta provided such a successful way of accounting for phenomena like the photoelectric effect. Most physicists would, however, probably have agreed that, even though radiation must have quantal features, these features appeared only when light was emitted or absorbed, and that the free propagation of radiation had to be described by the classical wave theory.[27]

It became impossible to maintain this separation of the domains in which the wave and particle theories of light were applicable after the discovery of the Compton effect.[28] Arthur Compton found that the wavelength of X rays increased when they were scattered by free electrons. Both Compton[29] and Peter Debye[30] independently worked out the now familiar equations for this Compton scattering by treating the X ray as a particle, a quantum of energy hv and momentum hv/c, and applying the conservation laws for momentum and energy to the collision between this quantum and an electron at rest. Since this approach led to a successful description of the Compton effect, most physicists considered the agreement between theory and experiment to be "definite evidence for the existence of light quanta," as J. H. Van Vleck put it.[31] Some were willing to go further. Arnold Sommerfeld, for example, wrote Compton that his discovery sounded "the death knell of the wave theory of radiation."[32] For now even freely propagating radiation seemed to have particle properties, as

27. See, for example, A. Sommerfeld, *Atomic Structure and Spectral Lines,* trans. by H. L. Brose from third German edition of 1922 (New York, 1923), p. 253.

28. A. H. Compton, "Secondary Radiations Produced by X-Rays, and Some of Their Applications to Physical Problems," *Bull. Nat. Res. Council, No. 20* (1922), p. 16.

29. A. H. Compton, "A Quantum Theory of the Scattering of X-Rays by Light Elements," *Phys. Rev., 21* (1923), 483.

30. P. Debye, "Zerstreuung von Röntgenstrahlen und Quantentheorie," *Phys. Z., 24* (1923), 161.

31. J. H. Van Vleck, "Quantum Principles and Line Spectra," *Bull. Nat. Res. Council, No. 54* (1926), p. 270.

32. Quoted by A. H. Compton in *J. Franklin Inst., 198* (1924), 70.

shown by the simple quantum theory of the Compton effect based solely on the conservation laws. The Compton effect was generally taken to be the kind of crucial experiment that Einstein had been looking for.

The quantal explanation of the Compton effect had other consequences. It made it possible to treat the old, perplexing, unsolved problem of the thermal equilibrium between free electrons and blackbody radiation. H. A. Lorentz and A. D. Fokker had worked on this problem a decade earlier and had been unable to construct a plausible theory that would account for a Maxwellian velocity distribution for the electrons and a Planck distribution for the radiation.[33] Wolfgang Pauli took up the problem again in 1923 and showed that Compton scattering provided a mechanism that would achieve this aim.[34] Pauli's method was a modification of the one used by Einstein in his 1916 theory of radiation. A few months after Pauli's paper appeared, Einstein and Ehrenfest showed that his argument could be clarified and made more intelligible if one looked at Compton scattering as a two-step process: the absorption of a quantum of frequency ν by an electron, and the emission of a quantum of frequency ν', both quanta being appropriately specified as to direction, and the entire process subjected to the energy and momentum conservation laws.[35] Once again Einstein's free light quanta seemed to be essential to the understanding of the whole effect.

Nevertheless, not quite everyone was persuaded that the Compton effect proved the existence of light quanta. The most important nonbeliever was Niels Bohr, and we must now try to see why he took up an opposing position.

5. Bohr's first famous series of papers, written in 1913, was concerned with the problem indicated by its title, "On the Constitution

33. See H. A. Lorentz, "Sur l'application au rayonnement du théorème de l'équipartition de l'énergie," in *La théorie du rayonnement et les quanta,* ed. P. Langevin and M. de Broglie (Paris, 1912), p. 35. See also W. Pauli, "Quantentheorie" in *Handbuch der Physik, 23, Quanten,* ed. H. Geiger (Berlin, 1926), p. 18. Reprinted in W. Pauli, *Collected Scientific Papers,* ed. R. Kronig and V. F. Weisskopf (New York, 1964), *1,* 288.
34. W. Pauli, "Über das thermische Gleichewicht zwischen Strahlung und freien Elektronen," *Z. Phys., 18* (1923), 272.
35. A. Einstein and P. Ehrenfest, "Zur Quantentheorie des Strahlungsgleichgewichts," *Z. Phys., 19* (1923), 301. See also M. J. Klein, *op. cit.* (note 11).

of Atoms and Molecules."[36] This research was not originally under-taken to explain atomic spectra, and Bohr's highly successful theory of the hydrogen spectrum, worked out only a few weeks before he sent the first paper of the series off for publication, seemed almost a distraction from what he described as the "main object of this paper —the discussion of the permanent state of a system consisting of nuclei and bound electrons."[37] As Bohr's ideas developed over the years, the analysis of spectra took on a more and more significant role, just because this analysis proved to be the best way to study atomic structure. Bohr's writings, and the old quantum theory generally, dealt with two basic problems: what is the nature of the stationary states of atomic systems, and how does the structure of the atom, de-fined through these stationary states, determine the physical and chemical properties of the corresponding element? (The character of the atomic spectrum was the most important of these physical prop-erties.)

In none of Bohr's writings before 1922, so far as I know, did he concern himself at any length with the problem of the nature of radiation.[38] He did make a remark on this problem in a lecture in Berlin in 1920, with Einstein present in the audience, but it was only to put it aside for the time being. "I shall not," he had said, "here discuss the familiar difficulties to which the 'hypothesis of light quanta' leads in connection with the phenomenon of interference, for the explanation of which the classical theory of radiation has shown itself to be so remarkably suited. Above all I shall not consider the problem of the nature of radiation."[39]

Although Bohr did not write on the radiation problem, he cer-tainly did think about it. He was deeply impressed by Einstein's new statistical derivation of the Planck distribution, which also supported his own characteristic postulate, and he incorporated the idea of transition probabilities, both spontaneous and induced, into his sub-

36. (a) N. Bohr, "On the Constitution of Atoms and Molecules," *Phil. Mag., 26* (1913), 1, 476, 857. (b) The three papers have been reprinted in a book with the same title (Copenhagen, 1963), with a long and informative introduction by L. Rosenfeld.
37. See Bohr, *op. cit.* (note 36[b]), p. 20.
38. The same comment was made by K. M. Meyer-Abich in his book, *op. cit.* (note 4[b]), p. 108.
39. N. Bohr, *The Theory of Spectra and Atomic Constitution* (Cambridge, 1922), p. 22.

sequent work. Bohr called particular attention to the fact that Einstein had made his basic assumptions in analogy to the classical theory of radiation.[40] The connection of new work to the classical theory was always a matter of concern to Bohr.

In his 1913 paper on the hydrogen spectrum Bohr had assumed that the frequency of a spectral line emitted by an atom was proportional to the difference between the energies of the initial and final stationary states of the atom. This, as Bohr said, was "in obvious contrast to the ordinary ideas of electrodynamics," since it destroyed the classical idea that the frequency emitted was the frequency of some internal motion in the atom. This assumption had been the most puzzling and disturbing feature of Bohr's theory. (Years later Erwin Schrödinger still described it as "monstrous" and "inconceivable."[41]) But Bohr had also shown that his new assumption led to frequencies that were completely consistent with what classical theory predicted for the emission in the long wavelength limit, the same limit in which Planck's distribution law reduced to the classical Rayleigh form.

Bohr took this result very seriously, so seriously indeed that it became a central theme and guiding principle in his work. In the region of large quantum numbers, where the stationary states are closely spaced and emitted wavelengths are long, the frequencies calculated from the quantum theory must agree with those calculated classically, and the transition probabilities are simply related to the amplitudes of the corresponding harmonic components in the motion. This principle of correspondence became a very powerful method for treating specific problems in the theory of spectra, but Bohr saw its real significance as being more than that: it made it possible, he wrote, "in a certain sense to regard this theory [the quantum theory of spectra] as a natural generalization of our ordinary ideas of radiation."[42] In his search for a new theory, the correspondence principle was one of the few sure guides; it gave Bohr a way of keeping in contact with the solid results of classical electro-

40. N. Bohr, "The Quantum Theory of Line-Spectra," *D. Kgl. Danske Vidensk. Selsk. Skrifter. Naturvidensk. og Mathem. Afd.*, (8) *4*, 1 (1918), 7.
41. E. Schrödinger, A. Einstein, M. Planck, and H. A. Lorentz, *Letters on Wave Mechanics*, ed. K. Przibram, trans. M. J. Klein (New York, 1967), p. 61.
42. N. Bohr, "The Effect of Electric and Magnetic Fields on Spectral Lines," *Proc. Phys. Soc., 35* (1923), 279.

magnetic theory, while seeking the quantum theory which would be its "natural generalization."

6. The third Solvay Congress, on "Atoms and Electrons," was held in April 1921. Bohr was to have been one of the principal speakers, but illness kept him from going to Brussels for the meeting. He never completed the report on the application of the quantum theory to atomic problems which he had planned to deliver, but a part of it was written. This was presented for him by Ehrenfest, and it appeared in the proceedings. This portion included a brief discussion of the radiation problem, emphasizing the importance of the correspondence principle.[43] Bohr devoted only a few sentences to the hypothesis of free light quanta, but one of his remarks is worth noticing here. "Such a concept," he wrote, "seems, on the one hand, to offer the only possibility of accounting for the photoelectric effect, *if we stick to the unrestricted applicability of the ideas of energy and momentum conservation.* On the other hand, however, it presents apparently insurmountable difficulties from the point of view of the phenomena of optical interference. . . ." The words I have emphasized suggest that Bohr was already considering the possibility that the conservation laws might *not* be universally valid.

That Bohr was indeed thinking along these lines is confirmed by Paul Ehrenfest's description of his friend's ideas some months later. In a postcard to Einstein which has already been quoted above, Ehrenfest wrote that he was very curious to know Bohr's reaction to Einstein's proposed crucial experiment. Ehrenfest had recently been in Copenhagen to visit Bohr, and he tried to give Einstein an idea of Bohr's current thinking. "If I am able to reproduce his opinion on these matters correctly, I might formulate it this way. He is much more willing to give up the energy and momentum theorems (in their classical form) for elementary atomic processes, and to maintain them only statistically, than to 'lay the blame on the aether.' "[44]

The idea that a restriction on the validity of the conservation laws might provide a way of reconciling the wave and quantum aspects of light seems to have occurred to several people in this period. As

43. N. Bohr, "L'application de la théorie des quanta aux problèmes atomiques," in *Atomes et électrons* (Paris, 1923), pp. 241–242.
44. P. Ehrenfest to A. Einstein, 17 January 1922.

early as the summer of 1919 Charles Darwin wrote to Bohr, sending him his general views on the problems of the quantum theory.[45] Darwin considered "the case against conservation quite overwhelming." He remarked that Frederick Lindemann had told him of a conversation with Einstein on this subject: Einstein "had tried without conservation," but found "it was no better than with."

Bohr immediately began to compose a long answer to Darwin's letter, in which he tried to formulate his own views on some of the basic questions of principle in the quantum theory. This proved to be too big a problem to be solved in one letter—it was Bohr's lifework —and the answer to Darwin was never sent.[46] The unfinished draft of Bohr's ideas was preserved, however, and it presents a remarkable sketch of "the scientific conscience (bad or good?) of a Quanticist," as Bohr described it almost three years later, when he finally did write to Darwin.[47]

Bohr started with some general remarks about the nature of scientific reasoning, presumably prompted by Darwin's criticism of the proofs of the necessity of a quantum theory, such as that given by Poincaré.[48] "All progress in science emphasizes difficulties," he wrote. "All progress in physics no proofs whatever but only simple connections of different conceptions." Bohr even went so far as to describe "most general reasoning in science as opportunistic." He agreed with Darwin that the photoelectric effect was "by far the central evidence" for the applicability of quanta to nonstatistical phenomena. Bohr also thought, however, that the "wonderful inverse of the photoelectric effect which we see in the phenomena of excitation of spectral lines" would "make the case very hard" for any attempt to explain such things statistically. He was inclined to accept the rather widespread view that the wave theory of light was valid for freely propagating radiation and "that all difficulties are concentrated on the interaction between the electromagnetic forces and matter." With respect to these interactions, however, Bohr was "inclined to take

45. C. Darwin to N. Bohr, 20 July 1919. (The documents referred to in notes 45–47 are part of the Archive for the History of Quantum Physics. They were pointed out to me by Professor Roger Stuewer.)
46. N. Bohr, undated draft of a letter to C. Darwin.
47. N. Bohr to C. Darwin, 14 February 1922.
48. H. Poincaré, "Sur la théorie des quanta," *Journal de Physique, 2* (1912), 5. See R. McCormmach, "Henri Poincaré and the Quantum Theory," *Isis, 58* (1967), 37.

the most radical or rather mystical views imaginable." He thought that conservation of energy was "quite out of question," and wondered if the frequency of the incident light were just "the key to the lock which controls the starting of the interatomic [intra-atomic?] process." Even the definition of energy was not a trivial problem in the quantum theory, and required the use of "the principle of mechanical transformability," Bohr's name for Ehrenfest's adiabatic principle. Only this kept it from being "so criminal as it looks at first sight to speak with such light heart of the fundamental difficulties touched upon above and still to attempt to be a serious worker in the present cribbled [crippled?] field of physics."

These comments were only part of the preliminary draft of a letter that was never sent, but Bohr expressed himself in much the same vein in a major paper several years later.[49] This work, completed in November 1922, was intended as the first of a series of essays on the application of the quantum theory to atomic structure. It was actually the only one of the series to be published, and it dealt with the fundamental postulates of the theory. Bohr tried to state and explain the principles which formed the basis for the applications of the theory to atomic structure, applications which had already included a wide range of phenomena and were constantly being extended. But even more than that, Bohr wanted to treat the question of "whether it is possible to present the principles of the quantum theory in such a way that their application appears free from contradiction."

It was in connection with this last question that Bohr finally confronted Einstein's hypothesis of light quanta, in the concluding chapter of his long paper. This chapter was entitled, "On the formal nature of the quantum theory," and Bohr emphasized that the hypothesis of light quanta should be considered as being only *formal*. The view that light propagated as localized and indivisible packets of energy had "placed certain classes of phenomena, such as the photoelectric effect, in a clear light in relation to the quantum theory," but it could "in no wise be regarded as a satisfactory solution." The light quantum hypothesis gave rise to "insuperable difficulties when applied to the explanation of the phenomena of

49. N. Bohr, "On the Application of the Quantum Theory to Atomic Structure, Part I, The Fundamental Postulates of the Quantum Theory," *Proc. Cambr. Phil. Soc. (Supplement)* (1924). (Originally published in German in *Z. Phys., 13* [1923], 117.)

interference," and it even "excluded in principle the possibility of a rational definition of the conception of a frequency v, which plays a principal part in this theory." One could not make an adequate picture of the processes involved, if one started with Einstein's hypothesis. In addition, the success of the hypothesis of light quanta in accounting for "certain aspects of the phenomena" supported the view that "a description of atomic processes in terms of space and time cannot be carried through in a manner free from contradiction by the use of conceptions borrowed from classical electrodynamics, which, up to this time, have been our only means of formulating the principles which form the basis of the actual applications of the quantum theory."[50]

Bohr also emphasized the great difficulties involved in trying to make a quantum theory of dispersion. The classical theory assumed that the illuminated atom produced secondary waves coherently related to the incident waves. The characteristic frequencies of the equivalent oscillators would, however, have to be those observed in the absorption spectrum, and one of the cardinal assumptions of the quantum theory was that the absorption frequencies were not the frequencies of any actual electronic motions in the atom.[51]

One conclusion could be drawn from all the difficulties: "A general description of the phenomena, in which the laws of the conservation of energy and momentum retain in detail their validity in their classical formulation, cannot be carried through." As a result, Bohr warned, "We must be prepared for the fact that deductions from these laws will not possess unlimited validity."[52] It was not the conservation laws but rather the correspondence principle and Ehrenfest's adiabatic principle to which Bohr looked for guidance. They were "suited, in a higher degree, to point out new ways for further extensions of the quantum theory of atomic structure," and they offered "a hope in the future of a consistent theory, which at the same time reproduces the characteristic features of the quantum theory . . . and, nevertheless, can be regarded as a rational generalisation of classical electrodynamics."[53]

Bohr's paper was published in the spring of 1923, so that his re-

50. *Ibid.*, pp. 34–35.
51. *Ibid.*, p. 38.
52. *Ibid.*, p. 40.
53. *Ibid.*, p. 42.

jection of light quanta appeared at about the same time as the works of Compton and Debye, works that showed how simply and naturally light quanta could be used to account for the Compton effect.

7. There were other physicists ready to consider giving up the strict validity of the conservation laws by this time. Sommerfeld commented on this idea in his influential treatise on atomic structure and spectra: "The mildest modification that must be applied to the wave theory is, therefore, that of disavowing the energy theorem for the single radiation phenomenon and allowing it to be valid only on the average for many processes." He considered this kind of change to be much less extreme than taking the light quantum hypothesis as valid, which would make Maxwell's equations for the field into "statistical approximations."[54] But even though some physicists were willing to deny the conservation laws, nobody was able to use this denial to construct a theory of radiation that could account for both the wave and the particle properties.

One of the people puzzling over these questions during the winter of 1923/24 was John C. Slater, who had just received his Ph.D. at Harvard and was spending the year in Europe on a traveling fellowship. Early in December 1923 Slater wrote from England to H. A. Kramers, who had been working with Bohr since 1916, to arrange the details of his arrival in Copenhagen later in the month.[55] Slater mentioned briefly that he thought he might have a way of getting a consistent explanation of dispersion and a variety of other problems by putting the emphasis on light quanta. One would have to construct an electromagnetic field that determined the motion of these quanta with the help of Poynting's theorem; the field would have the frequencies of the emission lines and amplitudes determined by the correspondence principle.

Slater explained his idea more clearly in a letter he sent to *Nature* from Copenhagen at the end of January.[56] He referred to the need for achieving consistency between those properties of light accounted for by waves and those accounted for by quanta. Although the dis-

54. A Sommerfeld, *Atomic Structure and Spectral Lines, op. cit.* (note 27), p. 253.

55. J. C. Slater to H. A. Kramers, 8 December 1923. (Archive for the History of Quantum Physics.)

56. J. C. Slater, "Radiation and Atoms," *Nature, 113* (1924), 307.

continuous side of the story was "apparently the more fundamental," Slater thought he could make progress "by associating the essentially continuous radiation field with the continuity of existence in stationary states, and the discontinuous changes of energy and momentum with the discontinuous transitions from one state to another."

He assumed that an atom in one of its stationary states was surrounded by "a virtual field of radiation, originating from oscillators having the frequencies of possible quantum transitions." This virtual radiation field would provide for statistical conservation of energy and momentum by determining the probabilities of the possible transitions. The virtual field of a given frequency produced by the atom itself would determine the spontaneous transition probability, while the virtual fields due to other atoms would determine the probabilities for induced emission or absorption, much as Einstein had suggested. When an atomic transition occurred, the virtual radiation field would have to change character abruptly, so that the frequencies it then contained would be those appropriate to the new stationary state of the atom.

Slater remarked that although his original goal had been to construct a field that would serve to guide light quanta, he had been persuaded by Kramers that his new approach really implied "a much greater independence between transition processes in distant atoms than [he] had perceived." Many years later Slater explained that the statistical version of the conservation laws was, in his words, "put into the theory by Bohr and Kramers, quite against my better judgment." He would have preferred to keep the light quanta "as real entities" and to have them satisfy the conservation laws exactly, but "Bohr and Kramers opposed this view so vigorously" that he went along with them "to keep peace and get the main part of the suggestion published."[57]

In his letter to *Nature,* Slater referred to a forthcoming paper, written jointly with Bohr and Kramers, for more details. Although the paper did appear under the names of all three men, Slater wrote to a friend that Bohr and Kramers had actually written all of it,[58] and the paper is certainly in Bohr's unmistakable style. Slater's vir-

57. J. C. Slater to B. L. van der Waerden, 4 November 1964. Quoted in B. L. van der Waerden, *op. cit.* (note 4[c]), p. 13.
58. J. C. Slater to J. H. Van Vleck, 27 July 1924. (Archive for the History of Quantum Physics.)

tual radiation field—associated with an atom in one of its stationary states and orginally intended as part of a theory of light quanta—had now become the core of a program for a new theory of radiation, a theory that would have no use for light quanta.

Bohr began with a long introduction, emphasizing both the difficulties in the basic ideas of the quantum theory and the progress that had nevertheless been made in atomic physics by using some of these ideas.[59] He pointed out once again that the quantum theory had only a "formal character" since it did not provide "a description of the mechanism of the discontinuous processes" that it used. Bohr mentioned Einstein's arguments based on the conditions for thermodynamic equilibrium between matter and radiation, and Einstein's conclusion from these arguments that light quanta must carry momentum hv/c in definite directions. He admitted that this conclusion was "considered as an argument for ascribing a certain physical reality to the theory of light quanta," and that it had recently been used with great success to explain the Compton effect and to clarify other problems.[60] Bohr's next paragraph, however, showed why he was not satisfied with this apparent progress. The quantum theory of atomic processes "must in a certain sense ultimately appear as a natural generalization" of classical electrodynamics. This principle, the correspondence principle in the broadest sense, was central for Bohr; it was his only guidance in the attempt to create a new physics. Einstein's light quanta could not be understood on correspondence terms, and for that reason the radiation theory would have to be constructed in some other way. Slater's idea seemed to provide the necessary starting point. This new approach did not "in any way remove the formal character of the theory"—it did not give the mechanism of transitions or avoid their probabilistic description—but it was "a definite advance" in its reinterpretation of radiation phenomena.[61]

An atom in a stationary state was to be thought of as "communicating continually with other atoms" through a virtual radiation field whose frequencies were those of all transitions from this state allowed by the earlier Bohr theory. In each stationary state the atom would

59. N. Bohr, H. A. Kramers, and J. C. Slater, "The Quantum Theory of Radiation," *Phil. Mag.*, 47 (1924), 785.
60. *Ibid.*, p. 789.
61. *Ibid.*, p. 790.

then be equivalent to a set of virtual harmonic oscillators, an approach already used a few years before by R. Ladenburg in his work on dispersion.[62] The frequency, intensity, and polarization of spectral lines would be related to the structure of the atom exactly as before, but the occurrence of transitions would be determined in quite a different way.

According to the Bohr, Kramers, Slater theory the occurrence of a transition in an atom would depend on the initial stationary state of that atom and on the states of those other atoms that produce the virtual radiation field at its location; it would not, however, depend on the occurrence of a transition in one of the latter atoms. This meant abandoning "any attempt at a causal connection between the transitions in distant atoms, and especially a direct application of the principles of conservation of energy and momentum, so characteristic for the classical theories."[63] Thus the light quantum theory would say that if an atom absorbed a quantum $h\nu$ in making a transition from state 1 to state 2 (where $h\nu = E_2 - E_1$), then some other atom must have previously emitted this quantum by making a transition from state 2 to state 1. The new theory, on the other hand, would say that the atom in state 1, which absorbs the radiation, must be subject to a virtual field of frequency ν, produced by another atom in state 2, but that no actual transition of this second atom would be required for the absorption to occur. This second atom could remain in state 2, and yet its virtual radiation field could produce transitions in which any number of other atoms gain the energy $E_2 - E_1$.

Although energy would not be conserved in the individual process of emission or absorption according to this point of view, there would still have to be conservation on the average over many such events. This was provided for by "the peculiarities of the interaction between the virtual field of radiation and the illuminated atoms."[64] The authors contemplated a mechanism much like that of the classical wave theory. An atom illuminated by virtual radiation would act as a source of secondary virtual radiation of the same frequency.

62. R. Ladenburg, "Die quantentheoretische Deutung der Zahl der Dispersionselektronen," Z. Phys., 4 (1921), 451. English translation in B. L. van der Waerden, op. cit. (note 4[c]), p. 139.

63. Bohr, Kramers, and Slater, op. cit. (note 59), p. 791.

64. Ibid., p. 793.

The amplitude of these secondary virtual waves would be large when and only when the incident frequency was very near the frequency of one of the virtual oscillators associated with the stationary state in which the atom happened to be. The relative phase of secondary and incident waves would determine whether the interference between the two would lead to a decrease or an increase in the intensity of the virtual radiation field. If this intensity were decreased, for example, the virtual field would be less capable of inducing transitions in other atoms. (The probability of such an induced transition, the probability that Einstein had introduced, was assumed to be determined by the intensity of the virtual radiation field at the frequency corresponding to the transition.) When the virtual oscillator corresponded to a transition that increased the atom's energy, the phase relations would have to be such as to decrease the intensity of the virtual radiation field, thereby ensuring the statistical validity of the conservation of energy. Similar remarks could be made for the momentum law, since a transition involving an energy change $h\nu$ would also produce a momentum change $h\nu/c$ in some direction.

Bohr, Kramers, and Slater discussed a variety of physical phenomena in their paper, showing how they were to be interpreted according to the new radiation theory. In some cases the new theory led to experimental consequences that differed sharply from those of the old. If, for example, one had a beam of atoms emerging from a luminescent discharge into a vacuum and one asked for the duration of the luminosity in the beam, the old and new theories gave different answers. According to the old theory, even if all the emerging atoms were to be in the same excited state and to have the same speed, the decay of the luminosity would vary as a superposition of exponential decays, one exponential for each different transition probability corresponding to each of the emitted lines. On the new theory, however, all of the spectral lines starting in the given state would decay at one and the same rate. The experimental data available did not allow one to distinguish between these two distinctly different predictions.[65]

Another experiment which Bohr and his collaborators could hardly avoid discussing was the Compton effect, especially as it was generally considered to be direct evidence for the existence of light quanta. Compton himself had already shown that the modified radiation with

65. *Ibid.*, p. 794.

27

its increased wavelength could be interpreted formally as secondary radiation coming from an imaginary recoiling source which produced a Doppler effect. This imaginary source could not be identified with the actual illuminated electron, since the velocity one had to ascribe to it differed from that of the electron. Bohr had to admit that this feature of the virtual wave interpretation was "strikingly unfamiliar to the classical conceptions," but he thought that this was no reason to reject it as inadequate, since it was only a "formal interpretation." Although the emission of scattered radiation of the proper modified wavelength could be accounted for in this way on the continuous virtual wave theory, Bohr and his collaborators still had to admit a discontinuous element into their analysis. They had to assume "that the illuminated electron possesses a certain probability of taking up in unit time a finite amount of momentum in any given direction."[66] The recoil of an electron and the production of scattered radiation of modified wavelength would be uncorrelated events on this view, a conclusion in sharp contrast to the results of the Compton-Debye analysis based on light quanta.

8.   No paper by Bohr would have been ignored in 1924, and certainly not a paper dealing with such fundamental issues as this one did. But it was never easy to grasp Bohr's meaning, and this time there was no structure of equations to help guide the reader through Bohr's dense and difficult prose. The complete mathematical content of the seventeen-page paper was, in fact, the single equation, $h\nu = E_1 - E_2$.

One of the first to react in print was Erwin Schrödinger.[67] He was much taken with the new way of looking at radiation, partly because it emphasized the continuous rather than the discontinuous aspects of the phenomena. Schrödinger liked this idea of a return to a wave theory which could dispense with light quanta. (He never even mentioned, however, that the radiation fields produced by atoms in their stationary states were only virtual fields.) What particularly attracted Schrödinger was the proposal that energy conservation is only a statistical law. He was familiar with this conjecture through the work

66. *Ibid.*, p. 799.
67. E. Schrödinger, "Bohrs neue Strahlungshypothese und der Energiesatz," *Naturwissenschaften, 12* (1924), 720. See the discussion in W. T. Scott, *Erwin Schrödinger. An Introduction to His Writings* (Amherst, 1967), pp. 30, 48.

of his teacher, Franz Exner, and had already speculated in his Zürich inaugural lecture about the possibility that the world was fundamentally acausal and that the conservation laws were only statistically valid. The Bohr, Kramers, Slater theory put these old conjectures and speculations "for the first time into a form one could grasp," and Schrödinger proceeded to work out some of its consequences, after giving a brief exposition of the theory in his own crisp style. Schrödinger emphasized one implication of statistical energy conservation: there must be corresponding fluctuations in the energy of an isolated system, "true" energy fluctuations in the sense that they were not due to contact with another system such as a heat bath. Although these intrinsic fluctuations would not have the disastrous consequences that a first inspection of the problem suggested (as Schrödinger demonstrated with the help of an example based on the runaway inflation of the times), they would have some peculiar features. While they would normally be much too small to be detected, they would necessarily grow arbitrarily large as the time of observation increased, a very puzzling result at best.

Other physicists attempted to work out the details of the program suggested by Bohr and his co-workers. Richard Becker in Berlin tried to construct a unified theory of absorption and dispersion on the basis of the new approach to radiation.[68] These phenomena had previously been treated quite separately, by the quantum and wave methods respectively. Becker tried to specify the nature of the spherical waves radiated by an atom in a stationary state in such a way that both phenomena could be understood together. In Amsterdam, J. D. van der Waals, Jr., pointed to an inconsistency in the Bohr, Kramers, Slater theory.[69] He saw no reason why one should not be able to treat the absorption of momentum continuously in this theory, just as one treated the absorption of energy. He thought this way of treating momentum was necessary for the consistency of the theory, and questioned the cogency of Einstein's old arguments that purported to make discontinuous momentum changes $h\nu/c$ into a thermodynamic necessity. This last point was also made at greater length by Pascual

68. R. Becker, "Über Absorption und Dispersion in Bohrs Quantentheorie," *Z. Phys.*, 27 (1924), 173.

69. J. D. van der Waals, Jr., "Remarques relatives à des questions du domaine de la théorie des quanta," *Arch. Néerl.*, 8 (1925), 300. This paper was pointed out to me by Professor Paul Forman.

Jordan in his Göttingen dissertation.[70] Jordan went on to suggest a generalization of Einstein's statistical arguments that would avoid the necessity for "needle radiation" but would still lead to Planck's radiation law. Einstein soon pointed out, however, that despite the "ingenious" nature of Jordan's reasoning, his generalization went too far; it would prevent one from defining an absorption coefficient at all, because Jordan did not treat the absorption of radiation coming from different directions as completely independent processes.[71]

Experimentalists also responded to the suggestions of Bohr, Kramers, and Slater. Within a month or so of the appearance of their paper, Walther Bothe and Hans Geiger proposed an experiment that would test one crucial prediction of the theory.[72] When one interpreted the Compton effect on the basis of the new theory, one had to conclude that the scattering of X radiation with an increase in wavelength was not necessarily correlated with the recoil of an electron. Bothe and Geiger announced that they were preparing to search for coincidences between Compton recoil electrons and scattered X rays (or, more precisely, the photoelectrons produced by the absorption of the scattered X rays) in an experiment using counters.

At about the same time Arthur Compton undertook an experiment, in collaboration with Alfred W. Simon, to test the same basic point.[73] Compton and Simon used the Wilson cloud chamber, rather than counters, and tried to check the relationship between the angle of scattering of the X rays (observed by means of secondary photoelectrons) and the angle of recoil of the Compton electron. The light quantum theory predicted a unique relationship between these two angles, while the new theory of Bohr, Kramers, and Slater called for no correlation between them. Both this experiment and Bothe and Geiger's were difficult, and their results were not available until the summer of 1925.

At Copenhagen the ideas of the Bohr, Kramers, Slater theory were developed in a variety of directions. Kramers took up the idea that

70. P. Jordan, "Zur Theorie der Quantenstrahlung," *Z. Phys., 30* (1924), 297. Similar criticisms were made by others. See J. H. Van Vleck, *op. cit.* (note 31), p. 269 for references. Also see W. Pauli, *op. cit.* (note 33), p. 16.

71. A. Einstein, "Bemerkung zu P. Jordans Abhandlung 'Zur Theorie der Quantenstrahlung,'" *Z. Phys., 31* (1925), 784.

72. W. Bothe and H. Geiger, "Ein Weg zur experimentellen Nachprüfung der Theorie von Bohr, Kramers und Slater," *Z. Phys., 26* (1924), 44.

73. A. H. Compton and A. W. Simon, "Directed Quanta of Scattered X-Rays," *Phys. Rev., 26* (1925), 289.

an atom in a stationary state acts like a set of virtual oscillators having the frequencies of all absorption and emission lines starting in that state.[74] The theory of dispersion that he developed on this basis made no use of the more controversial aspects of the joint paper. Kramers continued the work on dispersion in collaboration with Werner Heisenberg, and this work was the immediate predecessor of Heisenberg's new quantum mechanics.[75]

Bohr himself struggled to understand the full implications of his new ideas by examining their consequences for other kinds of atomic processes.[76] He made an analysis of the whole problem of collisions of atoms with charged particles. This problem dealt with a range of phenomena that showed the same kind of duality between classical continuity and quantal discreteness that one had in the domain of radiation. At one extreme was a process like the scattering of very fast alpha particles by atoms. It was here that a purely classical analysis had led Rutherford to the concept of the nuclear atom, on which all later developments in atomic physics were based. At the other extreme was the Franck-Hertz experiment, as clear a demonstration of the effects of discrete, quantized atomic states as one could ask for. In between one had a whole range of situations, including the loss of energy by charged particles passing through matter, a phenomenon Bohr had worked on years before.[77] Bohr argued that one could classify atomic collision phenomena according to whether or not they exhibited "reciprocity," by which he meant "a mutual coupling of the participating systems of such a nature that the collision would only be considered as completed, from the standpoint of either system, when the other system is brought into that stationary state which is taken to be the final result of the interaction."[78] Reciprocal collision processes were those that had inverse processes, like the

74. H. A. Kramers, "The Law of Dispersion and Bohr's Theory of Spectra," *Nature, 113* (1924), 673; "The Quantum Theory of Dispersion," *Nature, 114* (1924), 310. Both reprinted in van der Waerden, pp. 177, 199. There are also several unpublished manuscripts by Kramers from this period in the Archive for the History of Quantum Physics, which deal with possible extensions of the Bohr, Kramers, Slater work.

75. H. A. Kramers and W. Heisenberg, "Über die Streuung von Strahlen durch Atome," *Z. Phys., 31* (1925), 681. English trans. in van der Waerden, p. 223.

76. N. Bohr, "Über die Wirkung von Atomen bei Stössen," *Z. Phys., 34* (1925), 142.

77. N. Bohr, "On the Decrease of Velocity of Swiftly Moving Electrified Particles in Passing through Matter," *Phil. Mag., 30* (1915), 581.

78. N. Bohr, *op. cit.* (note 76), p. 143.

Franck-Hertz collisions. The absence of such inverse processes for nonreciprocal ones led, in Bohr's view, to difficulties in interpretation when one applied the exact conservation laws to them. Nonreciprocal collisions, such as those between a fast α particle and an atom, in which the duration of the collision is short compared to the periods of electronic motions within the atom, would be analogous to radiation processes: the conservation laws could be expected to be valid only on the average for such collisions.

Bohr proposed various ways in which his suggested distinction between reciprocal and nonreciprocal interactions might be tested experimentally. In this paper Bohr's characteristic expression, "We must be prepared to find . . . ," is used a number of times, warning the reader that nature is likely to be harder to understand and less adaptable to existing categories than he expects. Bohr concluded by emphasizing the tentative character of his proposed new viewpoint, although it did seem to offer a way out of the difficulties of maintaining both the quantum theory of atomic phenomena and the conservation laws.

9.   Bohr had been reluctant to accept light quanta. Einstein was even more reluctant to accept the alternative Bohr proposed. "Bohr's views on radiation interest me very much," he wrote to Hedwig Born, Max Born's wife, in April 1924. "But I shouldn't let myself be pushed into renouncing strict causality before it had been defended altogether differently from anything done up to now. The idea that an electron ejected by a light ray can choose *of its own free will* the moment and direction in which it will fly off, is intolerable to me. If it comes to that, I would rather be a shoemaker or even an employee in a gambling casino than a physicist. My attempts to give quanta a form one can grasp have failed again and again, it is true, but I am far from giving up hope."[79]

Einstein had some more specific things to say in a letter to Ehrenfest at the end of May:

> I reviewed the Bohr, Kramers, Slater paper at our colloquium the other day. This idea is an old acquaintance of mine, but I don't consider it to be the real thing. Principal reasons:

79. A. Einstein to H. Born, 29 April 1924. Quoted in M. Born, *Physik im Wandel meiner Zeit,* 4th ed. (Braunschweig, 1966), p. 294.

(1) Nature seems to adhere strictly to the conservation laws (Franck-Hertz, Stokes's rule). Why should action at a distance be an exception?

(2) A box with reflecting walls containing radiation, in empty space that is free of radiation, would have to carry out an ever increasing Brownian motion.

(3) A final abandonment of strict causality is very hard for me to tolerate.

(4) One would also almost have to require the existence of a *virtual* acoustic (elastic) radiation field for solids. For it is not easy to believe that quantum *mechanics* necessarily requires an electrical theory of matter as its foundation.

(5) The occurrence of ordinary scattering (not at the proper frequency of the molecules), which is above all standard for the optical behavior of bodies, fits badly into the scheme. . . .[80]

In the brief notes he wrote out for his colloquium talk, Einstein had included several other criticisms of the Bohr, Kramers, Slater idea.[81] He found the preordained harmony between the probabilities of absorption and emission and the intensities of the virtual radiation to be unsatisfactory. He wondered how the virtual field was to be arranged so that the return of a formerly free electron would correspond to a Bohr orbit, and considered this point "very suspicious."

A few months later Einstein mentioned the subject again in another letter to Ehrenfest, saying that Bohr and his collaborators had "abolished free quanta," but adding that free quanta "would not allow themselves to be dispensed with."[82] Ehrenfest, who would be more and more torn between the conflicting views of his two close friends as the years went by, wrote back: "If Bothe and Geiger find 'statistical independence' of the electron and the scattered light quantum it will prove *nothing*. But if they find a *correlation* it will be a triumph of Einstein over Bohr. This time, as an exception, I firmly believe you are right, and I would therefore be happy if the correlation were to be demonstrated."[83]

Even the newspapers were aware of the difference of opinion between Bohr and Einstein on a question of fundamental principle. At

80. A. Einstein to P. Ehrenfest, 31 May 1924.
81. Unpublished note in the Einstein Archive.
82. A. Einstein to P. Ehrenfest, 12 July 1924.
83. P. Ehrenfest to A. Einstein, 9 January 1925.

33

the end of October 1924 Einstein received a letter from Kurt Joel, a member of the editorial staff of the *Vossische Zeitung* in Berlin.[84] Joel wrote that reports and dispatches from Copenhagen indicated that there was a controversy over the nature of light and the conservation of energy, and that the outcome was likely to be decided by an experiment being performed by Geiger and Bothe in Berlin. He asked if Einstein would be kind enough to supply more information. Einstein had had some experience with the ways of the press by this time, and his answer was very brief.[85] Yes, there was a real difference between him and Bohr over the nature of light, but the reports Joel had forwarded were evidently from a not very well informed source. And, he added, there had been no written exchange of views with Bohr on this subject. (The two men never did correspond much with each other.)

When the results of the experiment by Bothe and Geiger were announced, after months of rumors, they seemed quite unambiguous. The observed counter coincidences between the Compton recoil electrons and the scattered X rays were orders of magnitude greater than the purely chance coincidences predicted by Bohr, Kramers, and Slater. "The experiments described are incompatible with Bohr's interpretation of the Compton effect," Bothe and Geiger wrote. The conclusion was clear: "One must therefore admit that the concept of light quanta possesses more reality than is supposed in this theory."[86]

Compton and Simon came to the same conclusion as a result of their cloud chamber test of the relationship between the angles of scattering and recoil. "These results do not appear to be reconcilable with the view of the statistical production of recoil and photoelectrons proposed by Bohr, Kramers, and Slater. They are, on the other hand, in direct support of the view that energy and momentum are conserved during the interaction between radiation and individual electrons." They also saw their results as directly supporting Einstein's picture of "directed quanta of radiant energy."[87]

84. K. Joel to A. Einstein, 28 October 1924.
85. A. Einstein to K. Joel, 3 November 1924.
86. W. Bothe and H. Geiger, "Über das Wesen des Comptoneffekts; ein experimenteller Beitrag zur Theorie der Strahlung," *Z. Phys., 32* (1925), 639. See also W. Bothe, "Absorption und Zerstreuung von Röntgenstrahlen" in *Handbuch der Physik, 23, Quanten,* ed. H. Geiger (Berlin, 1926), pp. 423–424.
87. A. H. Compton and A. W. Simon, *op. cit.* (note 73), p. 299.

Einstein remarked simply, in a letter to Ehrenfest: "We both had no doubts about it."[88]

10. The experimental refutation of the Bohr, Kramers, Slater theory did not solve any of the perplexing problems of radiation. The development of physics seemed to have produced insoluble difficulties. To one physicist, O. D. Chwolson in Leningrad, Bohr's proposal served as an instance of "what peculiar things the current efforts, one may well say the current desperate efforts, of physicists lead them to, as they strive to get physics out of the blind alley it is in now."[89] And J. H. Van Vleck, who was struggling with the problems himself, commented in a similar vein in the summer of 1925: "Modern physics certainly is passing through contortions in its attempt to explain the simultaneous appearance of quantum and classical phenomena; but it is not surprising that paradoxical theories are required to explain paradoxical phenomena."[90]

Bohr's response to the results of the Bothe-Geiger experiment came in a long "Postscript" that he added in July 1925 to his paper on atomic collisions, which was already in proof.[91] Bothe and Geiger had proved that the individual processes involved in the Compton effect were really coupled and not statistically independent as Bohr and his collaborators had proposed. The question now was, what did this mean? Which alternatives were now ruled out and which were still open? Bohr emphasized that the outcome of the Bothe-Geiger experiment "could not be looked at as simply distinguishing between two well-defined ways of describing the propagation of light in empty space, which would correspond to either a corpuscular or a wave theory of light." The problem lay deeper: what were the limits within which one could apply to atomic processes the kind of space-time picture that had previously served for the description of natural phenomena? The Bohr, Kramers, Slater theory had tried giving up the strict validity of the conservation laws just because there seemed to be no imaginable space-time mechanism which maintained the causal connections between individual atomic radiative processes and also managed to preserve a sufficiently close tie with the ideas of

88. A. Einstein to P. Ehrenfest, 18 August 1925.
89. O. D. Chwolson, *Die Physik 1914–1926* (Braunschweig, 1927), p. 392.
90. J. H. Van Vleck, *op. cit.* (note 31), p. 287.
91. N. Bohr, *op. cit.* (note 76), pp. 154–157.

classical electrodynamics. Now, despite the successful development of some of the ideas of Bohr, Kramers, and Slater in the theory of dispersion, the experimental results had closed off that way out of the difficulties. Since these results seemed to argue for the kind of corpuscular theory of light associated with Einstein's light quanta, Bohr warned that "one must be prepared to find that the generalization of classical electrodynamic theory that we are striving after will require a sweeping revolution in the concepts on which the description of nature has been based up to now."

That most critical of physicists, Wolfgang Pauli, agreed with Bohr's harsh conclusion.[92] He was convinced that light quanta must be assigned no less reality than electrons. Pauli thought that what needed thoroughgoing revision was not the energy and momentum laws but rather the classical concepts of force and motion, and especially the classical concept of the electromagnetic field. Pauli had been more than usually skeptical of the Bohr, Kramers, Slater theory anyway, and was happy to see it so quickly discredited by the experiments. He suggested to Kramers that it might otherwise have soon become a hindrance to the development of theoretical physics, particularly for those physicists whose sense of reality was not so strong as Bohr's.[93]

Einstein was as convinced as Bohr that there would be no easy answer to the riddle of radiation. He had never imagined that his light quantum hypothesis constituted a real theory, nor did he ever give up his efforts to construct such a theory, one that would unify the disparate concepts of particle mechanics and field electrodynamics. In December 1923—when the Compton effect finally persuaded many physicists that radiation did have the corpuscular features that Einstein had pointed out almost twenty years earlier, and when Bohr was ready to consider giving up causality, conservation, and detailed space-time descriptions of atomic phenomena—Einstein was pointing his researches in another direction. He read a paper to the Prussian Academy on the question, "Does field theory offer any possibilities for the solution of the quantum problem?"[94] This time it was Einstein who emphasized the "wonderful certainty" with which

92. W. Pauli, *op. cit.* (note 33), p. 86.
93. W. Pauli to H. A. Kramers, 27 July 1925. (Pauli Collection, Zürich. Professor Paul Forman was kind enough to provide me with a copy of this letter.)
94. A. Einstein, "Bietet die Feldtheorie Möglichkeiten für die Lösung des Quantenproblems?" *Berliner Berichte* (1923), p. 359.

the wave theory of light accounted for the complicated phenomena of optical interference and diffraction. No one who fully appreciated this "wonderful certainty" would find it hard to believe that a causal description in space and time by means of partial differential equations—a field theory—was well suited to do justice to the facts. Einstein was convinced that field theory did offer many unexplored possibilities which might allow one to put the quantum rules on a firm foundation, and that it would be unwise to abandon the goal of causal space-time description before these possibilities had all been explored. He had already begun to study overdetermined systems of equations in the hope that these would lead to laws that restricted the initial conditions in the manner of quantum conditions.

It was Einstein who was ready to devote himself to the exploration and development of the unheard-of idea that material particles should show wave properties, even as electromagnetic radiation showed corpuscular properties, when that idea was put forward by Louis de Broglie.[95] Einstein seized upon de Broglie's suggestion of matter waves, testing it, searching out its experimental consequences, and serving as its great advocate. Bohr's first comment in print on de Broglie's work came in his "Postscript" of July 1925, where he mentioned both de Broglie's thesis and Einstein's subsequent papers as examples of work that renounced the goal of space-time description.[96] One may doubt that de Broglie or Einstein viewed their work this way.

11. Werner Heisenberg once referred to the Bohr, Kramers, Slater theory as "the first serious attempt to resolve the paradoxes of radiation into rational physics."[97] That theory would certainly not have been recognized as falling under the label "rational physics" when it appeared. Even Bohr, who was striving to develop the new physics whose necessity he had been persuaded of for years, made no such claims for the work. He admitted that it had not "in any way removed the formal character of the [quantum] theory," so that even if it

95. For a detailed discussion see M. J. Klein, *op. cit.* (note 11).
96. N. Bohr, *op. cit.* (note 76), p. 157.
97. W. Heisenberg, "The Development of the Interpretation of the Quantum Theory," in *Niels Bohr and the Development of Physics,* ed. W. Pauli (London, 1955), p. 12.

were successful, it would serve only as an indication of a new line to follow.

Heisenberg's remark is also extraordinary for its suggestion that no one before Bohr, Kramers, and Slater had tried to resolve the wave-particle paradoxes into "rational physics." In April 1924 Einstein had described the current situation in an article on the Compton effect, written for a Berlin newspaper: "We now have two theories of light, both indispensable, but, it must be admitted, without any logical connection between them, despite twenty years of colossal effort by theoretical physicists."[98] Einstein's phrase "twenty years of colossal effort" was no exaggeration, though he was too modest even to hint that most of that effort was his own. He had been struggling to construct a "rational physics" that would resolve the paradoxes of radiation long before his colleagues recognized the existence of the problem, and he would go on with the struggle long after almost all of them were satisfied that the problem had been solved.

Einstein once wrote that what made Bohr "so marvelously attractive as a scientific thinker" was "his rare blend of boldness and caution."[99] (He could, of course, speak with some authority on these subjects.) Einstein's own blend of boldness and caution was complementary to Bohr's, to use the exactly appropriate term. He could never share Bohr's view that the new quantum physics constituted the long sought-for "rational generalization of classical physics," and he never stopped criticizing what he considered to be its inadequacies. Einstein had no illusions about the path he chose for himself. He knew that it subjected him to the accusation of "rigid adherence to classical theory," an accusation not always made, as he thought Bohr did make it, "in the friendliest of fashion." Einstein felt that his lonely efforts were demanded by "a coercion which I cannot evade,"[100] and he knew better than anyone else the price that they exacted. In 1951 he wrote to his old friend Michele Besso: "All the fifty years of conscious brooding have brought me no closer to the answer to the question, 'What are light quanta?'" But he added, "Of course today every rascal thinks he knows the answer, but he is

98. A. Einstein, "Das Comptonsche Experiment," *Berliner Tageblatt,* 20 April 1924, 1. Beiblatt.

99. A. Einstein, *The World as I See It,* p. 68.

100. A. Einstein, "Remarks Concerning the Essays Brought Together in This Cooperative Volume," in *Albert Einstein: Philosopher Scientist,* pp. 675–676.

deluding himself."[101] And when, just a few weeks before his death, Einstein wrote to Bohr to enlist his support for a public declaration warning the world about the hazards of an atomic arms race, he began with the remark: "Don't frown like that! This has nothing to do with our old controversy on physics, but rather concerns a matter on which we are in complete agreement."[102]

Bohr has written of the "deep and lasting impression" that his discussions with Einstein made on him. How deep and lasting they were was made clear by Abraham Pais, when he described the way in which Bohr would daily relive the struggles that went into the understanding of quantum mechanics. "This," Pais added, "I am convinced, was Bohr's inexhaustible source of identity. Einstein appeared forever as his leading spiritual sparring partner—even after the latter's death he would argue with him as if Einstein were still alive."[103]

ACKNOWLEDGMENTS

This work was supported in part by a grant from the National Science Foundation. Earlier and briefer versions of this article were given as papers to the History of Science Society (Dallas, December 1968) and the American Physical Society (New York City, February 1969). I am grateful to Professors Paul Forman and Roger Stuewer for pointing out a number of relevant documents to me. I want to thank Dr. Otto Nathan, Executor of the Estate of Albert Einstein, for granting me permission to quote from Einstein's unpublished letters, and Miss Helen Dukas for her generous bibliographic assistance in all matters concerning Einstein. I also thank Professor Aage Bohr for permission to quote from unpublished letters of Niels Bohr.

101. A. Einstein to M. Besso, 12 December 1951.
102. A. Einstein to N. Bohr, 2 March 1955. Reprinted in *Einstein on Peace*, ed. O. Nathan and H. Norden (New York, 1960), pp. 629–630.
103. A. Pais, "Reminiscences from the Post-war Years" in *Niels Bohr*, ed. S. Rozental (Amsterdam, 1967), p. 219.

Einstein, Lorentz, and the Electron Theory

BY RUSSELL MC CORMMACH[*]

When Albert Einstein began his career, H. A. Lorentz was Europe's foremost physical theorist.[1] His place was largely a consequence of the electron theory he had elaborated in the years from 1892 on. That theory made a deep impression on Einstein, and the field-theoretical objectives he pursued from an early point in his career were largely formed in the context of his study of Lorentz' work.

Einstein was personally close to Lorentz for nearly twenty years, until the latter's death in 1928, and he was intellectually close for an even longer time. He often tried out his latest ideas on Lorentz, whose opinion he valued above all others. Lorentz' critical grasp of any problem was immediate and deep, giving the impression he had already thought through every new idea. For Einstein, Lorentz was at once the most sympathetic and the most profoundly critical physicist of his day.

Lorentz had another trait Einstein greatly admired: it was his intellectual daring. Late in life Einstein noted that physics had absorbed so much of Lorentz' achievement that its unexpected, audacious character had been largely forgotten.[2] The novelty and continuing good standing of the relativity and quantum theories, both following closely upon the consolidation of the classical electron theory, have obscured the contemporary reputation of Lorentz' work.

* Department of History and Philosophy of Science, University of Pennsylvania, Philadelphia, Penna. 19104.
1. Einstein, "H. A. Lorentz, his Creative Genius and his Personality," *H. A. Lorentz. Impressions of his Life and Work,* ed. G. L. de Haas-Lorentz (Amsterdam, 1957), 5–9, esp. 5.
2. *Ibid.*

It is well known that Einstein and Lorentz disagreed profoundly over quanta and special relativity. In a certain sense, their disagreement is too well known; to correct facile assumptions about Einstein's borrowings from Lorentz, historians and philosophers have tended to stress only their differences. By ignoring an equally profound agreement, Einstein appears to lie outside all tradition, and much of the development of modern physics is scarcely intelligible as a result. It is not that Lorentz exerted a substantial ongoing influence on Einstein's research, but rather that Einstein extracted from Lorentz' electron theory the problem area he made characteristically his own. Lorentz' electron theory played a central role in posing the physics discipline's leading problems at the time Einstein began his career. The incompleteness that Einstein sensed in Lorentz' and others' attempted solutions to these problems did much to shape Einstein's approach to physical theory. The expectation, or vision, of physical reality that motivated his reform of classical theory was uniquely suited to the contemporary needs of classical theory. Even his revolutionary light-quantum hypothesis of 1905 was not intended as a new departure in twentieth-century physics so much as a partial resolution of the basic difficulties of nineteenth-century theory. (His physical vision and corresponding reform strategy for physical theory turned out not to be the most productive for problems stemming from early twentieth-century developments in atomic physics.) In what follows I have traced the relations of the theoretical ideals underlying Einstein's light-quantum hypothesis and his special and general relativity principles to the theoretical problems defined by the electron theory.

1. EINSTEIN'S PRE-1905 WORK

Two of Einstein's student notebooks are preserved in the Princeton Archive.[3] They contain his notes of Heinrich Friedrich Weber's lectures on theoretical physics delivered at the Zürich Polytechnic. Einstein's marginal comments on the lectures are the only contempo-

3. The heading of Notebook 1 is "Kollegium über Physik. Prof. Weber," that of Notebook 2, "Vorlesungen über Physik, Weber." I wish to thank Dr. Otto Nathan, Executor of the Albert Einstein Estate, for his kind permission to quote from the Einstein Collection. And I wish especially to thank Miss Helen Dukas for making these papers available and for her generous help in all bibliographical matters.

rary record of his thinking in his student years, 1896 to 1900. The first notebook contains lessons on the elementary properties of electricity and the liquefaction of gases. (There are only a couple of marginalia in this first book, but one of them is choice, revealing the critical, passionate nature of the notetaker. Beside an offending calculation, Einstein wrote "a true conjurer's trick.") In the middle of the second notebook an entirely new theme is introduced. The lecturer said that until now the "term 'heat' was used only as a sign for a quantity of an unknown something, without reflection on its real nature," an approach that was "fully satisfactory for the development of the phenomena, where heat has no connections with other physical magnitudes." He now intended to take up the "real nature" of heat, and he went on to introduce the kinetic theory. Einstein was clearly fascinated; he went over this section of his notes with critical care, adding a number of remarks in the margin, occasionally penciling in bold question marks. At one place, challenging an assertion Weber made about molecular forces, he wrote "investigate in the vacation," a directive it seems he followed.[4] This notebook strongly suggests that, for Einstein, the mechanical theory of heat was the most important physical subject taught at the Polytechnic.

A glance at Einstein's compulsory subjects at the Polytechnic corroborates his recollection that in his student days mechanics was presented as the core of physics.[5] Mechanics had fascinated him then for its extension into branches of physics, especially the kinetic theory of gases, that were not in any obvious way mechanical.[6] His fascination was excited by his outside reading as well as by his formal classwork at the Polytechnic. His favorite extracurricular authors were Hermann Helmholtz, Ludwig Boltzmann, Heinrich Hertz, James

---

4. In his first publication after leaving the Polytechnic, Einstein derived an expression for the inner energy of a liquid, and from it an expression for the internal work of expansion of the liquid against its molecular forces. Weber said that the latter expression has no theoretical foundation, and it is this assertion that Einstein intended to investigate in the vacation. This at least suggests that Einstein's first published research was in part provoked by his Polytechnic teacher's lectures. I do not mean to suggest that Einstein found Weber an inspiring teacher. He emphatically did not and had very negative things to say both about Weber as a person and as a physicist. All the same Einstein did take an interest in Weber's lectures on the kinetic-molecular theory of heat, and that is what I want to call attention to.

5. Einstein, "Autobiographical Notes," in *Albert Einstein: Philosopher-Scientist,* ed. P. A. Schilpp (Evanston, Ill., 1949), 19.

6. *Ibid.*

Clerk Maxwell, and Gustav Robert Kirchhoff, all fervent exponents of the view that mechanics is the basis of all physical science.[7]

The extension of the domain of mechanics was frequently coupled at this time with a theory of matter premised upon gravitationlike forces between particles. Einstein's first two publications incorporated this conventional theory of matter. The two papers he sent to the *Annalen der Physik* in 1900 and 1902[8] were an attempt to supply chemistry, not visibly a mechanical science, with a mechanical foundation. He distinguished the various chemical substances by the strength of their gravitationlike molecular forces. It was the unifying power of this theory of matter that he responded to, as he explained in a letter to his former classmate Marcel Grossmann in 1901.[9] Referring to his work on the "theory of the power of attraction of atoms," Einstein said that "it is a magnificient feeling to recognize the unity of a complex of phenomena which appear to be things quite apart from the direct visible truth." He looked forward to investigating the question of the "inner relationship of molecular forces to the Newtonian remote forces," and he thought of turning his researches into a doctor's dissertation. There is nothing especially novel about his research program. It points to his immersion in Newtonian-force physics and his acceptance of the long-familiar ideal of the unity of forces.

As early as 1897 his Polytechnic friend Michele Besso had alerted Einstein to Ernst Mach's critique of mechanics, and late in life Einstein recalled that this work had exercised a deep influence on him as a student.[10] He was impressed above all by Mach's scepticism

7. Philipp Frank, *Einstein, His Life and Times,* trans. S. Rosen, ed. and rev. S. Kusaka (New York, 1947), 20.

8. Einstein, "Folgerungen aus den Capillaritätserscheinungen," *Ann. d. Phys., 4* (1901), 513–523, and "Ueber die thermodynamische Theorie der Potentialdifferenz zwischen Metallen und vollständig dissociirten Lösungen ihrer Salze und über eine elektrische Methode zur Erforschung der Molecularkräfte," *Ann. d. Phys., 8* (1902), 798–814.

For an illuminating discussion of Einstein's use of thermodynamics in these and other early papers, see Martin J. Klein, "Thermodynamics in Einstein's Thought," *Science, 157* (1967), 509–516.

9. Einstein to Grossmann, 14 Apr. 1901, in Carl Seelig, *Albert Einstein,* trans. M. Savill (London, 1956), 53; quoted in Klein, "Thermodynamics in Einstein's Thought," *ibid.*

10. Seelig, *ibid.,* 33, and Einstein, "Autobiographical Notes," *op. cit.* (note 5), 21.

Gerald Holton has closely examined Einstein's relation to Mach in "Mach, Einstein, and the Search for Reality," *Daedalus, 97* (1968), 636–673.

toward the claim that mechanics is the basis of all physical science. Yet for several years after leaving the Polytechnic Einstein did not speak of any misgivings that Mach may have inspired in him; rather, he worked on problems falling wholly within the mechanical tradition. It was only after turning to the problem area of the electron theory that he spoke of his intention to seek a new world view superseding that of mechanics.

In the spring of 1901 Einstein told Grossmann of his continuing enthusiasm for molecular-force investigations.[11] He now believed that his theory could be extended from liquid to gas molecules. He did not, however, complete that extension, but instead began studying gas molecules from a different viewpoint. In another letter to Grossmann he said that he had become deeply occupied with Boltzmann's work on the kinetic theory of gases (probably his *Gastheorie,* which did not treat the foundations of statistical mechanics as thoroughly as his articles), and that he had written up a small exercise on the subject.[12] He sent his third paper to the *Annalen* that summer;[13] it was concerned with the kinetic theory of heat, the first fruit of his study of Boltzmann's statistical mechanics. He believed that Boltzmann and Maxwell had failed to show that mechanics is a "sufficient foundation" for the theory of heat, and he intended to fill this lacuna in the foundations of the subject. He showed that the laws of molecular mechanics and the probability calculus are sufficient to derive the laws of heat equilibrium and the second law of thermodynamics. His chief conclusion was that the second law appears as a "necessary consequence of the mechanical world picture." There is the hint in this that he did not have grave doubts about the mechanical world picture. If this is so, he was surely behind the times in 1901; but as he had not yet said anything about his views on light and electricity, there is no way of knowing precisely his opinion of the status of the mechanical world picture.

Einstein published one paper in each of the years 1903[14] and

11. Einstein to Grossmann, 14 Apr. 1901, *op. cit.* (note 9).
12. Einstein to Grossmann, summer 1901, Einstein Collection, Princeton.
13. Einstein, "Kinetische Theorie des Wärmegleichgewichtes und des zweiten Hauptsatzes der Thermodynamik," *Ann. d. Phys., 9* (1902), 417–433. This paper is discussed in Klein, "Thermodynamics in Einstein's Thought," *op. cit.* (note 8).
14. Einstein, "Eine Theorie der Grundlagen der Thermodynamik," *Ann. d. Phys., 11* (1903), 170–187. This and Einstein's 1904 paper are discussed in Klein, "Thermodynamics in Einstein's Thought," *op. cit.* (note 8).

1904,[15] continuing his study of the mechanical foundations of thermo-dynamics. His purpose in 1903 was to free these foundations from certain special assumptions drawn from kinetic theory. His main result in 1904 was a formula for the thermal mean-square fluctuation of the energy of a molecular system about its average energy. He showed that the fluctuation is proportional to Boltzmann's constant, and he accordingly characterized that constant as a measure of the thermal stability of the system. He noted that neither Boltzmann's constant nor the other quantities entering the fluctuation formula called to mind the assumptions underlying the theory. This gave him confidence to apply the formula to the thermal fluctuations of the energy of radiation, in full awareness that a space containing radi-ation cannot be regarded as housing a mechanical molecular system of the kind assumed in the derivation of the formula. He thought that in a space whose linear dimension is of the order of magnitude of a wavelength, the radiant energy fluctuation ought to be of the order of magnitude of the energy itself. Making use of his fluctuation formula and the Stefan-Boltzmann law for the total energy of black-body radiation, he derived the dependence of the linear dimension of that space on the temperature. This dependence was the same as the dependence of wavelength on temperature in the Wien displace-ment law, and the empirical constant entering the Wien law was close to that calculated from fluctuation theory. Einstein did not claim that he had provided a new derivation of Wien's law, but only that he had found a partial confirmation of the predicted fluctuations in energy. This confirmation, however, was enough to persuade him that the molecular-mechanical theory of heat really does apply to pure radiation.

Einstein's 1904 paper on the molecular theory of heat links his several better-known works of 1905. One of these deals with Brown-ian motion, a phenomenon he brought forward as evidence for the thermal fluctuation of particles about their average, thermodynamic behavior.[16] Brownian motion served as a test of the general correct-

15. Einstein, "Zur allgemeinen molekularen Theorie der Wärme," *Ann. d. Phys.,* *14* (1904), 354–362.

16. Einstein, "On the Movement of Small Particles Suspended in a Stationary Liquid Required by the Molecular-Kinetic Theory of Heat," *Ann. d. Phys.,* *17* (1905), 549–560, trans. R. Fürth in Einstein, *Investigations on the Theory of the Brownian Movement* (London, 1926), 1–18.

ness of his molecular heat theory, and it provided a way of determining atomic magnitudes. Einstein's other 1905 papers are concerned with quanta and relativity. In his 1904 article he said he had uncovered a "connection of the highest interest between the universal constants determined by the elementary quanta of matter and electricity and the order of magnitude of a wavelength of radiation without the help of special hypotheses."[17] One reason why this connection had the "highest interest" for him is that it pointed to a theoretical unity of molecular-mechanical and electromagnetic phenomena. To enlarge upon this unity was the express purpose of his next year's light-quantum and relativity studies. To understand why Einstein should have welcomed a connection between mechanics and electromagnetism, it is necessary to examine the place of electron theory in physics at the time.

2. LORENTZ' ELECTRON THEORY

Lorentz' was the most influential electron theory at the time Einstein began his career. There were other versions of the electron theory—notably Emil Wiechert's and Joseph Larmor's—but Einstein was not drawn to them. He never doubted that Lorentz' theory was the only one worth taking seriously. Before entering into his attempts at reforming Lorentz' theory, I will sketch the chief physical ideas of the theory and suggest the promise it held at the turn of the century.

Lorentz' great achievement was to unite the particulate view of electricity with Maxwell's ether-borne, contiguously acting electromagnetic forces. This achievement was founded on Lorentz' novel conception of the electromagnetic ether. The distinguishing features of this ether were that it occupies the same space as a charged particle and that its properties are in no way changed by its coextension with the particle. The ether completely permeates uncharged matter, too, and it is not displaced when material bodies move through it. The sole connection between ether and matter occurs through the spherical charged particles, or electrons, that are contained in ponderable molecules. The displacement of one charged particle produces a

17. Einstein, "Zur allgemeinen molekularen Theorie der Wärme," *op. cit.* (note 15), 354.

change in the state of the ether, and this change is propagated outward at the speed of light, influencing a second particle at a later time. Lorentz' ether is clearly not an ordinary mechanical substance, for it has no mechanical linkage with ponderable matter.

The entire mathematical basis of Lorentz' theory consists of five equations. The first four completely characterize the state of the ether and constitute Maxwell's electromagnetic theory of light. The fifth expresses the force of the ether on charged particles acting in the space occupied by the particles; it is Lorentz' own contribution, his fundamental law. The theory is founded on two sorts of mathematical expressions: partial differential equations relating the field to its particulate sources, and ordinary differential equations determining the motion of the charged particles under the influence of the field. Particle and field concepts appear side by side, and their irreducibility constitutes the essential dualism of Lorentz' theory.

The Lorentz force has two parts: an electric component that acts on a charge at rest in the ether (in a given state) and an additional electrodynamic component that acts on a charge only when it moves through the ether (in a given state) with a determinate velocity. I will not attempt to characterize the state of electrodynamics prior to Lorentz' theory beyond remarking that for nearly half a century electrodynamics involved motion-dependent forces, and that the principal Continental workers in the subject saw the outstanding problem as the need for eliminating such forces in favor of the motion-independent Coulomb force or potential. Lorentz now proceeded to ignore this tradition to found a theory upon a motion-dependent force, a force demanded by his notion of a stationary electromagnetic ether.

From the inception of his theory in 1892 to the end of the century,[18] Lorentz was primarily concerned to find equations for the motion of light in macroscopic dielectrics by averaging over the fields of numerous electrons. With the aid of a "local-time" variable, he

18. Lorentz' two principal statements of electron theory in this period are "La théorie électromagnétique de Maxwell et son application aux corps mouvants," *Arch. néerl.*, 25 (1892), 363, in *Collected Papers*, 2, 164–343, and *Versuch einer Theorie der electrischen und optischen Erscheinungen in bewegten Körpern* (Leiden, 1895), in *Collected Papers*, 5, 1–137.

The principal historical study of the formation of Lorentz' electron theory is Tetu Hirosige's "Origins of Lorentz' Theory of Electrons and the Concept of the Electromagnetic Field," *Historical Studies in the Physical Sciences, 1* (1969), 151–209.

explained the several positive effects of the motion of dielectric bodies through the ether: the aberration of light, the Doppler effect, and the Fizeau experiment. Especially pressing was the problem of reconciling a stationary ether with the repeated failure to detect optical and electric influences of the motion of the earth through the ether. For this purpose Lorentz defined transformations of the field variables which, together with his local-time transformation, preserved the form of Maxwell's equations when the latter were referred to moving coordinate frames. To account for the null-measurements of the Michelson-Morley experiment on the motion of the earth through the ether—measurements accurate to second-order quantities in the ratio of the earth's velocity to the velocity of light—he required an accessory hypothesis: he assumed that molecular forces propagate through the ether and that they transform exactly as the electric force, with the consequence that the dimensions of ponderable bodies contract in the direction of motion by the requisite second-order quantities. It follows that, to second-order quantities, the measured velocity of light is independent of the motion of the observer through the ether and so cannot serve as a means for detecting that motion.

Following the discovery of the empirical electron shortly before the turn of the century, Lorentz shifted his primary concern from the optics of macroscopic bodies to the mechanics of individual electrons. There were now means for directly testing the mechanics of high-speed electrons, and the electron theory became an instrument for the reform of molecular mechanics. Lorentz looked to this reformed mechanics as the cornerstone of an electromagnetic view of nature.

From 1899 to 1904,[19] Lorentz unified and simplified his theory by interpreting his dimensional contraction formula as a universal coordinate transformation. This interpretation entailed a number of departures from conventional dynamical concepts. First, Lorentz asserted that the masses of all particles, charged or not, vary with

19. See especially Lorentz, "Théorie simplifiée des phénomènes électriques et optiques dans des corps en mouvement," *Versl. Kon. Akad. Wetensch. Amsterdam,* 7 (1899), 507, in *Collected Papers, 5,* 139–155, and "Electromagnetic Phenomena in a System Moving with any Velocity Smaller than That of Light," *Versl. Kon. Akad. Wetensch. Amsterdam, 12* (1904), 986, in H. A. Lorentz, A. Einstein, H. Minkowski, and H. Weyl, *The Principle of Relativity* (New York, 1923), 11–34.

For recent discussions of these two papers see Hirosige, "Electrodynamics before the Theory of Relativity, 1890–1905," *Japanese Studies in the History of Science,* No. 5 (1966), 1–49, and Kenneth F. Schaffner, "The Lorentz Electron Theory and Relativity," *Amer. Journ. Phys., 37* (1969), 498–513.

motion, and that they vary in the same way. Second, he asserted that the mass of an electron is due to the electromagnetic self-reaction of the electron with its field, and that the mass has no constant, Newtonian component. Third, he supposed that a moving electron undergoes a physical deformation arising from the motion itself, and that all nonelectric molecular forces, whether between a ponderable particle and an electron or between two ponderable particles, are influenced by motion in the same way as the electrostatic force between electrons. Finally, he supposed that there exists in the velocity of light an upper limit to the velocity of all particles moving through the ether. Clearly, then, to accept Lorentz' electron theory was to accept the elements of a testable, non-Newtonian molecular mechanics. The theory originally took account only of electrons, but it came to transcend this restriction and make assertions about the physics of all matter. In problems in traditional mechanics the shapes, masses, and forces of bodies could be specified independently of their motion; in electron theory they had meaning only when expressed relatively to the motion of bodies through the stationary ether. The mechanical properties of matter were dependent upon the properties of the electromagnetic ether, intimating a unified physics of light and matter.

By means of his electron theory Lorentz resolved a number of difficult problems, e.g., normal and anomalous dispersion, the Zeeman effect, and Faraday's rotation of light. These specific accomplishments lent his work immense authority, and this authority enhanced the promise of electron theory as a theory of potentially universal scope.[20] At the turn of the century Lorentz,[21] Wilhelm Wien, Max Abraham,[22] and others indicated routes for deriving the laws of classical mechanics as special cases of the equations of electron theory. Their object was to eliminate mechanics as a separate domain of physics. All of molecular mechanics was to be contained within the

20. For a study of the place of electron theory in physics at this time see Russell McCormmach, "H. A. Lorentz and the Electromagnetic View of Nature," *Isis* (winter, 1970).

21. Lorentz, "Contributions to the Theory of Electrons," *Proc. Roy. Acad. Amsterdam, 5* (1903), 608, in *Collected Papers, 3,* 132–154.

22. W. Wien, "Ueber die Möglichkeit einer elektromagnetischen Begründung der Mechanik," *Arch. néerl., 5* (1900), 96–104, reprinted in *Ann. d. Phys., 5* (1901), 501–513; M. Abraham, "Prinzipien der Dynamik des Elektrons," *Ann. d. Phys., 10* (1903), 105–179.

laws of the continuous field, and the dualism in the foundations of physical theory was thereby to be removed. Lorentz expanded on the topic of the central place of electron theory in physical science, pointing to the probable incorporation into electron theory of spectroscopy, atomic structure, and chemistry;[23] and he devoted a considerable body of work to providing the thermodynamics of radiation with an electron-theoretical derivation.[24] And he showed that if all matter were constructed of charged particles the law of gravitation could be placed on electromagnetic foundations.[25] He foresaw a universe composed solely of ether and charged particles; he also recognized that this ultimately simple universe was not certain and that the search for its laws was at best only a heuristic program for physics.

The concept of a deformable electron proved to be an obstacle to the realization of this program. By 1904 Lorentz, following Abraham, recognized that such an electron contains an associated energy of deformation and that this energy is necessarily nonelectromagnetic.[26] A closely related diffculty stemmed from the requirement that to avoid infinite self-energies, the electron has to be finite in extent. To maintain a finite structure, there must be forces or rigid constraints that are themselves of nonelectromagnetic origin. These difficulties suggested that physics could not be reduced solely to electromagnetic constituents, at least not without a thoroughgoing alteration of the basis of electron theory. There was another formidable difficulty, one

23. Lorentz, "Elektromagnetische Theorien physikalischer Erscheinungen," *Phys. Zeit.*, *1* (1900), 498 and 514, in *Collected Papers, 8,* 333–352.

24. Lorentz, "The Theory of Radiation and the Second Law of Thermodynamics," *Versl. Kon. Akad. Wetensch. Amsterdam, 9* (1900), 418, in *Collected Papers, 6,* 265–279, and "Boltzmann's and Wien's Laws of Radiation," *Versl. Kon. Akad. Wetensch. Amsterdam, 9* (1901), 572, in *Collected Papers, 6,* 280–292.

25. Lorentz, "Considérations sur la pesanteur," *Versl. Kon. Akad. Wetensch. Amsterdam, 8* (1900), 603, in *Collected Papers, 5,* 198–215.

26. Abraham had observed that a moving deformed spherical electron with a surface charge distribution has an internal energy less than the internal energy of the electron at rest by an amount $1/6 \cdot c^2 \, \varepsilon^2 \, / \, R \, \{1 - (1 - \beta^2)\}$, where $\beta = v/c$, and v is the velocity of the electron, c the velocity of light, ε the charge, and R the electron radius. Lorentz accepted this calculation as a necessary and extremely important consequence of his theory. He said that should this change in internal energy not be found in experience, then the whole attempt to prove the complete independence of phenomena, such as Michelson's interference pattern, from motion through the ether has failed. Lorentz discussed this critical question in "Ergebnisse und Probleme der Elektronentheorie," in *Elektrotechn. Verein zu Berlin, 1904* (Berlin, 1905), in *Collected Papers, 8,* 76–124.

that arose in connection with one of the applications of electron theory. In 1903 Lorentz attempted to derive the formula for the energy density of blackbody radiation, employing the equipartition theorem and a mechanism borrowed from Paul Drude's 1900 electron theory of metals.[27] He was able to find only the long-wavelength limit of the energy spectrum, and he knew of no way to generalize his calculation. At the same time he recognized that Max Planck's 1900 quantum theory of blackbody radiation had succeeded where his electron theory had not and that Planck's energy quanta seemed irreconcilable with the assumptions of electron theory.

3. LIGHT QUANTA, RELATIVITY, AND THE ELECTRON THEORY

Although Einstein had not published on radiation before his statistical mechanical discussion of Wien's law in 1904, he had long thought about the fundamental problems of radiation. At age sixteen he had puzzled over the question of what light would look like to an observer moving with it.[28] And one of the extramechanical applications of mechanics that had most fascinated him in his student days was the theory of light as a wave motion in a quasi-rigid elastic ether.[29] In the same letter in 1901 to Grossmann in which he spoke of reading in Boltzmann's kinetic theory, he referred to his extensive, concurrent work on the principal difficulty of the light ether theory:

On the investigation of the motion of matter relative to the light ether a considerably simpler method has occurred to me again, which rests on the customary interference experiments. If inexorable fate only once would give me the needed time and peace to carry it through! If we see each other again, I will report to you on it.[30]

Einstein was clearly thinking about either Fizeau's 1859 interference measurements on light passing through moving water or—and this is more likely—Michelson's 1881 or Michelson and Morley's 1887 attempts to detect the earth's motion through the ether by interferometer methods. (Einstein's own testimony is conflicting as to when

27. Lorentz, "On the Emission and Absorption by Metals of Rays of Heat of Great Wave-Lengths," *Versl. Kon. Akad. Wetensch. Amsterdam, 11* (1903), 729, in *Collected Papers, 3,* 155–176.
28. Einstein, "Autobiographical Notes," *op. cit.* (note 5), 53.
29. *Ibid.,* 19.
30. Einstein to Grossmann, summer 1901, *op. cit.* (note 12).

he learned of the Michelson and the Michelson and Morley experiments.) He did not say which view of the ether he took most seriously, the view that it was a mechanical body or the view that it was a nonmechanical body. The former is more likely, given the nature of his other work and his fascination with the world-view implications of mechanics. It might be well to point out that he probably learned Maxwell's electromagnetic theory of light from the German translation of Maxwell and from the German texts of August Föppl[31] and Boltzmann;[32] both expositors, like Maxwell, took pains to present electromagnetic field theory as an application of mechanics. It is clear that Einstein was working desperately hard on the light ether problem in 1901 and that he had discussed it before with his classmate. In 1905 he dismissed the need for an ether in two sentences and merely mentioned in passing that attempts to detect the absolute motion of the earth had failed. These remarks give no hint of the years of effort he had devoted to the problem of motion relative to the ether. In 1905 he wrote the word ether in quotation marks; he did not in his 1901 reference, and there is no reason to think he disbelieved in the ether then.

The ether Einstein wrote about in 1905 was the stationary ether of the electron theory. The electron theory had not been taught at the Polytechnic, and Einstein had had to instruct himself in it. His earliest mention of it occurs in a letter to Besso at the beginning of 1903,[33] in which he said that after attacking the problem of molecular forces in gases he planned to make "comprehensive studies of the electron theory." It was no doubt then that he began poring over Lorentz' 1892 and 1895 treatises.[34] His critical appreciation of Lor-

---

31. August Föppl, *Einführung in die Maxwellsche Theorie der Elektrizität* (Leipzig, 1894).

Holton has suggested cogent reasons for Einstein's attraction to Föppl's text in "Influences on Einstein's Early Work in Relativity Theory," *American Scholar,* 37 (1967), 59–79.

32. Ludwig Boltzmann, *Vorlesungen über Maxwells Theorie der Elektricität und des Lichtes,* 2 vols. (Leipzig, 1891 and 1893).

33. Einstein to Besso, Jan. 1903, probably Jan. 26/27, Einstein Collection, Princeton.

34. In 1905, when he first published directly on the electron theory, Einstein was familiar with Lorentz' 1892 and 1895 works, but not with his later ones (Holton, "On the Origins of the Special Theory of Relativity," *Amer. Journ. Phys., 28* [1960], 627–636, esp. 635, ref. 33). It might seem strange that he did not follow Lorentz' work, given his admiration for it. However, it would have been difficult for Einstein at the patent office in Berne to obtain the *Proceedings of the Amsterdam Academy* where Lorentz' papers appeared (Holton, "Influences on Einstein's Early Work in Relativity Theory," *op. cit.* [note 31], 68–69).

entz' work would have been sharpened by his close reading of Poincaré's *Science and Hypothesis* that same year.[35] Poincaré generously praised Lorentz' theory as the most satisfactory theory in physics, and at the same time he was critical of the disharmony between its concepts and those of mechanics.

Except perhaps for the implicit reference to electron theory in the last section of his 1904 paper on fluctuation phenomena, Einstein's 1905 paper[36] on light quanta was the first public indication of his reading in the literature of the electron theory. He referred to Drude's application of electron theory to the theory of metals,[37] an application that would naturally have attracted his interest. Drude had united the electron theory with statistical mechanics, the subject Einstein had devoted himself to for the past three years. Einstein would have been interested in any case, since the electron theory of metals was then widely regarded as one of the most fruitful domains of application of the electron theory; Lorentz felt this way, for instance.[38] Einstein took special note of Drude's application of the statistical equipartition theorem to an assembly of molecules and electrons. He in turn applied the theorem in another physical context, one involving radiation as well as electrons and molecules. His purpose was to show that the accepted electron theory could not solve the problem of blackbody radiation. He imagined a space containing free electrons, gas molecules, and radiation. The space is enclosed by reflecting walls which themselves contain bound electrons, or "oscillators," capable of emitting and absorbing radiation. The equipartition theorem asserts that the average kinetic energy of a gas molecule, which is known from the kinetic theory of gases, equals the average kinetic energy of an oscillator electron. Einstein

35. In 1903 Conrad Habicht produced the "long-awaited" *Science and Hypothesis* for the small discussion group which Einstein belonged to (Seelig, *Albert Einstein, op. cit.* [note 9], 61).

36. Einstein, "Über einen die Erzeugung und Verwandlung des Lichtes betreffenden heuristischen Gesichtspunkt," translated by A. B. Arons and M. B. Peppard as "Einstein's Proposal of the Photon Concept—a Translation of the *Annalen der Physik* paper of 1905," *Amer. Journ. Phys., 33* (1965), 367–374.

See Klein's studies of Einstein's 1905 light-quantum hypothesis: "Einstein's First Paper on Quanta," *Natural Philosopher, 2* (1963), 59–86, and "Thermodynamics in Einstein's Thought," *op. cit.* (note 8).

37. Paul Drude, "Zur Elektronentheorie der Metalle," *Ann. d. Phys., 1* (1900), 566–613, and *3* (1900), 369–402.

38. Lorentz, "Ergebnisse und Probleme der Elektronentheorie," *op. cit.* (note 26).

related the equipartition energy of an oscillator electron to the radiant energy per unit volume of the cavity at a given frequency, using for this purpose an expression Planck had derived in 1900. The resulting formula for the energy density of blackbody radiation did not agree with experience. To clinch the argument that the accepted theory was in error, Einstein observed that his formula for the energy density implies that blackbody radiation has an infinite total energy. This means that radiation accounts for all of the energy, the electrons and molecules none. Einstein concluded that "in our model there can be no talk of a definite energy distribution between *ether* and matter" (my italics). Others had already shown this—e.g., Rayleigh in 1900—but Einstein did not know this at the time. He did not know either of Lorentz' 1903 paper on the radiation from metals, which would have given him additional support for his views.

It is a measure of Einstein's confidence in statistical mechanics that he did not single out the equipartition theorem as the source of the error. Others would probably have done so, since the status of the equipartition theorem was generally thought to be shaky. Einstein may well have believed that statistical mechanics was in difficulty, but the flaw in the electron-theoretical model that he drew attention to was its interpretation of light as a spatially continuous phenomenon. He did not question statistical mechanics, but confidently applied it to pure radiation, advocating on the basis of it a "very revolutionary" revision of electromagnetic theory (Einstein's words).[39]

The use of statistical mechanics in electron theory had yielded some promising results, for instance Drude's derivation of the ratio of thermal and electrical conductivities in metals. But the same principles failed to account for blackbody radiation in which light as well as electrons and molecules is involved. In Einstein's opinion this did not mean that statistical mechanics was inapplicable but that a new theory of light was called for. The equipartition theorem, which denied the possibility of an equilibrium energy distribution between ether and matter, did not deny the same possibility between light and matter if light, like discrete electrons and molecules but unlike the continuous ether, had only a finite number of degrees of freedom.

39. Einstein to Habicht, 1905, undated, in Seelig, *Albert Einstein, op. cit.* (note 9), 74, quoted in Klein, "Einstein's First Paper on Quanta," *op. cit.* (note 36), 59.

Einstein related his 1905 findings to a problem more basic than that of blackbody radiation. There was some recognition at the time that mechanics and electromagnetism had grown completely separate and that this was a highly unsatisfactory state of affairs. Wien, for example, gave this reason as the motivation for his 1900 paper on the electromagnetic foundations of mechanics, the earliest explicit statement of an electromagnetic program for physics (note 22, above). The light-quantum hypothesis was part of Einstein's response to this sense of a need for a unified foundation for physics. He objected to the "profound formal distinction" that existed between the concepts of the kinetic theory of gases and those of Maxwell's electromagnetic theory. The mathematical language of physics was not one but two languages, while physical reality was one. Einstein argued that the present physical concepts and the corollary mathematics could never lead to the unity that physicists sought. He called attention to the fact that the energy of a ponderable body is expressed by a discrete sum over the finite energies of individual atoms and electrons, while the energy of the electromagnetic field is expressed by a continuous spatial function. In his opinion it was the formalism of field physics, not that of particle physics, that was most in need of rethinking. The unwanted dualism could be removed by recognizing that the "theory of light which operates with continuous spatial functions may lead to contradictions with experience." He strongly hinted that the continuous ether of light, which in fact operates with continuous spatial functions, has to be eliminated, and he nearly said so, declaring that light is a "discontinuous medium." He was careful to emphasize that his arguments for the discontinuous structure of radiation were valid for light of high frequencies only, light for which Wien's distribution law is valid. Yet he presented the implications of this discontinuity for the accepted electromagnetic theory without any qualifications. The wording of the introduction of his 1905 paper suggests that he thought the formalism of electromagnetic theory would not be the formalism of field theory. By questioning the adequacy of a theory of light expressed in terms of continuous spatial functions, he seems to have anticipated a new electromagnetic theory modeled after the mechanical-molecular theory of gases. In the new theory, the wavelike property of light would apparently be an effect of the average dynamical behavior of aggregates of energy quanta. The analogy was of the relation of continuous thermodynamic functions to the average dynamical behavior of aggregates of material particles. Einstein offered an analysis of blackbody radiation to support his claim that light has discrete characteristics.

Einstein's analysis made use of the thermodynamics of radiation and Boltzmann's interpretation of entropy as the probability of state of a molecular system. He showed that the formal disparity between the theory of the continuous field and the dynamics of discontinuous matter does not exist in the molecular-thermodynamic description of light and matter. The entropy of radiation in an enclosure has the same formal dependence upon volume as does the entropy of an enclosed ideal gas. Because the entropy formula for an ideal gas depends upon the assumption that a gas consists of a finite number of independent, permanent point-particles, Einstein concluded that radiation in the high-frequency region behaves thermodynamically as though it too consists of a "finite number of [independent] energy quanta which are localized at points in space, which move without dividing, and which can only be produced and absorbed as complete units."

If light is really thus, then a new electromagnetic theory is needed to describe it. Einstein did not have that theory, and he spoke of light quanta as having heuristic value only. He did no more than hint at the nature of the future theory, a theory in which, presumably, the energy of light would no longer be expressed by continuous spatial functions. He seemed to imply that partial differential equations for the motion of light would be ruled out; for with such equations, a "finite number of parameters cannot be regarded as sufficient for the complete determination" of an electromagnetic state, and a finite number is required if an equilibrium distribution of energy between light and matter is to take place. The appropriate mathematics is, evidently, the ordinary differential equations of molecular mechanics. One likely reason why Einstein's light-quantum hypothesis received so little early support is that many physicists at the time were convinced that the route to the unification of physics was by means of continuous-field theory. Einstein's proposal looked like a throwback to a time when particle mechanics was the model for all physics.

Although Einstein did not refer explicitly to Lorentz' electron

57

theory, his light-quantum paper was in fact addressed to its major difficulty. The distinguishing feature of Lorentz' theory was its bifurcation of the electromagnetic world into two and only two entities, discrete electrons and the continuous ether. The unparalleled clarity of this theory appealed to Einstein, and it also prepared the way for the theory's downfall. The two halves of physics were discordant; Maxwell's equations of the continuous field, when combined with the mechanics of discrete particles, led to the wrong law of blackbody radiation, and the equations also led to difficulties in understanding emission-absorption phenomena, e.g., the production of photoelectrons. Einstein was convinced by 1905 that the practice, exemplified by Lorentz' theory, of admitting side by side as equally primitive concepts the discrete point-masses of mechanics and the continuous field represented an unsatisfactory stage in the development of theory, one which eventually had to go. He was never to lose this conviction, but he would soon reverse the prescription for the ills of a bifurcated physics that he had hinted at in his 1905 remarks; he would soon argue for partial differential equations and continuous functions as the unifying language of physics.

Einstein had long attempted to correct the faults of Lorentz' electron theory by direct, constructive approaches.[40] But by 1905 he had come to see that to succeed he must proceed indirectly, by means of some universal principle, and the model he had before him was thermodynamics.[41] He characterized thermodynamics as a "theory of principle," one based upon statements such as that of the impossibility of perpetual motion. He contrasted thermodynamics with the more common "constructive theory" built up from hypothetical statements;[42] Lorentz' electron theory was a theory of this latter type, as was the kinetic theory of gases. In 1905 Einstein refounded Lorentz' electron theory on a theory of principle, based on two universal postulates, the electromagnetic analogues of the laws of thermo-

40. Einstein, "Autobiographical Notes," *op. cit.* (note 5), 46 and 53.
41. *Ibid.*, 53.
42. Einstein, "What is the Theory of Relativity?" *London Times*, 28 Nov. 1919, reprinted in Einstein, *Ideas and Opinions*, ed. C. Seelig, trans. S. Bargmann (New York, 1954), 227–232, esp. 228.

Einstein's views on the nature of physical theory are discussed in Klein, "Thermodynamics in Einstein's Thought," *op. cit.* (note 8), and in Holton, "On the Origins of the Special Theory of Relativity," *op. cit.* (note 34).

dynamics. His postulates were kinematic, not electromagnetic, and so they applied equally to electrodynamics and mechanics, helping to formally unify these branches of physics. The first postulate, or the "principle of relativity," stipulated that the "same laws of electrodynamics and optics will be valid for all frames of reference for which the equations of mechanics hold good." The second postulate stated that light always moves with the same velocity in free space, regardless of the motion of the source. The principal experimental support for the first postulate was the "unsuccessful attempts [such as Michelson's] to discover any motion of the earth relatively to the 'light medium'." The second postulate preserved the one useful property of the ether: the velocity of light is independent of the velocity of the source. It was this postulate that physicists who desired a true emission theory found so difficult to accept; Ritz and others argued that in a true emission theory the velocity of light should depend on the velocity of the source, just as the velocity of a projectile does.

There was another basic reason not directly related to experiment that called for the relativity postulate. This had to do with Einstein's strong feeling for the internal perfection of theories. He felt that the accepted electrodynamics was markedly imperfect, and he illustrated the source of his dissatisfaction with an example. When a conductor is moved near a magnet at rest, a current is induced, and when a magnet is moved near a conductor at rest, a current is again induced; and the current is the same as long as the relative motion of the conductor and magnet is the same. Now although the effects are the same, the physical explanation in the two cases is different; and that is what troubled Einstein. According to the accepted theory, an electric field is produced in the surrounding space when the magnet is moved, and an energy is associated with this field. But when the conductor is moved, there is no electric field; instead there is an electromotive force in the space of the conductor, and, moreover, no energy is associated with this force. This electromotive force was proportional to the vector product of the velocity of the conductor through the stationary ether and the magnetic force. It was the so-called Lorentz force. Late in life Einstein recalled that what had led him directly to the principle of relativity was the conviction that the Lorentz force acting on a charged body moving in a magnetic field was "nothing else but an electric-field,"[43] a conviction very likely stemming from his prior doubts about the ether; without the ether the Lorentz force loses its fundamental status and needs reinterpretation as an electric field. The conviction that the Lorentz force was merely an electric field enabled him to resolve the asymmetry of magnet-conductor interaction, for it denied our ability to say whether it is the magnet or the conductor which is really in motion. The general expression of this conviction was the relativity postulate, which eliminated all unesthetic asymmetries endemic to ether-based electromagnetic theories.

Einstein brought his relativity ideas to fruition only shortly after writing out his ideas on light quanta. The timing was not a matter of coincidence, for the two sets of ideas were intimately related. He wrote his special theory of relativity in full awareness of its relation to his prior work on light quanta, a fact that is pointed up by his later remark that it was especially important for him that his relativistic kinematics did not depend on a knowledge of Maxwell's equations, since his study of blackbody radiation had convinced him that Maxwell's theory was only approximately true.[44] In his light-quantum paper he had concluded that, on the microscopic scale, the energy of light could not be described by continuous spatial functions, which would rule out the possibility of a continuous ether. His subsequent study of electrodynamics reinforced his doubts about the ether; he now declared that the ether was a "superfluous" concept in physics. The ether question was critical. Absolute translatory motions and continuous spatial energy distributions are unknown in particle mechanics, while they appear necessarily in electromagnetic theory if a stationary, continuous ether is presupposed. By discarding the ether the specialness of electromagnetic concepts can be partially eliminated, and the relativity and particle concepts of mechanics can be applied to the electromagnetic field as well. It seemed clear to other physicists at the time that Einstein's relativity principle favored mechanical over electromagnetic concepts. At the 1906 meeting of German Natural Scientists and Physicians, Abraham, W. Kaufmann, and A. Sommerfeld argued that relativity was

43. R. S. Shankland, "Michelson-Morley Experiment," *Amer. Journ. Phys.*, 32 (1964), 35.
44. M. Born, *Fünfzig Jahre Relativitäts-theorie*, ed. A. Mercier and M. Kervaire (Berne, 1955), 248.

merely a principle of mechanics. Believing that electromagnetic, not mechanical, concepts were those on which the future physics should be built, they looked upon Einstein's relativity postulate as a reactionary step.[45] By introducing mechanical concepts into electromagnetic theory in both his light-quantum and relativity studies, Einstein really did appear set on a reactionary course in 1905. His disinclination to grant an unquestioned precedence to electromagnetic concepts may be due in part to the fact that he had come to electromagnetism from the kinetic-molecular theory of heat and had worked solely with mechanical concepts up to that time.

As a student Einstein had been intrigued by the mechanical view of nature, according to which mechanics and electromagnetism were not the separate sciences they seemed to be in 1905. But by the time he began working on the electron theory, the program of reducing electromagnetic processes to motions in a mechanical ether had been pretty well abandoned in Europe. The prevalent alternatives to the mechanical program for unifying physics were the energetic and electromagnetic programs. In his student days Einstein had admired the chemistry of Wilhelm Ostwald, an energeticist. Ostwald denied the reality of an ether, asserting that radiation was energy existing independently in space; and he suggested, too, that mass was a secondary phenomenon of energy. These sound much like ideas Einstein was later to develop. Yet Einstein seems to have been little attracted to the energeticists' modes of thinking. Energetics was undoubtedly important for him, but not so much for its specific aims as for the locus of its concern. Energetics was an expression of the concern of physical scientists at the turn of the century with the unification and fundamentals of their field. The physics Einstein was nurtured in was permeated with a desire for unity, a desire he came fully to share.

Einstein was probably more sympathetic with the specific aims of the electromagnetic program than he was with those of energetics. By 1905, however, he had recognized that the electromagnetic program was inadequate, at least in its usual formulations. He could

45. See the discussion following Planck's talk in *Phys. Zeit.*, 7 (1906), 760–761. Hirosige discusses this aspect of the 1906 meeting in his important study of the fate of the ether after 1905, in "Theory of Relativity and the Ether," *Japanese Studies in the History of Science*, No. 7 (1968), 37–53, esp. 48–49.

not follow the reductionist strategy of referring all physics to Maxwell's equations and the ether, since he had found arguments for the inexactness of Maxwell's equations and the nonexistence of the ether. Although he fully endorsed the unifying motivation of the mechanical and electromagnetic world views, he could not subscribe to the program of either. He was forced to devise a research direction of his own.

Einstein recognized that not only electromagnetic concepts, but the mass and kinetic energy concepts of mechanics, too, had to be changed. Entirely in keeping with his goal of finding common concepts for mechanics and electromagnetism, he deduced from the electron theory elements of a revised mechanics. In his 1905 paper he showed that all mass, charged or otherwise, varies with motion and satisfies the formulas he derived for the longitudinal and transverse masses of the electron. He also found a new kinetic energy formula applying to electrons and molecules alike. And he argued that no particle, charged or uncharged, can travel at a speed greater than that of light, since otherwise its kinetic energy becomes infinite. He first derived these non-Newtonian mechanical conclusions for electrons only. He extended them from electrons to material particles on the ground that any material particle can be turned into an electron by the addition of a charge *"no matter how small."* It is curious to speak of adding an indefinitely small charge, since the charge of an electron is finite. Einstein could speak this way because he was concerned solely with the *"electromagnetic basis* of the Lorentzian electrodynamics and optics of moving bodies" (italics added). That basis was precisely Maxwell's equations with convection currents. Einstein recognized that the electron charge is foreign to Maxwell's equations, since these equations have nothing to say about why the charge has one value rather than another. He characterized an electron as a rigid body, with no further explanation of its stable structure. Again the reason is that the electromagnetic basis has nothing to say about what prevents an electron from bursting apart through the electrostatic repulsion of its parts. He needed to assume that the electron does in fact have parts, for the electromagnetic basis attributes infinite self-energy to point-electrons. Einstein later recalled that the need to assume that electrons are finite bodies and not point-particles underscored for him the unnaturalness of Lorentz' joint use of the ordinary differential equations of mass-points and the partial differential equations of the field.[46]

It might seem strange that Einstein did not mention light quanta in his relativity paper. But it would have been out of place to do so, since light quanta, like electrons, are not consequences of the electromagnetic basis. His light-quantum paper charted the limits of the accepted electromagnetic theory in its application to pure radiation; likewise, his relativity paper was concerned with the limits of that theory in its application to electricity. He recognized that light quanta are compatible with, though not consequences of, Maxwell's equations, if the latter are recognized as time-average descriptions of light-quantum behavior. Likewise, electrons are compatible with, though not consequences of, Maxwell's equations and the relativity principle; he showed this as follows. He added the customary convection current $\mathbf{u}\rho$ to Maxwell's equations, where ρ is the charge density and $\mathbf{u}$ is the velocity of the charge. He showed that the resulting equations have the same form in all material systems provided that $\mathbf{u}$ transforms as a velocity, as it must if Lorentz' basic assumption is correct that an electric current is the motion of electric particles. Einstein then showed that the invariance of Lorentz' equations entails a law of transformation of ρ, one differing from Lorentz', and he remarked on the "important law" that follows from it and from his transformation for $\mathbf{u}$: the total charge of a body is the same whether measured in a stationary or in a moving frame. This suggests that the charge of an electron, like the velocity of light, is a universal constant. The constancy of charge and the identification of $\mathbf{u}$ as a velocity are consequences of Einstein's theory, whereas in Lorentz' they are assumptions. The major significance is that Maxwell's equations are seen to imply certain properties of charged matter, another step toward the realization of the unity of field and particle. A related significance is that Einstein's formulation of electron theory has fewer *ad hoc* elements than Lorentz', an esthetic-epistemological consideration that was important to Einstein.

Writing to his close friend Conrad Habicht in 1905, Einstein said he would send four works in return for Habicht's thesis.[47] One con-

46. Einstein, "Physics and Reality," *Journ. Franklin Inst., 221* (1936), 349–382, reprinted in Einstein, *Ideas and Opinions, op. cit.* (note 42), 290–323, esp. 306.

47. Einstein to Habicht, 1905, undated, in Seelig, *Albert Einstein, op. cit.* (note 9), 74–75.

tained the germ of his relativity paper, and the rest were his finished papers on light quanta, Brownian motion, and the determination of atomic dimensions. It was an extraordinary catalogue of works, all falling within a single year, a year that was not yet over. Writing to Habicht a few months later, Einstein said that "there is not always a subtle theme to meditate upon[!] At least, not an exciting one. There is, of course, the theme of spectral lines, but I do not think that a simple connection of these phenomena with those already explored exists; so that for the moment the thing does not seem to show very much promise."[48] (This reference to spectra may be related to Einstein's immediate admiration[49] for Niels Bohr's work in 1913 and to his remark to Hevesy that year that long before he had had ideas similar to Bohr's but had not got far with them.[50]) He did have one new result to report, but wondered "whether the Good Lord does not laugh at it." The new result was the subject of a brief paper he published in the *Annalen* at the end of 1905;[51] here he showed that the mass (more precisely, inertia) of a body is a measure of its energy-content. This conclusion was based on his earlier "remarkable" discovery that the energy of light varies with the motion of the observer in the same way that frequency varies in the Doppler shift, the point where his relativistic electrodynamics implicitly supported his light-quantum hypothesis (according to which the quantum of energy is proportional to the frequency of the light). He found that if a body loses a quantity of energy by radiation or otherwise, then its mass decreases by that quantity divided by the square of the velocity of light. He observed that the exchange of radiation between bodies should involve an exchange of mass, a further implicit support for the kinetic-theory analogy underlying his light-quantum hypothesis: light quanta have mass exactly as do ordinary molecules. Einstein con-

48. Einstein to Habicht, 1905, undated, *ibid.*, 75–76.
49. Max von Laue, who saw Einstein frequently in Zürich in 1913, recalled that Einstein was excited at the time by Bohr's 1913 quantum theory (Seelig, *Albert Einstein, op. cit.* [note 9], 79). Einstein later wrote of his immense admiration, then and now, for Bohr's accomplishment (Einstein, "Autobiographical Notes," *op. cit.* [note 5], 47).
50. L. Rosenfeld quotes this remark in his introduction to Niels Bohr, *On the Constitution of Atoms and Molecules* (New York, 1963), xli–xlii.
51. Einstein, "Does the Inertia of a Body Depend upon Its Energy-Content?" *Ann. d. Phys., 18* (1905), 639–641, trans. in *The Principle of Relativity, op. cit.* (note 19), 69–71.

stantly regarded the equivalence of mass and energy as the most important consequence of his electrodynamics, even though he soon recognized that the variation of mass in available energy transformations was so small that there was no immediate hope of testing the prediction.[52] The enormous value he placed on this unifying principle is suggested by the parallel he drew between it and the outstanding nineteenth-century unification: "mass and energy are the great equivalents," he said, "comparable for example to heat and mechanical work."[53] For Einstein the great significance of his law of mass-energy equivalence was that it helped to bridge the concepts of mechanics and electromagnetism. The mechanical concept of mass lost its isolation, becoming a form of energy, characteristic of radiation as well as of ordinary matter.

In 1906 Einstein returned to his law of mass-energy equivalence, proving that a necessary and sufficient condition for its validity is the conservation of motion of the center of mass of a system in which electromagnetic and mechanical processes occur.[54] He gave two more special proofs the next year. But because the dependence of inertia on energy has such extraordinary generality he felt that a general proof should be found. In 1907 he felt that a general proof was still out of reach, however, because physics did not yet possess a "*complete world picture* corresponding to the relativity principle" (italics supplied).[55] This remark is revealing; within two years of his first writing on electrodynamics, he had come to envision a new world picture and to foresee that the relativity principle would condition its form. The relativity principle was profoundly important for him, serving as his chief heuristic support in his ongoing attempts at uniting mechanics and electromagnetism in a new world view.

Einstein worked at consolidating his world view from a number of directions. One was to continue his electron-theoretical studies, and in this connection a paper of his in 1907 has special interest.[56] It was

52. Einstein, "Le principe de relativité et ses conséquences dans la physique moderne," *Archives des sciences physiques et naturelles, 29* (1910), 5–28 and 125–144, esp. 144.
53. *Ibid.*
54. Einstein, "Das Prinzip von der Ehraltung der Schwerpunktsbewegung und die Trägheit der Energie," *Ann. d. Phys., 20* (1906), 627–633.
55. Einstein, "Über die vom Relativitätsprinzip geforderte Trägheit der Energie," *Ann. d. Phys., 23* (1907), 371–384, quotation on 371–372.
56. *Ibid.*

the first of his electrodynamic writings in which he mentioned light quanta. In it he explained why, in a sense, he was doing electrodynamics at all, since he knew that Maxwell's equations were inexact. He said that the present dualistic "electromechanical world picture" does not explain the entropy properties of radiation, or the emission and absorption of radiation, or the specific heats of bodies. It seemed to him that in periodic processes the energy can occur only in fixed quantities, or quanta, and that the manifold of possible processes is smaller than that permitted by conventional theory. In particular the idea of a continuous field conflicts with the quantum-theoretical demand that the instantaneous electromagnetic state is completely determined by a finite number of magnitudes. Not yet possessing a picture that satisfies this demand, Einstein said that it was proper to apply the present theory wherever entropy processes and transformations of elementary quantities of energy are not involved. To illustrate his meaning he referred to Brownian motion; that motion, he said, cannot be explained by mechanics and thermodynamics, and a profound change in the foundations of these subjects is correspondingly implied. Yet no one would hesitate to employ mechanics and thermodynamics so long as the problem does not concern the instantaneous states of a system in very small volumes. He appealed to Maxwell's theory with the same confidence and the same reservations, and, in this spirit, published fundamental studies on the Maxwell-Lorentz theory in the years after 1905. In these he developed, among other things, the laws of a "material point (electron),"[57] an expression that reflects his desire for a single mechanics for the material point and the electron.

At the same time that Einstein was working on the electron theory, he was trying to fit gravitation into a relativistic framework. This is not surprising, since in a complete world picture gravitation obviously must find its place. He had already freed gravitation from its exclusively mechanical context by his law of mass-energy equivalence, for radiation too has mass and should gravitate. What he now needed to do was revise the gravitational law so that it agreed with

57. See Einstein, "Über das Relativitätsprinzip und die aus demselben gezogenen Folgerungen," *Jahrbuch der Radioaktivität und Elektronik, 4* (1907), 411–462, esp. 431.

the demands of relativistic kinematics. He talked about this, about the direction of his early unpublished efforts at finding the new gravitational law, in his revealing "Notes on the Origin of the General Theory of Relativity."[58] From the start he sought a "field-law for gravitation, since it was no longer possible, at least in any natural way, to introduce direct action at a distance owing to the abolition of the notion of absolute simultaneity." That he applied his 1905 analysis of "simultaneity" to gravitation right after applying it to electrodynamics points up the universal significance he attributed to his relativity principle. It suggests too that his relativity principle may have been decisive in convincing him that action at a distance is untenable and that fields and not particles are the fundamental unifying concept. At first he retained the scalar potential of classical gravitation theory and merely generalized the Poisson equation by adding a second time-derivative term. The field equation then transforms correctly, and gravitation becomes a finitely propagated action. He did not get far with this approach, however. The difficulty was that according to the mass-energy law, the inertial mass of a body varies with its internal and kinetic energies, which means that the acceleration of free fall depends on these energies. This contradicts the notion, suggested by experience, and adopted by Einstein as a premise, that all bodies have the same gravitational acceleration regardless of their velocities and internal states. This was enough to persuade him that his 1905 principle of relativity was an inadequate basis for a gravitational theory and, hence, for a complete physics.

There was another way to approach the gravitational problem, one based on the recognition that the free fall of a body is independent of its energy if its gravitational mass varies with energy in the same way as its inertial mass. Now although there was no theoretical reason why the two kinds of mass should behave in the same way, Einstein did not doubt that they did so. Indeed, he made the strict equivalence of inertial and gravitational mass the key to a proper understanding of gravitation, and he developed this understanding in his first published statement on gravitational theory. In his survey article on relativity for the 1907 *Jahrbuch der Radioaktivität und*

58. Einstein, "Notes on the Origin of the General Theory of Relativity," *Mein Weltbild* (Amsterdam, 1934), trans. *Ideas and Opinions, op. cit.* (note 42), 285–290.

Elektronik,[59] he elevated the equality of inertial and gravitational mass, or, equivalently, the equality of the acceleration of the free fall of all bodies, to the status of an "equivalence principle." He expressed it this way: all bodies behave in a homogeneous gravitational field in such a way that this field is physically equivalent to a uniformly accelerated reference frame. By this principle Einstein could calculate motions in a gravitational field from the kinematic effects of uniform acceleration, even though he did not yet know the correct law of gravitational force. Accordingly he derived the influence of gravitation on the electromagnetic field and, in particular, the gravitational bending of light. The admission of accelerated reference frames and the resulting inconstancy of the velocity of light carried him outside the postulates of special relativity. It eventually led him to the general theory of relativity and to a profound investigation of the meaning of space and time measurements; but this was some years away in 1907. He sent his *Jahrbuch* article to Habicht at the end of 1907, informing his friend that he was now "busy on a relativistic theory of the gravitation law with which I hope to account for the still unexplained secular changes of the perihelion movement of mercury," and adding that "so far I have not managed to succeed."[60] He later wrote that the problem of extending relativity to accelerated frames had kept him busy from 1908 to 1911, and that he had attempted to draw special conclusions from this extension. Since he left no papers from this period that relate to the problem, it is not clear how he related it to his concurrent studies on the electron and quantum theories and to the unification problem in general. It may well be that when, around 1909, he attempted to construct a new mathematical basis for the electron theory, inspired by field-theoretic concepts, he did so partly as a result of conclusions he had reached in his continuing search for a field theory of gravitation and an extended relativity principle. There is simply no way of deciding this or the related questions of why Einstein began a concerted attack on the electron theory problem, which he did around 1909, and why he adopted the particular field concepts he did then.

59. Einstein, "Über das Relativitätsprinzip . . . ," *op. cit.* (note 57).
60. Einstein to Habicht, 24 Dec. 1907, in Seelig, *Albert Einstein, op. cit.* (note 9), 76.

4. EINSTEIN'S SEARCH FOR A NEW ELECTRON THEORY

A year after his light-quantum paper Einstein published on the radiation problem again.[61] Previously he had believed that Planck's theory was incompatible with his light-quantum hypothesis, but now he did not think so. He said that Planck's theory is based on the assumption that the energy of the elementary oscillators can have only values that are whole multiples of the elementary quantum and that the energy can change only discontinuously through the emission and absorption of light. In his derivation of the radiation formula, Planck had made use of Maxwell's theory, and that had been an improper use; Einstein explained that Maxwell's theory is able to give only the average, not the instantaneous, energy of a quantum oscillator. It was clear to Einstein now, as it had not been before, that the apparently incompatible foundations of Planck's theory meant that Planck had implicitly introduced a "new hypothetical element—the light-quantum hypothesis—into physics."

So far Einstein had discussed Planck's theory solely in connection with radiation. However, if the root problem of physical theory was the need for finding a common set of concepts for mechanics and electromagnetism, the implications of Planck's theory for mechanics needed exploring too. That exploration was the subject of Einstein's third paper on quanta, which he sent to the *Annalen*[62] at the end of 1906. Here he concluded that Planck's theory of radiation enforces a "modification of the molecular kinetic theory of heat," his first public acknowledgment that the theory that had guided his researches since 1901 needed revision. To show this need, he derived Planck's law in such a way as to bring into relief its meaning for mechanics. His key step was to recognize that the statistical weighting function is not constant in Planck's theory, as it is in the accepted molecular theory. Instead it is zero for all resonator energies except those close to integral multiples of the elementary quantum, and at each of these multiples it has an identical non-zero value. In his statistical analysis

61. Einstein, "Zur Theorie der Lichterzeugung und Lichtabsorption," *Ann. d. Phys., 20* (1906), 199–206.
62. Einstein, "Die Plancksche Theorie der Strahlung und die Theorie der spezifischen Wärme," *Ann. d. Phys., 22* (1907), 180–190.
The place of this paper in Einstein's thought and in the physics of the time is thoroughly explored in Klein, "Einstein, Specific Heats, and the Early Quantum Theory," *Science, 148* (1965), 173–180.

nothing depended on the fact that the oscillators were charged electrons rather than uncharged molecules. This encouraged him to associate the peculiar quantum features of the weighting function with molecular mechanics as well as electron theory. He put this conclusion in the form of a penetrating question and emphatic answer:

> If the elementary oscillators that are used in the theory of the energy exchange between radiation and matter cannot be interpreted in the sense of the present kinetic-molecular theory, must we not also modify the theory for the other oscillators that are used in the molecular theory of heat? There is no doubt about the answer, in my opinion. If Planck's theory of radiation strikes to the heart of the matter, then we must also expect to find contradictions between the present kinetic-molecular theory and experiment in other areas of the theory of heat. . . .[63]

Einstein went on to analyze a particular problem to support his contention that the same laws apply to molecules and electrons. He showed that the accepted kinetic-molecular theory leads to a formula for the specific heats of solids that contradicts experience. He then derived a more satisfactory formula by assuming that the average energy of a molecular oscillator of a solid is correctly described by Planck's formula for the average energy of an electron oscillator.

Einstein wrote in his autobiography that in his early work on the quantum theory his "major question" did not concern the detailed consequences of Planck's law, but rather the "general conclusions [which] can be drawn from the radiation-formula concerning the structure of radiation and even more generally concerning the *electromagnetic foundation of physics*" (italics added).[64] From the context of his recollection it is not clear just what period he was referring to, but I believe that he had in mind 1905 and the years immediately following. His publications and correspondence suggest that from at least 1909 to around 1911 he believed that the basic problem of physical theory, other than the problem of gravitation, would be solved by a new electromagnetics.[65] Basically he looked for an electron theory that would account for both electric and light quanta. Although he never found a theory that fully satisfied him, his work

63. The quotation is from Einstein, *ibid.*, 184; the translation is Klein's.
64. Einstein, "Autobiographical Notes," *op. cit.* (note 5), 47.
65. Klein has traced Einstein's arduous search for a new electromagnetics in "Thermodynamics in Einstein's Thought," *op. cit.* (note 8).

on the problem had great significance for his own development. In the course of it, he largely formulated the theoretical objectives he held the rest of his life. The theory he sought must be largely reconstructed from his unpublished work, since he made only a few qualitative remarks on it in his published writings.

It had undoubtedly been in his struggle to perfect Lorentz' theory that Einstein had come to recognize that the basic problem of physical theory was the separateness of mechanical and electromagnetic concepts. His 1904 and 1905 work was a first, partial response to this recognition, that of 1909 to 1911 was nothing less than an attempt to eradicate once and for all the dualistic error. In 1909 he began to argue that the only formalism capable of uniting physical theory was the continuous spatial functions and partial differential equations of field theory. His commitment was total; the field was the basic concept of his physics from this time on. In his autobiography he isolated the fundamental problem of Lorentz' theory:

> Kinetic energy and field-energy appear as essentially different things. This appears all the more unsatisfactory inasmuch as, according to Maxwell's theory, the magnetic field of a moving electric charge represents inertia. Why not then *total* inertia? Then only field-energy would be left, and the particle would be merely an area of special density of field-energy. In that case one could hope to deduce the concept of the mass-point together with the equations of the motion of the particles from the field equations—the disturbing dualism would have been removed.
>
> H. A. Lorentz knew this very well.[66]

I wish to call attention to the way that Einstein shifts imperceptibly from talking about the electron problem to talking about the problem of the material point; from at least 1905 he saw them as one and the same problem, though without assuming that all particles are electrons, as adherents of the electromagnetic view, such as Wien, had assumed. The connection of the particle and the field was, for Einstein, the central problem of physics, and he saw the whole of the problem contained in the connection of the electron and the electromagnetic field. The aim of deducing from the field equations the mass-point, or electron, together with its equations of motion was a cardinal point of his unifying program. Unlike the adherents of the

66. Einstein, "Autobiographical Notes," *op. cit.* (note 5), 37.

electromagnetic view, he did not take Maxwell's equations and the electron, conceived as some independently existing body, as given, but sought to deduce the electron from a revised version of Maxwell's equations. He later explained that there was nothing wrong in principle with his program for reforming electron theory. The whole difficulty was that it proved impossible to find the proper modification of Maxwell's equations without descending into "adventurous arbitrariness."[67] In 1909 Einstein believed that the special relativity principle was an adequate guide for finding the electromagnetic equations capable of describing particles. He was mistaken, a lesson he accepted after several years of arduous effort. Ultimately he looked to a more powerful universal principle—the postulate of general relativity—as offering the only hope of avoiding arbitrariness in the construction of field theories with particle-solutions.

In an early 1909 number of the *Physikalische Zeitschrift* Einstein revealed his deepest insight into the nature of radiation.[68] He showed that the formula for the energy fluctuations of blackbody radiation contains two terms. One he interpreted as referring to fluctuations caused by interfering waves, the other to fluctuations caused by changes in the density of light particles. He arrived at the same conclusion from the momentum fluctuations of a mirror suspended in a cavity containing an ideal gas and radiation. The meaning was clear: radiation contains two independent mechanisms, one wavelike and one particlelike. The problem was how to find a mathematical formalism that properly described both. Once that was found, the objectionable dualism of physical theory would be removed. The practical difficulty was that the fluctuation terms did not contain enough information to determine the new electromagnetic equations.

Einstein's 1909 paper was prompted in part by Lorentz' first article in a German periodical, in 1908, on the subject of quanta.[69] Lorentz had written that his long efforts to explain blackbody radiation by

67. *Ibid.*

68. Einstein, "Zum gegenwärtigen Stand des Strahlungsproblems," *Phys. Zeit.*, *10* (1909), 185–193.

Klein has discussed this together with another of Einstein's 1909 publications (see note 77, below) in "Thermodynamics in Einstein's Thought," *op. cit.* (note 8), and in "Einstein and the Wave-Particle Duality," *Natural Philosopher, 2* (1963), 59–86.

69. Lorentz, "Zur Strahlungstheorie," *Phys. Zeit.*, *9* (1908), 562, in *Collected Papers, 7*, 344–376.

combining the electron theory with the kinetic theory had been in vain. He now believed that the quantum theory was the only way and that the blackbody law could be deduced from the electron theory only after making far-reaching changes in that theory. Einstein agreed; in 1909 he set out to make those changes, to do nothing less than reduce the quantum theory to a revised electron theory. For this purpose he adapted a dimensional analysis of blackbody radiation by James Jeans. Einstein reasoned that in any theory the energy density ρ of blackbody radiation must involve the following quantities: c, the velocity of light; v, the frequency of light; ε, the electron charge; and a quantity RT/N proportional to the mean molecular energy, where R is the gas constant and N is Avogadro's number. Except for numerical factors, the dimensions of ρ fix the way these four magnitudes must combine: $\rho = \varepsilon^2/c^4 \cdot v^3 \cdot \psi$, where ψ is an undetermined function of the dimensionless quantity $N/R \cdot \varepsilon^2/c \cdot v/T$. Einstein thought that since numerical factors are unlikely to affect the order of magnitude, the coefficients in ρ should approximately equal those occurring in the quantum formula. Therefore, from Planck's law, $\rho = [8\pi h v^3/c^3] \, [\exp{(hv/kT)} - 1]^{-1}$, the following order-of-magnitude relations obtain: $h/c^3 = \varepsilon^2/c^4$, and $h/k = N/R \cdot \varepsilon^2/c$, from which $h = \varepsilon^2/c$ and $k = R/N$. Planck had derived the second one, and the first, too, was an old idea with him, though he had not developed it. By contemporary estimates, $h = 6 \cdot 10^{-27}$ and $\varepsilon^2/c = 7 \cdot 10^{-30}$; Einstein supposed that this three-decimal difference was due to the suppressed dimensionless factors. As he saw it, the significance of the derivation was that the light-quantum constant h is reduced to the elementary quantum ε of electricity. The strange discreteness of energy was seen then to be intimately related to the recognized discreteness of the electric quantum of electron theory. And the incompleteness of the Maxwell-Lorentz electrodynamics, in failing to explain the elementary quantum of electricity, the electron, is intimately related to the same theory's incompleteness in explaining light quanta. Einstein said that, to him, the relation $h = \varepsilon^2/c$ meant that the "same modification of the [Maxwell-Lorentz] theory that entails, as a consequence, the elementary quantum ε also will entail, as a consequence, the quantum structure of radiation." The long-familiar and most fundamental problem of the electron theory, namely, its inability to account, in electromagnetic terms, for the structure, sta-

73

bility, and charge of the electron, was now, for Einstein, the same problem as that which light quanta posed for the electron theory. Electrons and light quanta would be theoretically comprehended at one and the same time.

To this end, Einstein proposed replacing the basic equation of optics,

$$\frac{1}{c^2}\frac{\partial^2 \phi}{\partial t^2} - \left(\frac{\partial^2 \phi}{\partial x^2} + \frac{\partial^2 \phi}{\partial y^2} + \frac{\partial^2 \phi}{\partial z^2}\right) = 0 \,,$$

by one containing ε as a coefficient (probably occurring as ε^2, so that $h = \varepsilon^2/c$). This new equation, he thought, would complete the revision of Lorentz' theory he had begun in 1905; it would serve for the construction both of electrons and light quanta. He thought that the equation could not be linear and homogeneous; otherwise, as he explained elsewhere that year,[70] radiation would not have the proper statistical properties. In his autobiography he gave two other reasons for desiring nonlinear equations;[71] and, though I cannot be sure of this, I believe they were in his thoughts in 1909. First, only nonlinear equations can yield conditions of electron stability, for Maxwell's linear equations are powerless here. Second, if the equations are linear and homogeneous the sum of two solutions is again a solution, and it is therefore impossible to deduce an interaction between particles. In this event additional mechanical equations are required to describe the motion of particles under the influence of the field, an aspect of the unwanted dualism.

Einstein wrote to Lorentz from Berne in 1909,[72] enclosing a copy of his radiation paper. It was the beginning of an extraordinary correspondence that continued to Lorentz' death in 1928. This first exchange between the man who had raised classical physics to its nearly perfected state and the younger man who was determined to change it dealt, necessarily, with the character of the future physics. Einstein explained that, in his view, both molecular mechanics and Lorentz' electrodynamics were in contradiction with blackbody radiation. He wanted Lorentz to pay special attention to his formula for wave-particle fluctuations in radiant energy and to his discussion of the wave-particle equation he was seeking. He fervently hoped that

70. Einstein to Lorentz, 23 May 1909, Algemeen Rijksarchief, The Hague.
71. Einstein, "Autobiographical Notes," op. cit. (note 5), 77–78.
72. Einstein to Lorentz, 1909, undated, Algemeen Rijksarchief, The Hague.

Lorentz would point the way to the reform of electron theory or else show how the old principles could still be made to work.

Unknown to Einstein, Lorentz had already formed an opinion of light quanta; he had, in short, rejected them. He had explained his reasons at a meeting of Dutch physicists in Utrecht in April 1909 in what was, so far as I know, the first critical discussion of the light-quantum hypothesis.[73] As Lorentz cited only Einstein's 1905 paper, he had not been prompted to write on the subject by Einstein's recent paper in early 1909.

Lorentz' letter is lost. However, a draft reply in Dutch, dated 6 May 1909, has survived.[74] It consists of rough notes, incomplete sentences, fragments of calculations; yet there is enough to suggest the tenor of Lorentz' remarks, especially when read together with his and Einstein's papers and Einstein's second letter. Lorentz clearly took Einstein's latest arguments for quanta seriously, examining them minutely. On the main question Lorentz agreed completely with Einstein; an unchanged electron theory could not solve the radiation problem, and something like Planck's theory was needed. But he did not agree with Einstein that energy fluctuations imply light quanta. And he argued, as he had in his paper, that the optical phenomena of interference and the physiological phenomena of vision are incompatible with light quanta. He suggested instead that Planck's constant might be located in the ether, and that the needed reduction in the number of degrees of freedom of pure radiation might be attributed to a coupling of ethereal vibrations. Most important, he did "not agree that h is connected with electron theory." The three-decimal difference between h and ε^2/c was no small thing to him. And he did not find it satisfying that the equations of the ether should serve as the equations of motion of electrons. He said that he experienced the greatest difficulty over the suggestion that Maxwell's equations needed revision. Clearly Lorentz did not endorse Einstein's direction of electron-theory reform. At the same time he was not unsympathetic, remarking that his optical objections to light quanta were a "pity since the theory is good in magnitude."

73. This address was translated from the Dutch and published as "Die Hypothese der Lichtquanten," *Phys. Zeit.*, *11* (1910), 349–354.

74. Lorentz to Einstein, 6 May 1909, Lorentz Collection, Algemeen Rijksarchief, The Hague.

Einstein's long reply of 23 May 1909 contains a new and more detailed proposal for the revision of electron theory.[75] This letter goes further than any of his publications in throwing light on his thinking at this time. Lorentz had become an intimate to whom he revealed his most speculative thoughts. In the trial electron theory he sketched for Lorentz he fully exposed for the first time his field-theoretical ideal. His subsequent field theories, some of which he published and some not, were motivated by this ideal, as were his critiques of later developments of the quantum theory.

Einstein said that he too was displeased by the three-decimal difference in the relation of h and e. But he regarded the rough order-of-magnitude agreement as highly significant, observing that nature is sparing with her universal constants. Contrary to Lorentz he regarded the search for equations yielding electrons and light quanta as the most fruitful direction.

In his letter Einstein groped for a mathematics to fit the reality he intuited. That reality had changed since 1905 in an important respect, and Lorentz had failed to appreciate this. Lorentz reproduced in his letter the objections he had detailed in his talk on light quanta earlier in the year; in both places he spoke of light quanta as Einstein had spoken of them in 1905. In his talk Lorentz had introduced light quanta by quoting Einstein's 1905 observation that a theory of light operating with continuous spatial functions contradicts experience. The independence of light quanta is premised on the rejection of continuous spatial functions, and Lorentz thought that light quanta were still supposed to have the property of independence. Einstein, however, had rejected independent quanta and had couched his tentative theory in the language of continuous functions. He corrected Lorentz on this crucial point by carefully drawing the new physical image whose equations he was looking for. He explained that he now thought that light quanta are point-centers of extended vector fields and that they are capable of interacting through their fields. He thought that this should answer Lorentz' optical objections to light quanta, objections that had led Lorentz to conclude that light quanta cannot be point-particles but must occupy a finite volume. Einstein explained how a very natural extension of the classical electron theory might resolve the basic dualism of physical theory. His

75. Einstein to Lorentz, 23 May 1909, *op. cit.* (note 70).

EINSTEIN, LORENTZ, AND THE ELECTRON THEORY

model for a light quantum was no longer a neutral, or fieldless, gas particle, but an electron together with its field. The vector field of a light quantum, he supposed, cannot exist without an electron as its source; and, as in the case of an electron, the field of a light quantum falls off with distance. He explained to Lorentz the nature of the mathematics he thought might describe this unified conformation of field and matter.

Einstein was of a different mind than he had been earlier that year on the mathematical character of the field equations; he would now allow them to be linear and homogeneous. The key to the electromagnetic problem appeared to him to lie not in mathematical singularities but in the separability of the field equations (a separate equation for each field component), by which a finite quantity of energy travels without dispersion at the speed of light. He thought that with only a small change in Maxwell's equations he could show how a finite quantity of energy travels without dispersion at the speed of light, and he wrote out his most promising thoughts on the matter. His search for "new electromagnetic foundations" centered on equations of the type

$$\Delta\phi - \lambda^2\Delta\Delta\phi = 0 \, ,$$

where Δ stands for $\partial^2/\partial x^2 + \partial^2/\partial y^2 + \partial^2/\partial z^2$. A solution depending on $r = \sqrt{x^2 + y^2 + z^2}$ is

$$\phi = \varepsilon \, \frac{1 - e^{-r/\lambda}}{r} \, ,$$

where ε is an undetermined constant. This is the only solution with the following two properties necessary for representing electrons. As demanded by the physics of the situation, the field ϕ falls off with increasing distance from its source at $r = 0$, while, for large r, ϕ goes over to the Coulomb potential ε/r. And for $r = 0$, ϕ has the remarkable property of possessing no singularity, obviating the difficulty of the infinite self-energy of point-charges in Maxwell's theory. For Einstein a most important consequence was that a nonelectromagnetic, rigid structure for an electron is unnecessary. If his approach were correct, nothing would stand in the way of completing a purely electromagnetic electron dynamics, the minimum objective of all versions of the electromagnetic program for physics.

To make the equation dynamical and to insure its compatibility with the relativity principle, the operator Δ must be replaced by $\Delta -$

77

$1/c^2 \cdot \partial^2/\partial t^2$. Einstein envisioned four such transformed equations as a complete electrodynamic system. The other three would involve the components of the electromagnetic vector potential. No special equations for the motion of particles appeared; no ordinary differential equations marred the system. The chief difference between Einstein's and Maxwell's equations is that Einstein's are fourth-order rather than second-order, a revision demanded if singularities are not to occur for $r = 0$.

It is clear that Einstein wrote the field equation in the form he did as a result of a suggestion of Lorentz'. Lorentz had included in his letter a dimensional calculation for h that interested Einstein extremely. It led Einstein to recognize that it was no longer necessary to include ε in the equations, but that a constant λ with dimensions of length would work equally well. The electron charge ε was to be derived with the help of λ instead of entering the field equations as an arbitrary constant. A fundamental length, a quantity characteristic of waves and not of point-particles, was now primary, a proper field point of view.

Einstein said that a major innovation of his approach was that the field equations should be capable of determining the electric masses ε and current densities everywhere. He accordingly wrote the field equations so that the sources of the field ε do not enter them, but appear only in the solution ϕ. In his autobiography he was explicit on his "faith" that there are no arbitrary constants in nature; he said that his object was to find such strongly determinative field equations that if, say, the electron charge ε or the velocity of light c had different values the theory would be destroyed.[76]

The trouble is that, as it stands, the fourth-order linear equation does not fix the value of ε. Einstein recognized that he really still did need a nonlinear differential equation, and he wondered if one exists as an integral of the other. The second equation must lead to the same solution, ϕ, but it would now determine ε at the same time. If he could find this second equation, he said, it would be "very beautiful." He could not, however, for he was mistaken in thinking that a linear equation might be manipulated to yield both the sources of the field and their motion. The significant point is not the mathematical difficulties he encountered, but the type of mathematical

76. Einstein, "Autobiographical Notes," *op. cit.* (note 5), 63.

expression he chose to work with. A fourth-order equation elimi-
nates the infinite self-energy, his chief purpose in introducing it.
(The solution of the regular second-order equation, $\varepsilon/r \cdot \exp[-r/\lambda]$,
is singular at the origin.) He may also have had in mind the fact
that the fourth-order equation multiplies the number of solutions,
which would be desirable if both light quanta and electric quanta
are to be represented. Substituting a plane wave, $\phi = \exp(i[kx - \omega t])$,
into the fourth-order equation yields two dispersion relations instead
of one: $(-k^2 + w^2/c^2)(-k^2 + w^2/c^2 - 1/\lambda^2) = 0$; the first bracket would
apply to light quanta, the second, corresponding to a lesser velocity,
would presumably apply to electrons.

At the end of his letter Einstein begged Lorentz to tell him if his
hopes and efforts were empty. What Lorentz replied is not known.
He did not in any case persuade Einstein that his hopes were empty,
at least not immediately. In October 1909 Einstein submitted a second paper for publication in the *Physikalische Zeitschrift*,[77] one he
had delivered in Salzburg to a gathering of German scientists. The
subject once again was the radiation problem, and Einstein approached it firmly convinced that the next stage in the development
of physics would be a new electromagnetic theory. He predicted
that the theory would bring "a kind of fusion of the wave and emission [particle] theories" of light. He used his Salzburg address as an
opportunity to bring his thought closer to Lorentz'. He praised Lorentz' electron theory as the one ether-based theory that was useful
and clear in its foundations; and he showed that it was possible to
bring it into agreement with the principle of relativity. What remained to be done was to revise Lorentz' theory so that it agreed with
the facts of blackbody radiation.

In this paper Einstein stressed the significance of the relativity
principle for an understanding of radiation. According to relativity,
light is an independent inertia-bearing entity radiated into space, not
the state of a hypothetical medium. This is an important insight,
but it is not enough. Einstein told the Salzburg physicists that the
relativity principle cannot reveal the quantum structure of light, as
he had previously explained to Paul Ehrenfest that it cannot deter-

77. Einstein, "Über die Entwicklung unserer Anschauungen über das Wesen
und die Konstitution der Strahlung," *Phys. Zeit., 10* (1909), 817–826.

mine the structure of the electron.[78] The statistical properties of radiation afford some insight into its nature, but this guide too is extremely limited. As he explained the following year,[79] the probability can be calculated for a given particulate system, but should the probability alone be given, the corresponding particulate system cannot be uniquely deduced. All that he could hope for is that a statistical knowledge would rule out all theories incompatible with it. A more specific aid in understanding the properties of light was the electron theory's conception of the electron. He explained to Planck in the discussion following the reading of his paper that the precedent for light quanta was the electron theory's "molecularization of the carrier of the electrostatic field." He thought that "it is not out of the question that something similar will take place in radiation theory"; i.e., he hoped to introduce the older "particles" of light into Maxwell's field just as Lorentz had successfully reintroduced the older "atoms" of electricity into the field. He elaborated this parallel in the text of his address. His ideas are precisely those he had formulated in his correspondence with Lorentz:

> It appears to me at present that the most natural conception is that the existence of the electromagnetic field of light is bound to a singular point exactly as is the existence of an electrostatic field according to the electron theory. It is not ruled out that in such a theory the entire energy of the electromagnetic field could be seen as wholly concentrated in these local singularities, precisely as in the old action-at-a-distance theory. I conceive of each such singular point as surrounded by a force field which possesses the character of a wave and whose amplitude decreases with distance from the singular point. If many such singularities are present in a small space, the force fields will overlap and form an undulatory force field, which differs little from the wave field in the present electromagnetic theory.

It was a vivid physical image, and although Einstein was no closer to having the exact equations for describing it, he was confident of his understanding of the general properties the equations must have.

---

78. See Klein, "Thermodynamics in Einstein's Thought," *op. cit.* (note 8), 515–516.

79. Einstein, "Sur la théorie des quantités lumineuses et la question de la localisation de l'énergie électromagnétique," *Archives des sciences physiques et naturelles, 29* (1910), 525–528.

The solution to the "profound formal distinction" between electromagnetic and kinetic-theory concepts was to write physics in terms of continuous field quantities and nonlinear equations that yield singularity-free particle solutions. This was the permanent lesson Einstein took away from his early direct attempts to realize his unifying program for physics through a reformed electron theory. He was soon to find additional support for it in his general relativity theory, which required that field equations have just those properties he had decided upon on physical grounds.

With the help of general relativity Einstein derived gravitational equations that were nonlinear and inhomogeneous and could therefore account for particles and their interaction. In addition, these equations contained the laws of motion of particles, eliminating the need for particle mechanics in addition to field theory. This was precisely the kind of field theory he had come to seek in 1909 in the context of the electron-theory problem. (I do not mean to say that Einstein did not further refine his field-theoretical objectives after 1909. He did, of course, as a result of his experience with general relativity.)

5. GRAVITATION AND GENERAL RELATIVITY

Late in 1909 Einstein told his close friend Besso of a new idea for solving the radiation problem.[80] He observed that the expression for the electromagnetic energy density could be generalized by means of a certain transformation (the Lorentz gauge): if the scalar potential ϕ and the vector potential Γ_x, Γ_y, Γ_z, are replaced by $\phi - \partial\psi/\partial t$, $\Gamma_x + \partial\psi/\partial x$, etc., where ψ is any solution of $\Delta\psi - \partial^2\psi/\partial t^2 = 0$, the resulting modified energy density is compatible with Maxwell's equations. Einstein thought that the infinite number of energy distributions which are now possible might make it possible to introduce light quanta into the electromagnetic field. But nothing came of this either.

Again and again in his correspondence around 1909 Einstein spoke of his determination to solve the radiation riddle and his frustration with his slow progress. Everyone, he wrote J. J. Laub, ought to be

80. Einstein to Besso, 31 Dec. 1909, Einstein Collection, Princeton.

busying himself with the quantum-radiation problem.[81] For him its central importance lay in its promise of revealing how Maxwell's equations were to be revised to arrive at a relativistic world picture. The reformed Maxwell-Lorentz equations eluded him, however, and his expectation of solving the quantum problem in the near future gradually waned. In March 1910 he told Laub that the quantum theory is certain;[82] eight months later he wrote that he hoped to solve the radiation problem without quanta and that, in fact, he "no longer believed (at present)" in their reality.[83] Writing to Laub again in the summer of 1911, he did not once mention quanta but talked only of electrodynamics and of the extension of relativity to gravitation.[84]

This is not to say that Einstein abandoned his search for a theory of light quanta. Indeed, he was still plaguing himself with quanta in 1913, although feeling keenly the hopelessness of his continuing search.[85] He wrote again on the radiation problem in 1916[86] after having devoted himself almost exclusively to gravitation and general relativity for several years. He told Lorentz that same year that it was clear to him that the quantum difficulties applied to his new gravitational theory no less than they did to Maxwell's.[87] In 1918 he assured Besso that he believed in the reality of light quanta and that he was again struggling with the question, convinced that no one would take him seriously until he had a mathematical theory.[88] In 1924 he told Besso that he was still seeking to unify quanta and Maxwell's field.[89] That year he published on the quantum theory of an ideal gas, praising de Broglie's hypothesis of matter waves, a hypothesis that helped remove the formal difference between the concepts of field and matter.[90] Einstein in fact never ceased to be con-

81. Einstein to Laub, undated, in Seelig, *Albert Einstein, op. cit.* (note 9), 87. Klein has analyzed Einstein's correspondence on the radiation problem in "Thermodynamics in Einstein's Thought," *op. cit.* (note 8).
82. Einstein to Laub, 16 Mar. 1910, Einstein Collection, Princeton.
83. Einstein to Laub, 4 Nov. 1910, Einstein Collection, Princeton.
84. Einstein to Laub, 10 Aug. 1911, Einstein Collection, Princeton.
85. Einstein to Besso, 1913/1914, Einstein Collection, Princeton.
86. Einstein, "Zur Quantentheorie der Strahlung," *Phys. Zeit., 18* (1917), 121–128.
87. Einstein to Lorentz, 17 June 1916, Algemeen Rijksarchief, The Hague.
88. Einstein to Besso, 29 July 1918, Einstein Collection, Princeton.
89. Einstein to Besso, 24 May 1924, Einstein Collection, Princeton.
90. Einstein, "Quantentheorie des einatomigen idealen Gases," *Sitzungsberichte der Preussischen Akademie der Wissenschaften* (1924), 261–267.

cerned with the quantum problem, but his approach was less direct after 1911, and his concern was sometimes less evident to others.

In 1911 Einstein became deeply absorbed in the gravitational problem again.[91] That was also the year of the first Solvay Congress, which met to discuss the crisis in physics signaled by Planck's quantum theory. In 1913 Einstein and Grossmann published a first statement of the tensor theory of general relativity;[92] that same year Niels Bohr published his celebrated quantum theory of atoms and molecules, opening an extremely promising new domain of application of the quantum theory. Just when most physicists started to recognize the quantum problem as the central problem of physics, Einstein seemed to move away from it. There is, however, less of a change in the direction of his research in 1911 than first appears. His earlier preoccupation with electrons and light quanta does not conflict with his immersion in macroscopic physics. By 1909 Einstein had come to regard the particle aspect of nature as secondary to the field aspect. In his opinion it was futile to attempt to understand in any fundamental way the microscopic structure of radiation and matter until the macroscopic structure of the field was understood; when the latter was understood, quantum phenomena would be deduced from it.

From the early 1920's Einstein showed little enthusiasm for the direction the quantum theory took under Bohr's leadership. It was not that he grew conservative in his middle age, but that he remained convinced of the general correctness of the theoretical goal he had formulated in the years around 1909. The structures of gravitational and electromagnetic fields were the last consideration of the young quantum theorists, while for Einstein, whose thought was firmly rooted in the older electrodynamic tradition, these were the first consideration. The concepts of the accepted quantum theory were "largely taken over from classical mechanics," a retrograde step in Einstein's judgment. Long before, by 1909, he had persuaded himself that the unifying concepts were not mechanical, but field-theoretical; from then on he sought a field theory with quantum solutions, not a quan-

91. Einstein, "On the Influence of Gravitation on the Propagation of Light," *Ann. d. Phys., 35* (1911), 898–908, trans. in *The Principle of Relativity, op. cit.* (note 19), 99–108.

92. Einstein and Grossmann, "Entwurf einer verallgemeinerten Relativitäts-theorie und eine Theorie der Gravitation," *Zeit. für Math. u. Phys., 62* (1913), 225–244.

tum "mechanics." He emphatically disagreed with his colleagues' belief that it was impossible to describe the "atomic structure of matter and of radiation" by "continuous functions of space for which [partial] differential equations are valid."[93] In 1946, when he wrote his "Autobiographical Notes," he believed that such a description was possible as firmly as he had believed it in 1909. Quantum mechanics did acquire a formalism of continuous functions and partial differential equations, but its statistical interpretation seemed to require an abandonment of continuity and causality, cornerstones of late nineteenth-century thought that Einstein was profoundly reluctant to give up.

In 1911, the year in which he became deeply pessimistic over the prospect of an imminent solution to the electromagnetic problem, Einstein returned in print to his 1907 gravitational theory.[94] For the next several years he followed a new tack, looking to gravitation rather than electromagnetism as the starting point for the reform of physical theory. That tack proved immensely promising, so much so that by 1913 he could tell Lorentz that he now believed that the gravitational field was the framework which everything in physics depended on.[95] Two years later his gravitational theory had matured to the point where he could explain the anomalous motion of the planet Mercury, the critical test-problem whose solution had eluded him since 1907.[96]

Einstein recognized electromagnetic analogies in gravitational actions,[97] although, unlike Lorentz and others, he did not anticipate a reduction of gravitation to electromagnetism. The two force-fields were physically related, but had dissimilar mathematical structures. The conventional electromagnetic program postulated too narrow a base for the whole of physics, and Einstein extracted from it a more general field-theoretical ideal. He saw the foundation of a complete

93. Einstein, "Autobiographical Notes," *op. cit.* (note 5), 87.

94. Einstein, "On the Influence of Gravitation . . . ," *op. cit.* (note 91).

95. Einstein to Lorentz, 14 Aug. 1913, Algemeen Rijksarchief, The Hague.

96. Einstein, "Erklärung der Perihelbewegung des Merkur aus der allgemeinen Relativitätstheorie," *Sitzungsberichte der Preussische Akademie der Wissenschaften* (1915), 831–839.

97. Einstein showed that gravitation is a motion-dependent action: gravitating bodies in accelerated motion induce accelerations in other ponderable bodies. (Einstein, "Gibt es eine Gravitations-wirkung, die der elektrodynamischen Inductionswirkung analog ist?" *Vierteljahrsschrift für gerichtliche Medizin und öffentliches Sanitätswesen, 44* [1912], 37–40.)

physics as inhering in the combined electromagnetic and gravitational fields. He needed the combined fields to confront the classical problem of electron stability, which the electromagnetic field alone had been unable to solve.[98]

Einstein regarded his 1915 gravitational theory as his greatest achievement. Even so he saw it as merely a "preliminary formal frame for the representation of the entire physical reality."[99] It was nothing more than a theory of the pure gravitational field, which was isolated from the yet unknown total field. From 1916 on, Einstein worked at generalizing his theory of the pure gravitational field to a unified theory of the gravitational and electromagnetic fields. This was his ultimate response to the mechanical-electromagnetic crisis in physical theory he had first talked about in the opening of his 1905 light-quantum paper.

It was general relativity that brought Einstein and Lorentz into their closest scientific relation. While Lorentz admired the restricted principle of relativity, he never embraced it. The ether remained the center of his physics. To the end of his life he believed that absolute space and time are meaningful, even though a preferred reference frame and clocks at rest in it might never be detected. For him it was a matter of taste whether one adopted Einstein's original ether-denying interpretation of the equations of electron theory or his own; the known observable consequences were exactly the same upon either view. His attitude toward Einstein's extension of the relativity principle was entirely different. Lorentz published a number of important, constructive studies on the theories of general relativity and gravitation, and his correspondence with Einstein came to be almost exclusively concerned with this subject. He wrote to Einstein in 1916 saying that for months he had been occupied with general relativity theory and that now he believed he understood it in its "full beauty."[100] In the same letter he spoke of general relativity as seeming to agree with the ether hypothesis. Einstein did not dis-

---

98. Einstein, "Do Gravitational Fields Play an Essential Part in the Structure of the Elementary Particles of Matter?" *Sitzungsberichte der Preussischen Akademie der Wissenschaften* (1919), 349–356, trans. in *The Principle of Relativity, op. cit.* (note 19), 191–198.

99. Einstein, "Autobiographical Notes," *op. cit.* (note 5), 73.

100. Lorentz to Einstein, 6 June 1916, Rijksmuseum voor de Geschiedenis Natuurwetenschappen, Leiden.

pute the suggestion but only cautioned that the new ether must not conflict with relativity in the way that the earlier one did.[101] When Einstein accepted Lorentz' invitation to join the Leiden faculty in 1920, he proposed the ether as the topic of his inaugural address, seizing upon this opportunity to underscore the increasing harmony he felt between his and Lorentz' thought.[102] He explained in his lecture that to deny the ether is to deny that space has physical properties, and the theory of general relativity and its results are founded precisely on the physical character of space.[103]

Unlike Einstein, Lorentz did not oppose the direction that the quantum theory took; he was prepared to accept the axioms and to work out their consequences. Yet his heart was never really in the quantum theory. Writing to Einstein in 1923, he associated the great happiness he had known with the fact that in his day "we still lived in childlike innocence (knowing nothing of quanta), and physics was so beautiful."[104]

In the same letter Lorentz recalled Einstein's first visit to Leiden in 1911. It was the beginning of his instruction, Lorentz said; and the revolution that Einstein had brought about in their science had kept him young. That appreciation was deeply gratifying to Einstein, who venerated Lorentz as he did no one else. At the 1911 Solvay Congress Einstein learned little new about quanta, but he garnered rich impressions of the participants. He told a friend at the time that "H. A. Lorentz is a wonder of intelligence and fine tact. A living work of art! In my opinion he remains the most intelligent of the present theorists."[105] It was not just admiration for Lorentz' intelligence that Einstein felt; his admiration was inseparable from a deep affection. In a letter to Laub in 1909 he spoke in a profoundly personal vein: "I admire this man [Lorentz] more than any other. I might almost say I love him."[106] Much later, on the occasion

101. Einstein to Lorentz, 17 June 1916, op. cit. (note 87).
102. Einstein to Lorentz, 12 Jan. 1920, Algemeen Rijksarchief, The Hague.
103. Einstein, "Ether and the Theory of Relativity," an address delivered on 5 May 1920, in the University of Leiden, trans. by G. B. Jeffery and W. Perrett in Sidelights on Relativity (London, 1922), 3–24.
104. Lorentz to Einstein, 15 Sept. 1923, Rijksmuseum voor de Geschedenis der Natuurwetenschappen, Leiden.
105. Einstein to H. Zangger, 15 Nov. 1911, Einstein Collection, Princeton.
106. Einstein to Laub, dated 19, 1909, in Seelig, Albert Einstein, op. cit. (note 9), 124.

of the 100th anniversary of Lorentz' birth, Einstein said that Lorentz "meant more to me personally than anybody else I have met in my lifetime."[107]

ACKNOWLEDGMENTS: For their helpful criticisms of one or another draft of this article, I wish to thank Paul Forman, John L. Heilbron, Tetu Hirosige, and especially Martin J. Klein.

107. Einstein, "H. A. Lorentz, his Creative Genius and his Personality," *op. cit.* (note 1).

In Defense of Ether:
The British Response to Einstein's
Special Theory of Relativity,
1905–1911

BY STANLEY GOLDBERG[*]

1. INTRODUCTION

At the end of the nineteenth century, J. T. Merz, in his monu-
mental *History of European Thought in the Nineteenth Century*,[1]
expounded the thesis that the scientific spirit, which had begun in
France, had spread to Germany and then across the Channel to
Britain and had now become solidified and unified. He said that
"science at last had become international: isolated and secluded
centres of thought have become more and more rare. . . . National
peculiarities still exist, but are mainly to be sought in the remoter
and more hidden recesses of thought. . . . We can now speak of
European Thought, when at one time we should have had to distinguish between French, German, and English Thought."[2]

Of course one can cite features of the scientific enterprise that are
shared across national boundaries, such as canons of laboratory procedure, and the need for making researches public. But would one
have to search out the "remoter and more hidden recesses of thought"
to find significant national differences in science? My skepticism on

* Department of Physics, Antioch College, Yellow Springs, Ohio 45387.
1. J. T. Merz, *A History of European Thought in the Nineteenth Century*, 4
vols. (Chicago, 1904–1912).
2. *Ibid., 1,* 16–20.

this point was a major motivation for this study. I believe that national differences have persisted in science after Merz's time, and that they have involved more than linguistic difficulties. An analogy may be drawn to music: while the notes are a universal language, the interpretation of the notes is a highly cultural and personal affair between the conductor, musicians, and audience. In physical science the notes become the mathematical notation, which in every case must be interpreted.

Gerald Holton has analyzed the process of scientific inquiry and has suggested that there is more in it than empirical observation and analytic process. He insists on a third component, which he calls "themata": "The strong hold that certain themes have on the mind of the scientist helps to explain his commitment to some point of view that may run exactly counter to all accepted doctrine and to the clear evidence of the senses."[3] Are national scientific characteristics an important component in shaping the themata Holton has spoken of? It was this question that led me to study the response to Einstein's theory of relativity in four countries: France, Germany, Britain, and the United States.[4] I chose to study the period between 1905, the year of Einstein's first paper on relativity, and 1911, the year of the first Solvay Conference, when the thoughts of many early contributors to the theory of relativity turned to the quantum theory. During this time, over ninety-nine percent of the literature on Einstein's theory emanated from these four countries.[5] The present paper is a portion of this larger study, an examination of the British response to relativity. This was a unique response in at least two respects: it was far more uniform than the responses in other countries; and British scientists were more outspoken on the subject than were their Continental colleagues. Before turning to a detailed consideration of the British response, it will be helpful to sketch the context in which Einstein's theory was first advanced, and to evaluate briefly how it was received elsewhere.

3. G. Holton, "The Thematic Imagination in Science" in Holton, ed., *Science and Culture* (Boston, 1965), pp. 88–108, quotation on p. 100.

4. S. Goldberg, *Early Response to Einstein's Special Theory of Relativity, 1905–1912: A Case Study in National Differences* (unpubl. Ph.D. diss., Harvard University, 1969).

5. M. Lecat, *Bibliographie de la relativité* (Brussels, 1924), pp. 201–202.

2. THE THEORY OF RELATIVITY

There has been much confusion and difference of opinion over the originality of Einstein's contribution to the theory of relativity. On the one hand, it has been argued that the theory was largely the creation of H. A. Lorentz and H. Poincaré, and that Einstein's 1905 formulation added little that was new. This is the line that Whittaker took,[6] and more recently Keswami put forward the same view.[7] On the other hand, Holton and I have argued that Einstein's theory of relativity has little more than the formalism in common with the electron theory of Lorentz and Poincaré.[8]

There are three innovative features of Einstein's theory that set it apart from the work of his predecessors. First, the theory is derived from two postulates, that of relativity, and that of the invariance of velocity of light. Second, the theory is one of measurement, and in this sense, it is a kinematical, not a dynamical, theory; it says nothing about the structure of electrons or of matter in general, but only of the way events are measured in different inertial frames of reference. Finally, the theory does not depend in any way on the existence of a medium for the propagation of electromagnetic waves.

In the introduction to his first paper on special relativity, Einstein poses a paradox: when a conductor and a magnet are moved relatively to one another, the current induced in the conductor does not depend on which of the two is moved. Yet, according to the theory, the motion of a magnet gives rise to an electric field, while the motion of a conductor does not. Coupling this paradox with the

6. E. T. Whittaker, *A History of the Theories of Aether and Electricity*, 2 vols. (New York, 1960), 2, 27–77, esp. p. 44.

7. G. H. Keswami, "Origin and Concept of Relativity," *Brit. J. Phil. Sci., 15* (1965), 286–306.

8. G. Holton, "On the Origins of the Special Theory of Relativity," *Amer. J. Phys., 28* (1960), 627–636; "Influences on Einstein's Early Work in Relativity Theory," *The American Scholar, 37* (1968), 60–79. S. Goldberg, "Henri Poincaré and Einstein's Theory of Relativity," *Amer. J. Phys., 35* (1967), 933–944; "The Silence of Poincaré and Einstein's Relativity," *Brit. J. Hist. Sci.,* in press; "The Lorentz Theory of Electrons and Einstein's Relativity," *Amer. J. Phys.,* in press.

G. Holton has recently analyzed the foundations of relativity in his valuable "Einstein, Michelson, and the 'Crucial' Experiment," *Isis, 60* (1969), 133–197. See especially pp. 160–167. My analysis was written prior to seeing Holton's paper; however, there is substantial agreement in our points of view. I take this opportunity to acknowledge a special debt of gratitude to Holton, since my analysis owes much to conversations we have had over the past ten years.

futility of attempts to measure the motion of the earth through the ether, Einstein was led to suggest that the "phenomena of electrodynamics as well as mechanics possess no properties corresponding to the idea of absolute rest."[9] Without further consideration he stated his intention to "raise this conjecture . . . to the state of a postulate,"[10] the principle of relativity. Einstein never questioned this postulate or appealed to experience to make it seem more reasonable. This contrasts with the attitude of Poincaré, who saw the relativity principle as a generalization from experience, one that required only one contrary instance to render it inoperative.[11]

It was not until after Einstein's 1905 paper that Lorentz considered the principle of relativity as a basic concept rather than as a result of the electron theory. Whereas Lorentz' theory of 1892 predicted no first-order effects due to motion through the ether, it did predict second-order effects; that is, phenomena depending on the square of the ratio of the velocity of the apparatus through the ether to the velocity of light in free space. Second-order experiments, such as the Michelson-Morley experiment, however, failed to yield the expected results. The equations of the electromagnetic field appeared to be covariant in practice while the theory predicted otherwise. In the face of this puzzle, Lorentz was content to work out the transformation equations between inertial frames of reference that rendered the equations of the electromagnetic field covariant. By an accomplished piece of inductive detective work, he found, in 1904, results that were formally identical to those that Einstein published a year later.[12] However, in Einstein's paper the transformation equations are *derived* from considerations based on his two postulates.

Einstein's second postulate was that "light is always propagated in empty space with a definite velocity c which is independent of the

9. A. Einstein, "On the Electrodynamics of Moving Bodies," *Ann. d. Phys., 17* (1905), 891–917, trans. W. Perret and G. B. Jeffrey in Einstein, *et al., The Principle of Relativity* (New York, 1923), p. 37.

10. *Ibid.,* p. 38.

11. H. Poincaré, "Sur la dynamique de l'électron," *Rev. gen. sci. pures et appliquées, 19* (1908), 386–342. See S. Goldberg, "Henri Poincaré and . . . Relativity." Poincaré's continuing doubts about the principle were brought into sharp focus by Kaufmann's 1906 claim that his experiments on the specific charge of high-speed electrons were not in accord with either the Einstein or the Lorentz theories. For his part, Einstein maintained a significant silence.

12. H. A. Lorentz, "Electromagnetic Phenomena in a System Moving with Any Velocity Less than That of Light," *Proc. Acad. Sci. Amst., 6* (1904), 809, repr. in A. Einstein, *et al., The Principle of Relativity,* pp. 11–34.

state of motion of the emitting body."[13] In his "Autobiographical Notes," written approximately forty years after the publication of his first paper on relativity, Einstein reminisced about the source of his second postulate: "After ten years of reflection . . . a [general] principle resulted from a paradox upon which I had already hit at the age of sixteen: If I pursue a beam of light with the velocity c (velocity of light in a vacuum), I should observe such a beam of light as a spatially oscillatory electromagnetic field at rest. However, there seems to be no such thing, whether on the basis of appearance or according to Maxwell's equations."[14]

Again Einstein had leaped from a suggestive thought experiment to a heuristic postulate. In 1916 he published what is perhaps still the most charming and the clearest, if the most terse, popularization of the theory of relativity. In it he asks how the postulate of the invariance of the velocity of light could be confirmed experimentally. He proceeds to show that since a one-way measurement of the velocity of light is impossible, it is necessary to make a stipulation: "that light requires the same time to traverse the path . . . [AM] as the path. . . . [BM, M being the midpoint of the line AB] is in reality neither *a supposition nor a hypothesis* about the physical nature of light, but a *stipulation* which I can make of my own free will."[15] Such a stipulation is necessary in any inertial frame of reference if one is to arrive at an intuitive definition of simultaneity. And such a stipulation is equivalent to the postulate of the invariance of the velocity of light. It *follows* that observers in different inertial frames disagree on judgements of simultaneity. Though such results deduced from the postulates are counterintuitive, there is no contradiction between the postulates.

As early as 1898, Poincaré had also considered the problem of synchronizing clocks using light signals; and he had stated that the invariance of the velocity of light was a fact about which there could be no question. However, in his view, it was an empirical fact which could be disproved by experiment. In any event, Poincaré never saw the need for making the invariance of the velocity of light

13. A. Einstein, "On the Electrodynamics of Moving Bodies," *op. cit.* (note 9), p. 38.
14. A. Einstein, "Autobiographical Notes" in P. A. Schillp, *Albert Einstein: Philosopher-Scientist* (New York, 1959), p. 53.
15. A. Einstein, *Relativity, The Special, The General Theory* (Chicago, 1951), p. 18.

a postulate and, indeed, in his analysis of the process of synchronizing clocks, he showed clearly that he believed that the invariance was due to a strange compensation, an effect of the Lorentz-Fitz-Gerald contraction. With Lorentz, he would have agreed that there was only one frame of reference in which the velocity of light was really constant—the frame of the ether. In all other frames, it only appeared to be constant. Lorentz shared Poincaré's view on this matter.

As Einstein pointed out in his first paper on relativity, electro-dynamic theory is based on "the kinematics of the rigid body, since the assertions of any such theory have to do with the relationships between rigid bodies (systems of coordinates), clocks, and electro-magnetic processes. Insufficient consideration of this circumstance lies at the root of the difficulties which the electrodynamics of moving bodies at present encounters."[16] His paper is accordingly prefaced by a kinematical part in which the postulates are presented. Here Einstein deduces the relativity of simultaneity, the Lorentz transformation equations, the "Lorentz-FitzGerald" contraction, time dilation, and the composition rule for velocities. In the sequel dynamical part of the paper, these results are applied to specific circumstances in electrodynamics.

For example, with regard to the contraction of lengths in the direction of motion, Einstein may be interpreted as follows: when one makes a measurement of length, one must either have the object in his frame of reference or, failing that, make simultaneous observations of the ends of the object (or some other equivalent procedure). But since observers in different frames of reference will disagree about judgements of simultaneity, it follows that they will disagree as well with regard to judgements of length. The disagreement is not due to any dynamical change in the object; indeed, the construction and nature of the object are not even considered. The disagreement is an artifact of the measuring process itself and may be traced ultimately to Einstein's two postulates. Einstein maintained this kinematical view in his subsequent work. Thus when Paul Ehrenfest pleaded in 1907 that Einstein supply more information about his "theory of electrons" so that a decision could be made on certain paradoxes of electron rigidity Ehrenfest had pointed out, Einstein's

16. A. Einstein, "On the Electrodynamics of Moving Bodies," *op. cit.* (note 9), p. 38.

reply was simply to deny that he had a theory of electrons. Rather, he said, his was a theory concerned with the relationship between rigid rods, perfect clocks, and light signals.[17]

On the other hand, the theory of Lorentz and Poincaré was not restricted to the problem of measuring events by systems in relative motion. It was their hope to be able to subsume all physical phenomena under the assumptions of one theory. Thus within the electron theory, one attempted to explain such seemingly diverse phenomena as heat conductivity, electrical conductivity, energy radiation, reflection, refraction, and dispersion of light, elasticity, and strength of materials. Newtonian forces were not taken as fundamental; they were to be understood in terms of the interactions of electrons with other electrons through the absolutely fixed, space-filling ether. The problem of the electrodynamics of bodies in motion became important only because it was a troublesome area, an area which resisted, until Lorentz' 1904 paper, every assault.

Lorentz assumed the contraction hypothesis first as a way of explaining the Michelson-Morley and other second-order experiments. He later independently postulated the same kind of contraction for electrons. Prior to 1905 he never derived the contraction effect from first principles via his own transformation equations. To Lorentz the contraction was a real phenomenon in the sense that things actually became shorter in the direction of motion as viewed from different inertial frames of reference. Furthermore, in Lorentz' view, not only was the contraction independent of judgements of simultaneity, but discrepancies in such judgements were not significant and were themselves only disagreements on paper. The time dilation which ensues from the transformation equations was only an "aid to calculation."

The Lorentz theory was a dynamical theory throughout. Lorentz even tried to account for contraction by hypothesizing an alteration in the electrical forces between electrons in materials in motion. And when Abraham challenged the deformable electron of Lorentz' on the grounds that such an electron would be unstable, Poincaré

17. A. Einstein, "Bemerkungen zu der Notiz von Herrn Paul Ehrenfest: 'Die Translation deformierbarer Elektronen und der Flächensatz,'" *Ann. d. Phys., 23* (1907), 206–208. See M. J. Klein, "Thermodynamics in Einstein's Thought," *Science, 157* (1967), 509–516, esp. pp. 515–516, and S. Goldberg, "Henri Poincaré and Einstein's Theory of Relativity," *op. cit.* (note 8), 938.

suggested that the stability of the electron was maintained by an external force exerted by the ether.[18] Such considerations are completely absent from the Einstein theory.

In his 1905 paper on relativity, Einstein makes only one mention of the concept of the ether—not, as is often assumed, to say that the ether does not exist, but rather to say that in working out the theory, "the introduction of a 'luminiferous ether' will prove to be superfluous. . . ."[19] In the Lorentz theory, while the ether does not play an active role (disregarding the "Poincaré pressure" which maintained the deformable electron), it is vital. True time, according to Lorentz, is the time measured in the frame of reference at rest in the ether. And it is only in the ether frame that the velocity of light is really the constant c. Lorentz did eventually recognize that one could indeed arrive at the same formal results as he had without using the concept of an ether, but he always preferred a theory which utilized the concept. Then, too, it must be recalled that the theory of electrons required the ether to play the role of mediator and transmitter of electromagnetic fields; the fields of electrons interacted with the fields of other electrons in the ether. Lorentz would have agreed with Poincaré: "Beyond the electrons and the ether, there is nothing."[20]

3. THE RESPONSE TO EINSTEIN'S THEORY OF RELATIVITY IN FRANCE, GERMANY, AND THE UNITED STATES

Outside of scientists at German-speaking universities, the response to Einstein's theory was anything but immediate. In fact, in France, there was no response at all, and one can hardly find any mention even of Einstein's name in the context of problems associated with the theory of relativity prior to Einstein's 1910 visit to France.[21] During the 1905–1911 period, by far the most outstanding figure in France actively working on the problem of the electrodynamics of

18. For discussions of the points raised here, see G. Holton's and S. Goldberg's articles cited in note 8, and also S. Goldberg, "The Abraham Theory of the Electron: Symbiosis between Theory and Experiment," *Archive for History of Exact Sciences,* in press.

19. A Einstein, "On the Electrodynamics of Moving Bodies," *op. cit.* (note 9), p. 38.

20. H. Poincaré, *Science and Method* (New York, Dover, n.d.), p. 209.

21. See A. Einstein, "Le principe de la relativité et ses conséquences dans la physique moderne," *Arch. sci. phys. et nat., 29* (1910), 5–28.

moving bodies was Henri Poincaré. Poincaré never once referred to Einstein in this context. Since we have no private papers of Poincaré to guide us, this puzzle may never be solved with certainty. Recently I have analyzed Poincaré's attitudes towards electrodynamics and towards theories in general[22] and have suggested that he may have held the view that Einstein's work was only a small and rather insignificant part of a much larger theory that Lorentz and he had developed. Apparently Poincaré's influence was responsible for the fact that almost no one else in France had much to say about Einstein's theory prior to Poincaré's death in 1911. Even as late as 1955, the complaint was heard that the syllabus for the *certificat de mécanique rationnelle* contained not one lesson on the kinematics of high velocities.[23]

The response to the theory of relativity in Germany stands in sharp contrast to that in France. All during the period under investigation, Einstein's theory was discussed, criticized, elaborated upon, and defended. While I intend to make a detailed analysis of the response in Germany another time,[24] I want to mention several features of that response here. It was only in Germany that the theory was elaborated upon: physicists such as Max Planck, Max Born, Max von Laue, and Jakob Laub published lengthy studies of various implications of the theory, including generalized mechanics, relativistic thermodynamics, the concept of rigidity, and relativistic optics.[25]

Many German physicists opposed Einstein's theory, but it is only in Germany that its opponents understood it. Thus Max Abraham and Wilhelm Kaufmann both opposed the theory on the grounds that experiments on the specific charge of high-speed electrons gave results which were not in accord with the theory; yet they exhibited a firm grasp of the principles on which Einstein based his analysis.

22. S. Goldberg, "The Silence of Poincaré . . . ," *op. cit.* (note 8).
23. H. Arzelies, *La Cinématique Relativiste* (Paris, 1955), p. vii.
24. I am preparing four papers on the German response to Einstein's theory. The first is on the role of experiment in judgements about the theory; the second is on the debate on the nature of rigid bodies; the third is on the development of generalized relativistic mechanics; and the fourth is on relativistic thermodynamics.
25. Two contemporary reviews of the theory of relativity serve as excellent guides to the literature: A. Einstein, "Ueber das Relativitätsprinzip und die aus demselben gezogene Folgerungen," *Jahrb. Rad. u. Elek., 4* (1907), 411–462; J. Laub, "Ueber die experimentellen Grundlagen des Relativitätsprinzips," *Jahrb. Rad. u. Elek., 7* (1910), 405–463.

They explicitly rejected those principles.[26] Fritz Hasenöhrl also rejected the theory, even though he recognized that, using it, Planck had given a very satisfactory rendition of generalized mechanics and the physics of moving black bodies. Hasenöhrl preferred, he said, Lorentz' theory because it explicitly introduced the concept of the ether.[27]

Regardless of their attitudes towards the theory, German physicists took the theory seriously. This may sound like a strange comment to make as a characteristics of a national response, but the German response stands in sharp contrast to the response in other countries in just this particular. It was the seriousness of the German response, in my view, which ultimately led to the acceptance of relativity, for it insured that the theory would be examined, criticized, and elaborated upon. Once that process began, and the heuristic power of the theory was revealed, it was only a matter of time before it came into general use.

When one turns to the response to relativity in the United States, one finds little evidence that the theory was taken seriously or that any commentator had anything but a dim notion of Einstein's program. The first to deal with the theory itself were G. N. Lewis and R. C. Tolman who, in 1909, published a highly original exposition of the theory.[28] But, like most American scientists, Lewis and Tolman emphasized the practical aspects of the theory. The difference between them and other American scientists was this: they considered the theory to be a practical and useful one with experimentally proven postulates, while most of the others did not. Indeed, the tendency in America was to ridicule the theory of relativity as being totally impractical and absurd. For example, in his Presidential Address to the American Association for the Advancement of Science in 1911, W. F. Magie was moved to proclaim that "I do not believe that there is any man now living who can assert with truth that he can conceive of time which is a function of velocity or is willing to go to the stake for the conviction that his 'now' is another man's 'future' or still another man's 'past.' "[29] Magie concluded that

26. S. Goldberg, *Early Response* . . . , *op. cit.* (note 4), Chap. 2.
27. *Ibid.*
28. G. N. Lewis and R. C. Tolman, "The Principle of Relativity and Non-Newtonian Mechanics," *Phil. Mag.,* *18* (1909), 510–523.
29. W. F. Magie, "Primary Concepts of Physics," *Science, 35* (1912), 281–293; passage cited is on p. 292.

Einstein's theory could not be fundamental, since a fundamental theory must be intelligible to all people. And there were few scientists in America willing or able to dispute him.

4. THE BRITISH RESPONSE TO EINSTEIN'S THEORY

If there is a word to characterize British physics in the nineteenth century it is "ether." "Ether mechanics," as I will term the investigation of the properties of the ether, was not a study restricted to Britain. However, its effect on British physics in the early twentieth century was far more profound that it was on the physics of any other country in Europe. And when one examines the sources of ether theory *construction* in the nineteenth century, the preponderance of British authors is easily discernible.[30] One can trace the concept of the ether in Britain at least as far back as the time of Newton. However, the year 1800 is a convenient bench mark. In that year Thomas Young published his first paper on light as a wave phenomenon. British concern with the existence and nature of the ether grew steadily[31] until, at the end of the century, there were those who believed that the time was close at hand when the true structure of the etherial medium would be revealed to man.[32]

In a sense, it is misleading to speak of a reaction to the theory of relativity in Britain. More accurately, the British were reacting to what they perceived to be an attack on the ether; for, in fact, many British scientists were quite ignorant of the details of the theory of relativity. As late as 1923, N. R. Campbell could write of the average British physicist that he was "still ignorant of Einstein's work and not very much interested in it. Physicists of great ability, who would

30. See the first volume of Whittaker, *op. cit.* (note 6).
31. The role that the ether played in nineteenth-century British physics has been analyzed by several authors. See E. T. Whittaker, *A History of the Theories of Aether and Electricity, op. cit.* (note 6); J. T. Merz, *A History of European Thought in the Nineteenth Century, op. cit.* (note 1); C. C. Gillispie, *The Edge of Objectivity* (Princeton, 1959); Mary Hesse, *Forces and Fields* (New York, 1959). Work on the ether in Britain was concentrated in what has become known as the "Cambridge School," a group of mathematical physicists who studied and worked at Cambridge, and who included Green, Stokes, and Kelvin.
32. The most confident hopes for the imminent solution of the problem of the ether were expressed by Oliver Lodge in his *Modern Views of Electricity*, 1st ed. (London, 1889), p. ix. "[What is the Aether?] is *the* question of the physical world at the present time. But it is not unanswerable. It is in my belief, not far from being answered" (emphasis in original). This statement was repeated in the 2nd edition in 1892, but was deleted from the 3rd edition in 1907.

be ashamed to admit that any other branch of physics is beyond their powers, will confess cheerfully to a complete inability to understand relativity."[33] Although there were other factors involved, such as the direction Rutherford and his experimental school gave to early twentieth-century British physics, the comparative ignorance of and disinterest in relativity on the part of British physicists was primarily due to the unique character of their theoretical commitment to the ether.

Pervasiveness of the Concept of Ether in Britain

Perhaps no figure of nineteenth-century physics contributed more to ether mechanics than William Thomson, Lord Kelvin.[34] Section A, Mathematical Physics, at the British Association for the Advancement of Science in 1907 witnessed Kelvin's last public appearance before his death. At this meeting, he read a paper on the motions of the ether. The view he put forward (his views of the ether were in constant flux) was that the ether must be an "elastic, compressible, nongravitational solid."[35] He cited a paper by Oliver Lodge[36] which had concluded that "every cubic millimeter of the universal ether of space must possess the equivalent of a thousand tons and every part must be squirming with the velocity of light.[37] Lodge based this conclusion on the fact that the presence of transverse waves in the interior of a fluid can only be explained using gyrostatic principles, and on the fact that the internal circulatory motion of a fluid must be comparable with the speed with which such waves are propagated, which, in this case, was the velocity of light.

The significance of Kelvin's and Lodge's views is not that they are based on the assumption of an ether—many physicists in Europe still clung to the belief in the ether in 1907. Rather, what is significant

33. N. R. Campbell, *Relativity* (Cambridge, 1923), p. v.

34. On the influence of Kelvin in this regard, see S. P. Thomson, *The Life of William Thomson, Baron Kelvin of Largs*, 2 vols. (London, 1910); E. T. Whittaker, *op. cit.* (note 6), *1*, Chap. 9, "Models of the Aether"; J. T. Merz, *op. cit.* (note 1), *2*, Chap. 2.

35. Anon., "Mathematics and Physics at the British Association, 1907," *Nature, 76* (1907), 457–462, p. 457.

36. Oliver Lodge (1851–1940) studied at the Royal College of Science and at University College, London. In 1881 he became Professor of Physics at Liverpool, and in 1900 he was appointed first principal of the new university at Birmingham.

37. Anon., "Mathematics and Physics at the British Association, 1907," *op. cit.* (note 35), p. 459.

about these views—views that were shared by many leading British physicists—is that they regard the ether as a mechanical object. The fact that the ether must exhibit properties unlike any substance within human experience had driven Lorentz, for example, to divest the ether of all mechanical properties. In fact, to Lorentz the ether ultimately represented nothing more than the frame of reference in which absolute time was to be measured, a frame that was, by his own admission, undetectable in principle.[38]

It is easy to trace the source of Kelvin's 1907 mechanical ideas on the ether; he had spent a major portion of his professional life building a host of mechanical models which might represent the ether. Nor is it hard to find the source of Lodge's conception of the ether. His major research interests were in the propagation of electromagnetic waves; and if one did away with the ether, in his view, it would be impossible to account for the fact of wireless transmission.[39] But the ether meant much more than that to him. His biographers said that the "concept of the ether of space dominated his whole life. In his youth as a physicist he was well in the forefront of the science of his day, following the steps of Faraday and Maxwell by concentrating his attention on the properties of the medium surrounding electrified bodies."[40] Late in life, in the preface to a book in which he summed up his philosophical outlook, he wrote that the "Ether of Space has been my life study. . . . I always meant some day to write a scientific treatise about the Ether of Space; but when in my old age I came to write this book, I found that the Ether pervaded all my ideas, both of this world and the next. I could no longer keep my treatise within the proposed scientific confines; it escaped in every direction and now I find has grown into a comprehensive statement of my philosophy."[41]

38. H. A. Lorentz, "Alte und Neue Fragen der Physik," *Phys. Zs., 11* (1910), 1234–1257. See S. Goldberg, "The Lorentz Theory of Electrons . . . ," *op. cit.* (note 8).

39. O. Lodge, *Relativity: A Very Elementary Exposition,* 3rd ed. (London, 1926). The first edition was published in 1925.

40. R. A. Gregory and A. Ferguson, "Oliver Lodge," *Obituary Notices of Fellows of the Royal Society, 3* (1941).

41. O. Lodge, *My Philosophy: Reporting My Views on the Many Functions of the Ether of Space* (London, 1933), Preface. Lodge had a great interest in psychical phenomena, conducting experiments on extrasensory perception and consulting with some of the major mediums of his day on their work. He thought that besides the role that the ether played in the physical world, the ether was the agent responsible for the transference of thoughts in extrasensory phenomena (A. Ferguson, "Oliver J. Lodge," *DNB 1931–40*).

From an early date Lodge's clear ideas on the ether played an integral part in his work. Subsequent editions in 1892 and 1907 of his *Modern Views of Electricity* reveal that his etherial conception of electricity did not change. In his 1907 preface he remarked that "it is noteworthy that few actual *corrections* have had to be made; showing that new discoveries are of a supplementary rather than a revolutionary character. The doctrine expounded in this book is the etherial theory of electricity. Crudely, one may say that as heat is a form of energy, or mode of motion, so electricity is a form of ether or a mode of etherial manifestation."[42] He thought that nonphysicists "may possibly be surprised to see the intimate way in which the ether is now spoken of by physicists and the assuredness with which it is experimented on. They may be inclined to imagine that it is still a hypothetical medium whose existence is a matter of opinion. Such is not the case. The existence of an ether can legitimately be denied in the same terms as the existence of matter can be denied, but only so. . . . The evidence for ether is as strong and direct as the evidence for air."[43] In an appendix to *Modern Views of Electricity,* based on a lecture given in 1882, Lodge elaborated upon his conception of the ether. He noted that the idea of an ether was not a new one. The idea, he said, arose as soon as it became clear just how enormous was the extent of interstellar space. As to what was in this space, metaphysics had given two different answers. Some thinkers believed a vacuum could not possibly exist, while others believed that emptiness was necessary to provide room for motion. If metaphysics was to have any weight or validity for Lodge, it had to provide an unconscious appeal to common sense.

> If a highly developed mind or set of minds, find a doctrine about some comparatively simple and fundamental matter absolutely unthinkable, it is evidence and is accepted as good evidence that the unthinkable state of things is one that has no existence. . . .
>
> .   .   .   .
>
> Now if there is one thing with which the human race has been more conversant from time immemorial than an ether and concerning which

42. O. Lodge, *Modern Views of Electricity,* 3rd ed., pp. v–vi. Emphasis in original. Whether or not Lodge was referring to the theory of relativity here is not clear, since there is no reference to the theory in this edition.

43. O. Lodge, *Modern Views of Electricity,* 1st ed., pp. vii–viii. These statements also appear in the prefaces to the second and third editions.

more experience has been unconsciously accumulated than about almost anything else that can be mentioned, it is *the action of one body on another*. . . . Every activity of every kind that we are conscious of may be taken as an illustration of the action of one body on another. Now I wish to appeal to this mass of experience, and to ask, is not the direct action of one body on another across empty space, and with no means of communication whatever, is not this absolutely unthinkable? I think that whenever one body acts by obvious contact, we are satisfied and have a feeling that the phenomenon is simple and intelligible; but that whenever one body apparently acts on another at a distance, we are irresistibly impelled to look for the connecting medium.[44]

"The medium," Lodge claimed, "is now accepted as a necessity by all modern physicists."[45] He characterized the "modern view" of the ether as "one continuous substance filling all space; which can vibrate as light; which can be sheared into positive and negative electricity; which in whirls constitutes matter; and which transmits by continuity and not by impact, every action and reaction of which matter is capable."[46] Though Lodge first expressed these ideas in 1882, he repeated them in 1892, 1907, and, indeed, to the end of his life.[47] Lodge's belief in the ether was not simply an unthinking adherence to British tradition. It stemmed from a simple and straightforward metaphysics, an inability to conceive of two bodies affecting each other without some connection between the two.

In a new chapter[48] added to the 1907 edition of *Modern Views of Electricity*, Lodge again detailed his opinions on the constitution of the ether. He said it must be continuous[49] and frictionless, and at the

44. O. Lodge, "The Ether and Its Functions," delivered at the London Institution, 28 Dec. 1882; see *Modern Views of Electricity*, 1st ed. pp. 327–358, quotation on pp. 328–332.
45. *Ibid.*, p. 338.
46. *Ibid.*, p. 358.
47. O. Lodge, *My Philosophy, op. cit.* (note 41), *passim*.
48. This chapter was also published as a paper, "Modern Views of the Aether," *Nature*, 75 (1907), 519–522; and it served as the basis for Lodge's 1907 address to the British Association.
49. The property of continuity is one on which Lodge had much to say. In his Presidential Address to the British Association in 1913, entitled "Continuity," Lodge commented on some recent trends in physics:

Indeed, a whole system of non-Newtonian mechanics has been devised having as its foundation the recently discovered changes which must occur in bodies moving at speeds nearly comparable with that of light. . . . But I do not consider it so revolutionary as to overturn Newtonian mechanics. After all, a variation of mass is familiar enough and it would be a great mistake to say that Newton's second law breaks down

same time it must be rigid and perfectly elastic. Following Kelvin, Heaviside, FitzGerald, and Larmor, Lodge assumed a magnetic field to be a circulation of etherial fluid. The direction of flow was along the lines of magnetic induction, and the magnetic intensity at any point was proportional to the speed of flow. Assuming that the circulation of ether around the equator of an electron was equal to the velocity of the electron's forward motion and that the mechanical energy and the magnetic energy of the field were equal, Lodge arrived at a relationship between known electromagnetic quantities and the unknown density of the ether. From this he calculated the density to be of the order of magnitude of 10^{12} grams per cubic centimeter.[50] All of this supports the view that, to Lodge, the ether was a real substance capable of carrying electromagnetic energy, of transmitting all known forces, and of serving as the Cartesianlike source of charge and of matter. In Lodge's hands, the ether had a substantiation which one was not likely to find in France or Germany.

To a modern scientist, reading Lodge's work is similar to falling to the bottom of Lewis Carroll's rabbit hole. But it would be too easy to dismiss Lodge's work as the fantastic creation of an overimaginative individual. I think it is important that Lodge's assessment of the place and importance of the ether be taken quite seriously as a barometer of the thinking of many of his British colleagues, and as a key to the kinds of issues that were considered important in physics in Britain near the turn of the century. For while Lodge's passion for the ether was more vocal and more obvious than was that of most of his colleagues, he is more typical than aberrant. For example, while not everyone agreed with Lodge's calculations concerning the density of the ether, the calculations were taken seriously

merely because mass is not constant. . . . In fact, variable masses are the commonest, for friction may abrade any moving body to a microscopic extent (*Brit. Assoc. Rep., 1907*, pp. 13–14).

. . . .

I urge that we remain with or go back to, Newton. I see no reason against retaining all Newton's laws discarding nothing, but supplementing them in the light of further knowledge" (*ibid.*, p. 15).

Lodge believed that the main trend of the controversy over relativity was over continuity vs. discontinuity (*ibid.*, p. 22), and he declared himself an "upholder of ultimate continuity and a fervent believer in the Ether of Space" (*ibid.*, p. 32).

50. O. Lodge, "Modern Views of the Aether," *op. cit.* (note 48).

and did arouse considerable comment in both Britain and America. C. W. Richardson, professor of physics at Princeton University, thought that the flow of ether should be identified with the Poynting vector. This, he argued, would seem most natural.[51] C. V. Burton, on the other hand, eschewed any program that identified the flow of ether either with the magnetic or the electric fields or with the Poynting vector. Rather, Burton argued, the ether was "a-phenomenal," by which he meant that it was "without influence on the senses of the observer or upon any instrumental test or measurement which he can make."[52]

For Ebenezer Cunningham,[53] the ether "is in fact not a medium with an objective reality but a mental image which is only unique under certain conditions."[54] Just what those conditions were, Cunningham did not specify. But he was willing to ascribe to the view that each frame of reference has its own ether, each ether having a uniform velocity corresponding to the motion of its frame with respect to absolute space.[55] In other words, there would have to be an ether corresponding to every degree of uniform motion. In this way, an observer in any frame of reference would be unable to detect the ether associated with his frame. But what of the ethers associated with all the other frames of reference? Cunningham had no comment. Presumably, since he urged that the ether be taken as a "mental image," he was not bothered by such a question.

Cunningham drew inspiration for this position from Joseph Larmor,[56] though Larmor himself could not concur with any theory that ascribed "a uniform motion to the whole of the ether because there is no conceivable means of producing or altering such a motion."[57] Larmor believed that "an infinitely extended aether postulates absolute motion as a fact in the only real sense of that term, namely, motion relative to the remote quiescent regions of the aether: since

51. O. W. Richardson, "Structure of the Aether," *Nature, 76* (1907), 78.
52. C. V. Burton, "The Structure of the Aether," *Nature, 76* (1907), 150–151.
53. Ebenezer Cunningham (1881–   ) attended St. John's College, Cambridge, where he graduated senior wrangler in 1902. He was Smith's Prizeman in 1904. He was lecturer in mathematics at the University of Liverpool between 1904 and 1907; he later taught at University College, London. He returned to St. John's College in 1911, where he remained in active work until 1946.
54. E. Cunningham, *The Principle of Relativity* (Cambridge, 1914), Chap. 15.
55. E. Cunningham, "Structure of the Aether," *Nature, 76* (1907), 222.
56. *Ibid.*
57. J. Larmor, "The Aether and Absolute Motion," *Nature, 76* (1907), 269–270.

that determination is made, arguments from relativity of motion must lapse."[58] Larmor,[59] whose influence on British physics was great, rejected any substantial view of the ether. He expressed himself on this subject most forcefully in his book *Aether and Matter*,[60] where he said that an understanding of the general dynamical and physical relationships of matter must rest on the assumption that matter consists of atomic aggregates immersed in the ether. But "all that is known (or perhaps need be known) of the aether itself may be formulated as a scheme of differential equations deriving properties of a continuum in space, which it would be gratuitous to further explain by a complication of structure."[61] Eddington reported that Larmor believed the ether was not a material medium, and that he saw no point in trying to explain its properties as though it were. Larmor's ether was only to be defined by mathematical equations, and in this regard it resembled the ether of Lorentz. Larmor, like Lodge, believed that the ether was something other than a material medium. But Lodge believed that it was at least a detectable entity; his mysticism, his penchant for physical models,[62] and his experience with electromagnetic radiation led him to the belief that the ether, while not material, was substantial. Larmor, however, trained from the start as a theoretician, embraced a mathematical, abstract view of the ether.

   J. J. Thomson, like Lodge, was an experimentalist,[63] and his views

58. *Ibid.*
59. Joseph Larmor (1871–1942) attended St. John's College, Cambridge, where he graduated first wrangler in 1880. He succeeded G. G. Stokes as Lucasian Professor at Cambridge.
60. J. Larmor, *Aether and Matter: A Development of the Dynamical Relations of the Aether to Material Systems on the Basis of the Atomic Constitution of Matter* (Cambridge, 1900). This book, which was based on several earlier papers, won the Adams Prize in 1900.
61. *Ibid.*, p. 78.
62. Lodge devoted a good deal of time to constructing models similar to those of Clerk Maxwell. His model for electricity was an elaborate gear and pinion affair. "If we think of electricity in the several molecules of the insulating medium connected like so many cogwheels gearing into one another and also into those of the metal, it is easy to picture a sideways spread of rotation brought about by the current just as a moving rack will rotate a set of pinions gearing into it and into each other." (O. Lodge, *Modern Views of Electricity*, 1st ed., p. 177. The same model is presented in all later editions.
63. J. J. Thomson (1856–1940) was Cavendish Professor of Experimental Physics at Cambridge from 1894 to 1919. According to Rayleigh, Thomson's whole life was devoted to experimental physics. (Rayleigh, "Thomson, J. J.," *Obituary Notices of Fellows of the Royal Society, 3* [1941].)

on the ether were similar to those of Lodge. He expressed these views most clearly in his Adamson Lecture at the University of Manchester in 1907.[64] His concern here was with the degree of generality of Newton's third law. He noted that a system in which the third law did not apply could not be imitated by any mechanical model, and that there did seem to be situations in which the third law was not valid. Consider, he said, two electrical bodies A and B in rapid motion. The electromagnetic forces these two bodies exert on each other are not equal and opposite, unless the two bodies are moving with the same speed in the same direction. As a result, the momentum of the system composed of A and B does not remain constant. If one were to conclude from this that electrified bodies are not subject to the third law, it would mean "giving up the hope of regarding electrical phenomena as arising from the properties of matter in motion." But, Thomson noted with relief, such a conclusion was not necessary. The solution lay in hypothesizing an "invisible system," connected to the system containing A and B, which possesses mass and momentum; and any momentum gained or lost by the system containing A and B could be ascribed to the invisible system.[65] He identified the invisible system with the ether. Thomson might be interpreted as suggesting that the ether was simply a device to save and extend the applicability of Newton's third law; but the ether meant more to him than this. His Presidential Address to the British Association at Winnipeg in 1909 leaves no doubt about his position: "The aether is not a fantastic creation of the speculative philosopher; it is as essential to us as the air we breathe." It is the "seat of electrical and magnetic forces," and it is the "bank in which we may deposit energy and withdraw it at our convenience."[66]

I have presented a sampling of opinion in Britain during the first decade of this century on the nature of the ether. This opinion was not a response to Einstein's theory, since most British physicists had no knowledge of the theory at this time. It would be the easiest thing in the world to sweep early twentieth-century British considerations of the ether aside. They were not heuristic. They did not lead any-

64. J. J. Thomson, *On the Light Thrown by Recent Investigations on Electricity and on the Relation between Matter and Ether; The Adamson Lecture Delivered at the University on November 4, 1907* (Manchester, 1908).

65. *Ibid.*, pp. 7–8.

66. J. J. Thomson, "Presidential Address to the British Association at Winnipeg, 1909," *The Electrician, 63* (1909), 776–779, quotation on p. 778.

where, but rather represented something of a last breath for a theory that was nearly moribund. But they were also unique. In Germany, physicists, even those physicists committed to the ether, were engaged in a searching analysis of Einstein's theory. While one or two leading German physicists can be cited who ignored the theory and concentrated on ether mechanics, this was not typical, nor was their work given prominent display or attention.[67] In France, there was little attention paid to Einstein's work, but then, too, there was little active concern with the nature of the ether. In America, though there was also visible concern with the ether, physicists followed British work more than striking out in independent directions.

British Model Building and the Ether

In J. T. Merz's view the differences in scientific attitude and methodology between countries of Western Europe were more pronounced at the beginning than at the end of the nineteenth century. At the end of the century the "intercourse of the different nations," he said, had "done much to destroy these national peculiarities."[68] However, when one compares British and German physics in 1905, there is little evidence that national peculiarities had been destroyed. The comments of the French philosopher-physicist Pierre Duhem, writing on British physics in 1906, are illuminating on this point.[69] The French, he said, are astonished by the way the English always accompany the exposition of their theories with models. For instance, the "French or German physicist conceives, in the space separating two conductors, abstract lines of force having no thickness or real existence; the English physicist materializes these lines and thickens them to the dimensions of a tube which he will fill with vulcanized rubber." Lodge epitomized for Duhem the Englishman's commitment to the purely mechanical explanation of physical phenomena.

---

67. See W. Kass, "Die Natur des Äthers," *Das Weltall, 8* (1908), 134–136; P. Lenard, "Äther und Materie," *Sitzgb. Heidelberg. Ak. Wiss.* (1910), 37 ff.; W. Ritz, "Du rôle de l'éther en physique," *Riviste di Scienza, 3* (1908), 260–274; H. Witte, "Ätherfrage," *Jahrb. Rad. u. Elek.,* 7 (1910), 205–261.

68. J. T. Merz, *A History of European Thought, op. cit.* (note 1), *1,* 252.

69. P. Duhem, *The Aim and Structure of Physical Theory* (New York, 1962), pp. 69–72.

When we open Lodge's *Modern Views of Electricity,* we find "nothing but strings which move around pulleys, which roll around drums, which go through pearl beads, which carry weights; and tubes which pump water while others swell and contract; toothed wheels which are geared to one another and engage hooks. We thought we were entering the tranquil and neatly ordered abode of reason, but we find ourselves in a factory." Duhem observed that "understanding a physical phenomenon is, therefore, for the physicists of the English school, the same thing as designing a model imitating the phenomenon; whence the nature of material things is to be understood by imagining a mechanism whose performance will represent and simulate the properties of bodies. The English school is completely committed to the purely mechanical explanation of physical phenomena."

To Duhem, the construction of mechanical models was not a sufficient characteristic in itself to set off British physics from the physics of the Continent. It was rather the particular form that model building in Britain assumed that was distinct. In continental Europe, he declared, "the sense for abstraction may lapse, but it never falls asleep completely,"[70] as it had done in England.

> When an English physicist seeks to construct a model appropriate to a group of physical laws, he is not embarrassed by any cosmological principle, and is not constrained by any logical necessity. He does not aim to deduce his model from a philosophical system nor even to put it into accord with such a system. He has only one object: to create a visible and palpable image of the abstract laws that his mind cannot grasp without the aid of this model. . . . The English physicist does not therefore ask any metaphysics to furnish the elements with which he can design his mechanisms. He does not aim to know what the irreducible properties of the ultimate elements of matter are.[71]

While much of what Duhem says is valid, he is wrong on two counts. First of all, there were Englishmen, such as Larmor, who were perfectly willing to let the concept of the ether rest on nothing more than the mathematical equations that describe the electromagnetic field. Second, it is simply incorrect to say that the British penchant for model building overrode or ignored all cosmological principles

70. *Ibid.,* p. 73.
71. *Ibid.,* p. 74.

and that it did not stem from any uniform metaphysics. The model building itself can be construed as representing a metaphysics. The metaphysics of Oliver Lodge is quite plain: he could not imagine a world in which action at a distance could be operative. It would be quite natural for him to rely on some medium or system of particles to intervene between bodies interacting electrodynamically. J. J. Thomson wanted to preserve the conservation of momentum in all physical interactions and to leave open the possibility of a mechanical explanation of all phenomena; this, again, is certainly a metaphysics.

The case of J. J. Thomson and the vortex atom offers a counter-example to Duhem's further claim that the English scientist was prone to adopt inconsistent models for different phenomena.[72] The study of vortex motion, especially in connection with the motion of the ether, was a favorite subject in Britain in the second half of the nineteenth century. J. J. Thomson submitted an essay for the Adams Prize in 1882 on the subject of the interaction of two vortex rings. He was attracted to the subject by Kelvin's suggestion that matter might consist in vortex rings in an ideal fluid.[73] The "spartan simplicity" of this suggestion appealed to him, as did its promise to lead to a "theory more fundamental and definite than any that had been advanced before." Though Kelvin's enthusiasm for the explanatory power of the vortex ring faded, J. J. Thomson became more enthusiastic with the passing of time.[74] "I regard," the latter once commented, "the vortex-atom explanation as the goal [sic] at which to aim."[75] In his 1907 *Corpuscular Theory of Matter,* he wrote

72. There is some truth in Duhem's claim. Kelvin was probably the greatest offender in the building of inconsistent models. There are plenty of such examples in his 1884 Baltimore lectures. It was in these lectures that he made his famous admission that "I never satisfy myself until I can make a mechanical model of a thing. If I can make a mechanical model, I understand it" (*Lectures on Molecular Dynamics and the Wave Theory of Light* [Baltimore, 1884]).

73. J. J. Thomson, *Recollections and Reflections* (London, 1936), pp. 94–95.

74. Kelvin finally abandoned the model: "I am afraid it is not possible to explain all the properties of matter by the vortex-atom theory alone. . . . We may expect that the time will come when we shall understand the nature of an atom. With great regret I abandon the idea that a mere configuration of motion suffices" (quoted in J. T. Merz, *A History of European Thought, op. cit.* [note 1], 2, 182). On the other hand, J. J. Thomson said: "With reference to the vortex-atom theory, I do not know of any phenomenon which is manifestly incapable of being explained by it; and personally, I generally endeavour (often without success) to picture to myself some kind of vortex-ring mechanism to account for the phenomenon with which I am dealing" (*ibid.,* p. 183).

75. *Ibid.*

that he considered the vortex theory to be "fundamental."[76] The point is that in all of his work Thomson was faithful to the vortex model. Thomson, like Lodge, eventually accepted the results of relativity, but not the spirit. His commitment to the ether remained unspoiled; he could account for the results of relativity his way. Maxwell's equations and the electrical theory of matter, without the aid of relativity, had explained the contraction of moving bodies and the variation of mass with velocity. He thought that it was "reasonable to regard Maxwell's equations as the fundamental principle rather than that of relativity, and also to regard the ether as the seat of mass, momentum, and energy of matter."[77]

Duhem's thesis must be qualified in yet another way. He implies that British model building was part of a commitment to a completely mechanical view of nature. One can certainly find this commitment in Kelvin who, failing to find an adequate mechanical model of electromagnetic phenomena, eventually despaired of being able to understand electromagnetism. But by the turn of the century British physicists were beginning to abandon their frustrating attempts to provide a mechanical model of electromagnetism. Rather than trying to show that all electrical phenomena were fundamentally mechanical, the British now undertook to show that all mechanical phenomena were fundamentally electrical. Larmor's *Aether and Matter* was written from this point of view. J. J. Thomson, while clinging to the vortex theory of the atom, was of the opinion that the experimental results of electrodynamics could be accounted for on the basis of the electrical theory of matter. He was also of the opinion that the entire mass of the electron was electrical,[78] and that, since the entire mass of the atom is due to positive and negative charges, the entire mass of the atom was electrical.[79]

The change in allegiance from a mechanical to an electrical view of the universe was partly due to a feeling of failure and discouragement. It also owed much to the influence of the work of Lorentz,

76. J. J. Thomson, *The Corpuscular Theory of Matter* (London, 1907), pp. 1–2.
77. J. J. Thomson, *Recollections and Reflections, op. cit.* (note 73), pp. 432–443.
78. J. J. Thomson, *The Corpuscular Theory of Matter, op. cit.* (note 76), pp. 301–302.
79. *Ibid.*, pp. 1–2. Thomson's adherence to an electrical view of matter must be qualified in at least one regard. While holding that the entire mass of the atom should be considered as electrical, he did cling to the hope that electrical phenomena could be understood in terms of the properties of bodies in motion.

111

work which many British physicists saw as paralleling that of Larmor.[80] Lorentz' attempts at constructing first- and second-order theories of moving bodies were widely interpreted, with justification, as assuming an electrical view of matter. In the eyes of most British physicists, Lorentz' attempts had been successful, and they saw this as tending to confirm the electrical view of matter. To them the Lorentz theory was not *ad hoc;* rather, it represented a type of theory they were quite familiar with, a type that involved the tailoring of a model to fit the data.

Lodge viewed the null result of the Michelson-Morley experiment as confirmation of the contraction effect, which Lorentz and Fitz-Gerald independently hypothesized to account for that experiment. It was while Lodge was discussing the difficulties of the Michelson experiment with FitzGerald that FitzGerald first hit upon the contraction idea. Lodge later recalled their conversation: FitzGerald suggested that " 'perhaps the stone slab is affected by the motion.' I rejoined that it was a 45 degree shear that was needed. To which he replied, 'Well that's all right—a simple distortion.' *And very soon he said, 'And I believe it occurs and that the Michelson experiment demonstrates it.'* And is such a hypothesis gratuitous? Not at all: *in the light of the electrical theory of matter such an effect ought to occur.*"[81]

When J. J. Thomson said that the fact that momentum was not conserved in the interactions of moving charges *proved* that the invisible universe of the ether really existed, he was expressing a limited view of physical explanation similar to that of Larmor, FitzGerald, and other British physicists. Thomson argued that theories were not intended to explain anything in an ultimate sense: "From the point of view of the physicist, a theory of matter is a policy rather than a creed; its object is to connect or coordinate apparently diverse phenomena, and above all to suggest, stimulate and direct experiment."[82] But if that is the case, why were British physicists unwilling to give an ear to Einstein's theory, a theory which could certainly "coordinate apparently diverse phenomena"?

80. In a recent personal communication with the author, Cunningham referred to the Lorentz transformations as "commonly known as the Larmor-Lorentz transformations."

81. O. Lodge, "Continuity," *op. cit.* (note 49), pp. 56–57. Emphasis added. See also O. Lodge, *My Philosophy, op. cit.* (note 41), p. 69.

82. Thomson, *The Corpuscular Theory of Matter, op. cit.* (note 76), pp. 1–2.

The answer is that the theory could not be accepted in Britain until it could be made consonant with the existing British ideas about the ether. First, British physicists had to implicitly deny relativity the status of a theory; second, they had to make it a subordinate principle, modifying the electron theory of matter; and third, they had to redefine the ether so as to render it undetectable in principle and describable by equations agreeing with those of the theory of relativity.

The Contributions of Ebenezer Cunningham

Lodge's observation in 1907[83] that new theories were supplementary to the electron theory of matter rather than revolutionary provided the stepping-stone to the introduction of relativity theory in Britain. Ebenezer Cunningham was the first British physicist to give serious consideration to relativity. He wrote about it first in 1907 in a paper that dealt with Abraham's objection to the Lorentz electron.[84] Abraham had argued that there was no external source of energy available for deforming the Lorentz electron and that, furthermore, the deformable electron would be unstable.[85] Cunningham demonstrated that Abraham's conclusions were true only because Abraham had imposed his own rigid electron[86] on Lorentz' equations for the deformable electron. He claimed that if one allowed a change in dimensions in the direction of motion, both energy and momentum would be conserved and the electron would be stable.

Cunningham said that "Maxwell's equations represent equally well the sequence of electromagnetic phenomena relative to a set of axes moving relative to the aether as relative to a set of axes fixed in the aether."[87] This was his statement of the principle of relativity, from which he deduced "how a light wave traveling outward in all directions with a velocity c relative to one observer A, may at the

83. Lodge, "Continuity," *op. cit.* (note 49).
84. E. Cunningham, "On the Electromagnetic Mass of a Moving Electron," *Philosophical Magazine, 14* (1907), 538–547.
85. M. Abraham, *Theorie der Elektrizität* (Leipzig, 1905), *2,* 205 ff.
86. See M. Abraham, "Prinzipien der Dynamik des Elektrons," *Annalen der Physik, 10* (1903), 105–179.
87. E. Cunningham, "On the Electromagnetic Mass of a Moving Electron," *op. cit.* (note 84), p. 538.

same time be traveling outward in all directions with the same velocity relative to an observer B moving relative to A with velocity v."[88] By the technique of equating the expressions for two spheres of light, he derived the Lorentz transformations. When A. H. Bucherer challenged this derivation on the grounds that he was "not aware that such a requirement [invariance of the velocity of light] . . . [was] necessary to explain any known fact of observation,"[89] Cunningham replied[90] that he was not making the claim that the invariance of the velocity of light was a fact of experience; rather, he took it to be a statement derivable from the principle of relativity. His interpretation of relativity stemmed from his conviction that the ether was not a substantive entity and that an ether had to be defined for each inertial frame of reference. In each such frame the ether would be at rest, and the velocity of light would be the same in all directions.

Cunningham's eclectic attitude can be further illustrated by two papers he published in 1909. One of these, "The Motional Effects on the Maxwell Aether-Stress,"[91] represents an effort in pure ether mechanics. The task that Cunningham set himself in this paper was to bridge the gap between Maxwell's ether-stress and Newton's third law. In the second paper Cunningham came back to the problem of relativity. He wrote that the absence of detectable effects of the earth's motion relative to the ether had been "fully accounted for by Lorentz and Einstein *provided* the hypothesis of electromagnetism as the ultimate basis of mater can be accepted so that the only available means of estimating the distance between two points is the measuring of the time of propagation of the effects between the bodies, such propagation taking place in accordance with the equa-

88. *Ibid.*
89. A. H. Bucherer, "On the Principle of Relativity," *Philosophical Magazine, 15* (1907), 318. Bucherer had developed his own theory of the electron which contained a transformation equation for the mass of a moving electron and for the dimensions of the electron. The theory relied on the assumption that the volume of the electron was an invariant to all observers, and the contraction occurred in such a way as to maintain a constant volume. Subsequently Bucherer relinquished his theory in favor of Einstein's as a result of experiments he performed on the specific charge of the electron.
90. E. Cunningham, "On the Principle of Relativity and the Electromagnetic Mass of the Electron: A Reply to Dr. A. H. Bucherer," *Philosophical Magazine, 16* (1908), 423–428.
91. E. Cunningham, "The Motional Effects on the Maxwell Aether-Stress," *Proceedings of the Royal Society,* Ser. A, *83* (1909), 109–119.

tions of the electron theory."[92] Cunningham saw the "foundations" of the theory of relativity as the Lorentz transformation.[93]

In 1911 Cunningham was asked by the Cambridge University Press to write a book on the theory of relativity.[94] The book came out in 1914, and in it he gave his view of the place of relativity in physics at the time. He believed that *"there is a real place for it as a hypothesis supplementary to and independent of electrical theory owing to the limitations to which that theory is subject"*[95] (emphasis added). And he believed that it

> holds the same place in the physical thought of today, that the [Newtonian] principle of dynamical relativity held in the time when the laws of dynamics were considered as ultimate and all embracing. It consists in the general hypothesis, based on a certain amount of experimental evidence that *the problem of determining in a physical sense the absolute velocity of a body is one that can no more be solved uniquely by the help of optical and electrical phenomena, than it could by means of dynamical observations.* If we speak of a "fixed aether" as the background of electrical activity, it is the hypothesis that *the velocity of any piece of matter relative to the aether is unknowable.* To put it more precisely, it states that we *neither have nor expect to have any experimental evidence of the uniqueness of the framework which we call "the aether" but that if there is one, there is an infinite number of such frames of reference, any one of which has, relative to another, a uniform velocity of translating without rotation, the velocity being of arbitrary magnitude.* . . . [Relativity] is an *empirical* principle, suggested by an observed group of facts, namely the failure of experimental devices for determining the velocity of the earth relative to the luminiferous aether, and would make it a *criterion* of theories of matter that they should give an account of this failure, and it suggests modifications where the theory is insufficient to do so. But like all physical principles, it is to be probed by further experience.[96]

In one fell swoop, Cunningham had reduced the "theory of relativity" to the first postulate, "the principle of relativity." He had

92. E. Cunningham, "The Principle of Relativity in Electrodynamics and an Extension Thereof," *Proceedings of the Mathematics Society of London, 8* (1909), 77–98, quotation on p. 77.
93. *Ibid.*, p. 78.
94. E. Cunningham, personal communication.
95. E. Cunningham, *The Principle of Relativity* (Cambridge, 1914), p. v.
96. *Ibid.*, pp. 7–8.

made it supplementary to the electron theory of matter, and he had reasserted his position of the undetectability of the ether. He viewed the second postulate of relativity as a consequence of the first. The first postulate itself was, in his view, a generalization from experience about the ether—not a "metaphysical dogma." His consignment of the theory of relativity to the status of a principle supplementary to the electron theory was intimately related to his view of that principle as an empirical generalization. After reviewing some experiments on the motion of the earth through the ether, he explained that "it is because the experimental evidence extends into regions where existing electrical theory is insufficient, that the principle of relativity becomes of importance as a supplementary and independent hypothesis."[97]

Cunningham differentiated between the work of Lorentz, Larmor, and Einstein on the one hand, and the work of nineteenth-century mechanists on the other, terming the latter work "experimental relativity."[98] By this he meant that the failure to find evidence of absolute motion relative to the ether had incidentally led to a revision of the concepts of space and time.[99]

> It may be said that the arguments differ *only* in the point of view, but for the moment it is the new point of view that is vital. The older point of view was the outcome of the strivings of the mechanical school of physicists after an objective mechanical elastic ether and their adoption of a metaphysical concept of space and time. But when mechanics is placed in a derivative place and electrical theory in the fundamental position, when we are prepared to admit that many of the properties of matter are, when analysed, found to be compounded of electrical action, it becomes very necessary to consider with an open mind the possibility that some of what seem the most obvious modes of thought may require revision, and among other things to avoid the vicious circle of ideas involved in likening the aether to a species of matter with molar properties which are an idealization of the properties of matter as we imperfectly know them.[100]

There was no doubt in Cunningham's mind that the mechanists' search for what he called an "objective aether" had failed. But he

97. *Ibid.,* p. 41. Emphasis in original.
98. *Ibid.,* p. 45.
99. *Ibid.*
100. *Ibid.,* p. 50.

did not want to abandon the ether as a concept. He recognized that the ether would have to be "subject to the Einstein transformation of space and time when the frame of reference is altered."[101] But while he was cautious about the ether, he devoted a whole chapter of his book to "Relativity and an Objective Aether." He argued that he could make the principle of relativity and the concept of an ether compatible if he were allowed to make the ether *a priori* undetectable. This is the sense in which he conceded to the ether an "objective reality." Thus the program that Cunningham had outlined as early as 1907 was culminated.[102] Despite the efforts of Cunningham to make a reconciliation between the theory of relativity and the ether, most British physicists simply ignored relativity and the work of Einstein and maintained an unquestioning commitment to an ether-based physics.[103] In the entire period between 1905 and 1911, only one physicist seems to have questioned the validity of the concept of the ether; that physicist was Norman R. Campbell.

The Contributions of Norman R. Campbell

Campbell[104] challenged the ether in a series of three papers beginning in 1910.[105] The concept of the ether, he said, seemed unsatisfac-

101. *Ibid.*, p. 52.
102. See E. Cunningham, *Relativity and the Electron Theory* (London, 1915). This book is an abbreviated version of *The Principle of Relativity*. In the 1915 publication Cunningham repeated his claim that the theory of relativity was supplementary to the electron theory. But he noted with some force the *ad hoc* nature of the Lorentz contraction (*ibid.*, p. 25).
103. During the period 1905–1911, publications on ether mechanics continued unabated. Typical titles are C. V. Burton, "The Sun's Motion with Respect to the Aether," *Philosophical Magazine, 19* (1910), 417–423, and J. J. Thomson, "Light Thrown by Recent Investigations on Electricity and the Relation between Matter and Ether," *Smithsonian Reports, 1909,* pp. 233–244.
104. N. R. Campbell (1880–1949) was trained at Trinity College, Cambridge, becoming a fellow there in 1904. He worked in the Cavendish Laboratory under J. J. Thomson. In 1913 he was awarded the Cavendish research fellowship at Leeds, where Bragg was then Cavendish Professor. In 1919 he joined the staff of the research laboratories of General Electric; he was instrumental in establishing the freedom of the research worker in an industrial laboratory. He was deeply interested in philosophy and wrote several tracts on the philosophy of physics, including his classic *Physics: The Elements,* which has been reprinted as *The Foundations of Science* (New York, 1957).
105. N. R. Campbell, "The Aether," *Philosophical Magazine, 19* (1910), 181–191; "The Common Sense of Relativity," *Philosophical Magazine, 21* (1911), 502–517; "Relativity and the Conservation of Mass," *Philosophical Magazine, 21* (1911), 626–630.

tory in modern physics, even though some authors still considered it of supreme importance. He noted that one of the founders of the atomic theory of radiation, J. J. Thomson, had recently devoted his entire Presidential Address at the British Association to the properties of the ether. Campbell suggested that perhaps it was the shyness of scientists in discussing the "essential foundations" of their studies that explained their reluctance to attack or defend the concept of the ether. He did not share their reluctance, and he produced an example to demonstrate the untenability of the ether and the need for relativity.

> Consider the case of two or more electrically charged bodies moving with different uniform velocities relative to some observer. Round each body is distributed electrical energy localized in the aether; the positions of the portions of the aether which contain stated amounts of energy (belonging to one and the same body), relative to each other or to the charged nucleus, are not changed by the motion. It seems obvious and simple to identify points in the aether . . . by the amounts of energy contained in them. Then the velocity of the aether relative to the observer would be different according as one or the other of the charged bodies was considered [since the charged bodies are moving relative to each other] and would be in each case the same as the velocity of the corresponding charged body relative to the observer.
>
> Such, I think, is the simple and obvious view leading to the principle of relativity.[106]

But such an example did not necessarily lead one to the theory of relativity. Faced with the kind of contradiction Campbell proposed, Cunningham had simply redefined the ether in such a way that the contradiction would make no difference. Campbell himself recognized that one could do this simply by refusing to identify the ether as the locus of electromagnetic energy. But in that case, the ether became undetectable and therefore physically meaningless.

Cunningham said that the ether apologists had tried a new tack: "They now said that the 'difference between the velocities relative to the aether of any two bodies was equal to their relative velocities, but that the velocity relative to the aether of any body was uncertain to the extent of a constant.' And then they tried to show that this constant was far beyond the range of experiment right now." This

106. N. R. Campbell, "The Aether," *ibid.,* pp. 184–185.

seemed utterly unconvincing to him. He conjectured that the "future historian of physics will be astounded that the vast majority of physicists should accept a system of such bewildering complexity and precarious validity rather than abandon ideas which seem to have their sole origin in the use of the word 'aether' and reject those to which so many lanes of thought point insistently." He saw no point in trying to save the ether concept, since it "has never been the source of anything but fallacy and confusion of thought"; it should be relegated to the "dust-heap where 'phlogiston' and 'caloric' are now mouldering."[107] His position was that the concept of the ether should be given up since there was no way to test its existence. He insisted that all the apparent paradoxes of relativity will disappear "if attention is kept rigidly fixed upon quantities which are actually observed."[108]

Campbell's assertions about the meaninglessness of the concept of the ether were unique in Britain at this time. They are similar to the strictures of Duhem. Like Duhem, Campbell believed that metaphysics and science were completely separate: "one of the chief characteristics which distinguishes science from metaphysics, and the feature which makes men of science so averse from the latter, is that in science, but not in metaphysics, it is possible to obtain universal assent for conclusions, and to present results which do not lose their value because when they are presented, they are so obvious as to be indubitable."[109] But to claim that science is without metaphysics, or to claim indubitability of scientific results, was to express a metaphysics. It was this positivistic strain in Campbell that made him blind to the value that the ether concept had had in physics. It also prevented him from understanding the basis of the British physicists' commitment to the ether.

According to Campbell, the reason for the lack of interest in relativity theory and for the persistence of the concept of the ether was that the man in the laboratory "has been offered profound mathematical investigations, which are intensely important and interesting but tend to obscure the fundamental points at issue in the mind of one who thinks physically rather than mathematically. And on the other hand, he has been offered collections of apparently paradoxical

107. *Ibid.*, p. 190.
108. N. R. Campbell, "The Common Sense of Relativity," *ibid.*, p. 515.
109. N. R. Campbell, *The Foundations of Science, op. cit.* (note 104), p. 10.

conclusions deduced from the principle which are sometimes elegant and entertaining, but more often fallacious."[110] Campbell was mistaken. What stopped the British physicists was not primarily the mathematics or the paradoxical conclusions, which, in any case, were present in Lorentz' and Larmor's theories too, but the foundations and ether-denying implications of relativity. Campbell's "measurement metaphysics" was so intense that he preferred to use it to deduce reasons for the British physicists' difficulties with the theory rather than look for the actual reasons.

In spite of Campbell's emphasis on measurement and performable experiments, and his objection that the concept of the ether was meaningless, he was not averse to the use of models. In reply to the objections to relativity theory that it had no physical meaning and that it "destroyed utterly the old theory of light based on an elastic aether and puts nothing in its place,"[111] he maintained that a physical theory of light could be produced which was consistent with the principle of relativity. Campbell was of the opinion that a model would be an essential part of the construction of such a theory.[112]

5. CONCLUSIONS

It needs emphasizing that there was little work done in Britain on relativity in 1905–1911. The British contribution to the relativity literature represents less than ten percent of the total in this period.[113] Cunningham notes that he "was surprised to be asked to speak to the British Association in 1910" on the subject of relativity, because he "did not think that there was much interest in the matter at that time."[114] Indeed, even those British scientists who, in one way or another, became aware of the theory seemed to have difficulty in understanding it. For example, in 1909 the astronomer F. C. Searle wrote to Einstein to thank him for sending a paper on rela-

110. N. R. Campbell, "The Common Sense of Relativity," *op. cit.* (note 105), p. 502.
111. *Ibid.,* p. 515.
112. *Ibid.* See also N. R. Campbell, *The Foundations of Science, op. cit.* (note 104), p. 130. For a discussion of the conflict over the use of models between Duhem and Campbell, see Mary Hesse, *Models and Analogies in Science* (London, 1963).
113. This estimate is based on M. Lecat, *Bibliographie de la relativité, op. cit.* (note 5).
114. Personal communication.

tivity: "I have not been able so far to gain any really clear idea as to the principles involved or as to their meaning and those to whom I have spoken in England about the subject seem to have the same feeling."[115] Searle thought that it would be a good idea if Einstein were to write a short account of relativity for an English journal to help the English gain a better grasp of the ideas. When, in 1911, the British Association finally devoted a session to the "Principle of Relativity,"[116] there was little sign of any change in British attitudes toward relativity. The *Nature* reporter at the meeting noted C. V. Burton's "satisfaction that no one had confessed a disbelief in the aether."[117] Through 1911 the British rejection of relativity was founded on a tenacious adherence to an ether-based physics.

While some British physicists like Cunningham argued that experiment would be the final arbiter in deciding the worth of the theory of relativity, the bulk of those physicists concerned with the problems of electrodynamics were not concerned with experimental issues. It is simply incorrect to state, as Campbell did, that the major difficulty the British physicist encountered in dealing with relativity was with the consequences of the theory. In fact the British physicist was unwilling to accept the premises and assumptions on which the theory was based. But since Campbell did not believe that scientific matters involved metaphysics, or more accurately, since he believed that these were two separate and separable areas and that he and his colleagues were doing "science," there was no reason for him to look for assumptions or presuppositions as such. But the presuppositions were there. For example, many physicists shared Lodge's inability to conceive of a theory which did not account for "apparent" action at a distance in terms of pushes and pulls.

Toward an Understanding of the British Response to Relativity

There are, no doubt, many features of British culture and society that could aid in an understanding of the nearly uniform British

115. Letter from F. C. Searle to Albert Einstein, 20 May 1909, Einstein Archives, Princeton University. I wish to thank Dr. Otto Nathan, executor of the Albert Einstein Estate, for permission to quote from this letter.
116. Anon., "Mathematics and Physics at the British Association, 1911," *Nature*, 87 (1911), 498–502.
117. *Ibid.*, p. 500.

neglect of Einstein's theory and the concomitant insistence on retaining the concept of the ether. The feature I wish to bring out here—and one I feel is instrumental—is the structure of British education in physics. A British theoretical physicist interested in electrodynamics was trained to do ether mechanics; it is what he had to learn, and it is what he knew best.

Most British physicists, experimental and theoretical, were trained at Cambridge University. The Mathematical Tripos at Cambridge was the most competitive arena for mathematicians and mathematical physicists in Britain. The importance for students of developing skills in solving mechanical problems of the ether is suggested by questions appearing in the 1901 Tripos:

> Obtain the energy function of an isotropic elastic medium and assuming that waves of dilation are propagated through the medium with an indefinitely great velocity and that the difference between different media is one of density only, find the intensities of the reflected and refracted waves when plane waves are incident on a plane interface separating two media.

> Waves of light are incident on a face of a uniaxial crystal cut perpendicularly to its axis. Find on MacCullagh's theory the intensities of reflected and refracted waves (1) when they are polarized in the plane of incidence, (2) when they are polarized perpendicularly to it.[118]

Presumably, one would have needed to commit MacCullagh's theory to memory. This theory, which had been created in 1839, required an elastic-solid ether that exhibited the property of "rotational elasticity." No known elastic solid had ever exhibited this property, and there had been considerable early skepticism about the theory for this reason. The 1901 Tripos also placed heavy emphasis on fluid mechanics, a subject closely related to the concerns of ether physicists. A typical question was: "Give an account of Helmholtz' and Lord Kelvin's theory of vortex motion." A few more sample questions on elastic-solid theory and fluid mechanics from the 1903 and 1905 examinations point to the ether-mechanics competency expected of Cambridge graduates:

> Investigate Helmholtz' expression for the velocity of a fluid in vortex motion in terms of a vector potential.

118. *Mathematical Tripos, Part II, Thursday, May 30, 1901* (Cambridge, 1901).

Assuming that the spin at any point within a circular vortex ring of finite cross-section is proportional to the distance of the point from the axis of the ring, prove that the vector potential at any point P is directed at right angles to the axial plane through P, and is equal in magnitude to the value at P of the gravitational potential of the ring, supposed to have at any point Q a density proportional to the projection on the axial plane through P of the perpendicular from Q to the axis of the ring.

Deduce expressions for the stream-function in the case of Hill's spherical vortex, and express the velocity of translation in terms of the strength of the vortex.[119]

Discuss generally evidence for the conclusion that the velocity of the earth's motion does not sensibly affect observable optical phenomena. Show however that if two independent sources of light could interfere, a simple experiment would at once reveal the earth's motion in space.[120]

The potential energy of an elastic medium is a homogeneous quadratic function of the components of strain; prove that, if the directions of transverse vibrations are always in the wave front, the wave surface is Fresnel's.[121]

Such questions may be found in almost any of the Tripos examinations through 1910. It is important to note as well that these were timed examinations and that the student had to be extremely dexterous with the mathematical manipulations.

J. J. Thomson, who studied ("coached") with Routh at Cambridge, recalled that

Routh's system certainly succeeded in the object for which it was designed, that of training men to take high places in the tripos; for in the thirty-three years from 1855 to 1888 in which it was in force, he had 27 Senior Wranglers and he taught 24 in 24 consecutive years. . . . Until quite near the time when he gave up "coaching," candidates for the Mathematical Tripos were expected to be acquainted with the whole of the wide range of pure and applied mathematics included in the examination. . . . [Routh] was Senior Wrangler in the year when Clerk Maxwell was second. Perhaps no other man has ever exerted so much influ-

119. *Mathematical Tripos, Part II, Saturday, June 6, 1903* (Cambridge, 1903).
120. *Ibid.*
121. *Mathematical Tripos, Part II, Wednesday, May 31, 1905* (Cambridge, 1905).

ence on the teaching of mathematics; for about half a century the vast majority of professors of mathematics in English, Scotch, Welsh, and Colonial universities, and also the teachers of mathematics in the larger schools, had been pupils of his and to a very large extent adopted his methods. In the textbooks of the time old pupils of Routh would be continually meeting with passages which they recognized as echoes of what they had heard in his classroom or seen in his manuscripts. . . . Routh like Maxwell studied mathematics under Hopkins, the great "coach" at that time, who had taught Stokes and William Thomson and scored 17 Senior Wranglers before he retired.[122]

There existed then a natural filter for processing mathematicians and mathematical physicists in Britain. It is understandable why so many of them acted with such uniformity when confronted with the theory of relativity. They had studied under Hopkins, or under one of his pupils, such as Routh. They had been trained to master the partial differential equations of waves moving through "froths, jellies and vortices." It is understandable also why a person like Lodge, who had not been trained at Cambridge, might have had a similar outlook, since English theoretical physics was dominated by Cambridge. (What is difficult to understand is how Campbell, who was trained at Cambridge, was able to escape the prevailing ether orientation of British physics.)

Cunningham states that the Tripos he took in 1902 hardly dealt with electrodynamics, and that Maxwell's work had not yet filtered down to the point of being used as a textbook. With regard to the theory of relativity, he recalls that it was he who introduced systematic teaching of it in 1911. He says that this time was "a period of transition, which culminated in 1910 when the order of merit in the Tripos was abolished and teaching in all subjects took a much more theoretical turn, while the Cavendish Laboratory moved on into atomic theory. . . . Modern physics was in its infancy. I had to find for myself the work of Lorentz, Planck, etc. after I had graduated. Eddington, Jeans and others were still to come."[123] Cunningham's remarks have a familiar ring. Others[124] have noted the length of

122. J. J. Thomson, *Recollections and Reflections, op. cit.* (note 73), pp. 38–40.
123. Personal communication.
124. See E. Ashby, *Technology and the Academics: An Essay on Universities and the Scientific Revolution* (London, 1959), Chap. 2; J. T. Merz, *op. cit.* (note 1), *1*, 233 ff.

time required for scientific innovations to filter down to the classrooms and examinations of Cambridge—from Leibniz' "d's" through Young's wave theory of light to Einstein's relativity.

ACKNOWLEDGMENTS

I am indebted to the following people for comments and criticisms of earlier drafts of this paper: Professor Paul Forman, University of Rochester; Professor L. K. Nash, Harvard University; Professor E. L. Yates, University of Zambia. I am particularly grateful to Professor Russell McCormmach, University of Pennsylvania, who commented extensively on the manuscript, and to Professor Gerald Holton, Harvard University, under whose guidance and direction this study was first conceived and carried out.

The Rise of Physical Science at Victorian Cambridge

BY ROMUALDAS SVIEDRYS *

COMMENTARY BY ARNOLD THACKRAY **

REPLY BY ROMUALDAS SVIEDRYS

The creation of the chair of engineering, the rise of physiology, and the establishment of the Cavendish Laboratory at Cambridge University in the 1870's marked, no doubt, a milestone in the development of a university formerly preoccupied with classics and sterile mathematical competitions. These changes were a dramatic illustration of the rise of the University and its professorial body at the expense of the Colleges and the College tutors. And it exposes a profound conflict between the advocates of science and those who claimed, with William Whewell, that the progressive sciences such as chemistry and mineralogy were amorphous, contradictory, constantly in a state of flux, and therefore unable to contribute anything valuable to the mission of providing a liberal education for the Cambridge gentleman.[1] A study of the rise of science at Cambridge Uni-

* Department of Social Sciences, Brooklyn Polytechnic Institute, Brooklyn, N.Y. 11201.

The invited paper and commentary were delivered at the History of Science Society meeting in Dallas in December 1968. The reply was added later.

** Department of History and Philosophy of Science, University of Pennsylvania, Philadelphia, Penna. 19104.

1. The ideal of liberal education as it was expounded by William Whewell is discussed in the unpublished doctoral dissertation of Robert G. McPherson, *The Evolution of Liberal Education in Oxford and Cambridge 1800–1877* (Johns Hopkins University, 1957), 139–153. Forty years later that ideal changed completely, as can be seen in the theory of higher education advocated by another fellow of Trinity College. McPherson devotes the seventh chapter of his dissertation to Henry Sidgwick, who took part in the internal reform and sided with the professorial body, 268–324.

versity reveals, at a more fundamental level, the pattern of scientific institutional change during the Victorian era at Cambridge. In this paper, I shall examine first the mechanism of this change by studying the rise of the chemical and engineering laboratories at Cambridge, and then discuss the extent to which the Cavendish Laboratory was characteristic of the pattern of growth at Cambridge and of the organization of science in British educational institutions at the time.

By mid-nineteenth century, Cambridge was still not a University but rather a collection of autonomous Colleges. Student loyalties were to the latter and not to the University. Indeed, Cambridge was a financially poor University composed of rich Colleges. The Colleges derived their incomes from student fees, bequests, and inherited estates, the major part of which was spent on fellowships that provided comfortable and lifelong leisure to the dons. The goal of a Cambridge education was the production of a social type, and the tutors accordingly infused their students with the values and beliefs that characterized the British gentleman. These tutors and the senior fellows, by controlling the seventeen Colleges, were effectively in control of the University. By contrast, the University professors, unlike their counterparts at German universities, were few in number and powerless.[2] Their lecture material was never placed on the examinations controlled by the fellows, who, moreover, encouraged undergraduates to stay away from professorial lectures. Those students who tried attending the lectures found them too advanced. Consequently, professorial lectures in natural science, astronomy, and mathematics were unattended by the majority of Cambridge's talented students. The emphasis upon forming gentlemen for the governing class necessarily precluded specialized professional training. Critics pointed out that the Scottish University taught a man how to make a thousand a year; Cambridge, how to spend it. Scholarship standards were surprisingly low, except for those few students who aspired to mathematical honors and who

2. There were twenty Cambridge professorships in 1850, eight of which were in scientific subjects. Some professors lectured sporadically, and some did not lecture at all, during the first half of the nineteenth century. The salary of many professors was less than one hundred pounds a year, while a few received over three hundred pounds. (D. A. Winstanley, *Early Victorian Cambridge* [Cambridge, 1940], 175–178 and 181–186.)

were trained by private mathematical coaches.

The period from 1860 to 1883 was one of the greatest periods of expansion and change at Cambridge University. The number of students doubled; the University's statutes were streamlined (1877–1883); religious dissenters were admitted to fellowships (1871); and new fields were recognized to which students could be admitted for examinations and degrees, loosening the monopoly of pure mathematical studies. In 1882, Colleges were compelled to contribute a small portion of their annual incomes to the University for the support of professorial chairs and other teaching facilities. Thirteen chairs were established during the period, six of which were in scientific subjects.[3] The professors of natural science, mathematics, and physical science gained substantial control over the natural and mathematical tripos examinations, while the staff of University-paid lecturers and demonstrators expanded significantly, rivaling the private mathematical coaches and increasingly taking over the teaching duties that once were entirely in the hands of tutors. The new professors managed to increase their power, prestige, and income vis-à-vis both the tutors and the better-endowed professors of divinity, law, and ancient languages. The tutors opposed the introduction of new chairs; they argued against the dangers of specialization, but they really opposed the increasing rise of the professorial power, aware that they were not prepared to handle the new scientific subjects. Tutors and fellows of Colleges, together with sundry supporters of liberal education, defended what they considered worthy of preservation at Cambridge, namely, their way of life. The professors advocated their way, based on recognition of research as an academic activity, together with professional training. I propose to examine the nature of the conflict that ensued by examining the manner in which the chemical, engineering, and physics laboratories were established at Cambridge.

The first British educational institutions to establish laboratories fully devoted to training chemists were the University of Glasgow

3. The new scientific chairs were in pure mathematics (1860), zoology (1866), experimental physics (1871), mechanism and applied science (1875), physiology (1883), and pathology (1883).

(1829) and University College, London (1829).[4] By mid-century, every major institution of higher education offered courses in chemistry and had at least a modest chemical laboratory[5]—every major institution except Cambridge, that is. It was only in 1865 that Liveing,[6] the professor of chemistry at Cambridge, announced a course of practical instruction in chemistry, culminating almost fifteen years of struggle for the recognition of the subject. Even then, Liveing's chemical laboratory was an old anatomical museum, and he would not have had even that if the professor of anatomy had not lent his own money to the University to construct new facilities

4. Thomas Thomson was the first Regius Professor of Chemistry at Glasgow from 1818 until 1852; in 1820 he began giving practical training to his students. In 1829, Glasgow University spent five thousand pounds to provide a new chemical laboratory for him. (James Coutts, *A History of the University of Glasgow, from its Foundation in 1451 to 1909* [Glasgow, 1909].) In the early 1830's, another chemistry laboratory was established in Glasgow by Thomas Graham, who was teaching at the Anderson's College, the first purely technical institution established in Britain (1796). Chemistry had a long tradition at Edinburgh University, and under Charles Hope, the professor of chemistry from 1795 until 1844, the teaching of practical chemistry began in 1823. (Alexander Grant, *The Story of the University of Edinburgh* [London, 1884], 2, 398.) In London the new educational institutions catered to the needs of the rising middle classes and made provision for the teaching of chemistry from the start. London University (later, University College) gave the first laboratory course in practical chemistry in the spring of 1829; for a program of its chemical instruction see *Second Statement by the Council of the University of London explanatory of the Plan of Instruction* (London, 1828), 62–65. On the chemistry laboratory and the first professor, see Hugh Hale Bellot, *University College London 1826–1926* (London, 1929), 124–127. The other institution in London to pioneer chemical education was King's College. (F. J. C. Hearnshaw, *Centenary History of King's College, London 1828–1928* [London, 1929].)

5. University College created a second chair, devoted to analytical and practical chemistry, in 1845, and it erected a new chemical laboratory at a cost of two thousand five hundred pounds to accommodate the increasing number of students. The first school teaching chemistry exclusively was the Royal College of Chemistry, also established in 1845. In 1851 a chemistry laboratory was built at the Royal School of Mines, and in 1852 the trustees of Owens College raised ten thousand pounds to provide a laboratory for Edward Frankland.

6. George D. Liveing (1827–1924) entered St. John's College in 1847, graduating in 1850. He then read for the newly established natural science tripos; at the first examination held in 1851 he placed first, with distinction, in chemistry and mineralogy. Later he studied with Karl Rammelsberg at Berlin, and, upon his return to Cambridge in 1852, the professor of medicine asked him to provide practical instruction in chemistry for his medical students. In 1861 Liveing was elected to the chair of chemistry, succeeding the Reverend James Cummings, who had held that position since 1815 and had lectured sporadically. (*Dictionary of National Biography*, supplement 1922–1930, 510–512.)

for himself.[7] More than a decade earlier, the fledgling Owens College in Manchester could boast of far better facilities than Cambridge. In 1852, the trustees of Owens raised ten thousand pounds for a chemical laboratory on the appeal of their chemistry professor, and they provided him with salaried assistants. Liveing raised the same demands at almost the same time at Cambridge and got nowhere. His first "laboratory" had been a cottage, equipped at his own expense, and he had no paid assistants until 1870. His laboratory for many years was an institution of low prestige in the eyes of the dons, attended mainly by medical students, few of whom were interested in research. Some Colleges established their own chemical laboratories rather than strengthen the University's.[8] Liveing had a lackluster career, overworked in his teaching and his directing of laboratory courses for undergraduates. His first laboratory in which he finally gave University-approved courses came in 1885, some thirty years after the innovation of laboratory instruction in chemistry was accepted at most other institutions.

A similar situation developed at Cambridge with the introduction of the engineering chair and the engineering laboratory. To circumvent the high costs of engineering apprenticeship, some institutions began offering courses in engineering subjects in the early 1830's.[9] The railroad construction boom in 1837–1848 overstrained both the supply and the working capacity of existing engineers to

7. These facilities were part of the Museum that was finished in 1866. Since the time of the establishment of the natural science tripos (1848), the University of Cambridge was under obligation to provide accommodations for the teaching of the subjects comprising the examination, including chemistry. A proposal to build lecture rooms and laboratories for the science professors was presented to the Senate, but rejected in 1854. Trinity College was the only College willing to contribute freely four thousand pounds to the total estimated cost of twenty-three thousand pounds. The project was revived in 1860, just before the natural science tripos became a degree-granting examination, but chemistry and anatomy were dropped to reduce cost. The plan was rejected once more in May 1862, but it passed the following year. Construction of the Museum began in late autumn, 1863. For details see Robert Willis and John Willis Clark, *The Architectural History of the University of Cambridge* (Cambridge, 1886), *3*, 157–181.

8. On the Colleges' chemical laboratories at Cambridge, see F. G. Mann, "The Pace of Chemistry at Cambridge," *Proceedings of the Chemical Society, 1* (1957), 190–193.

9. Engineering instruction began in 1828 at University College, in 1831 at King's College, in 1837 at the University of Durham, and in 1843 at Queen's College, Belfast. (Thomas J. N. Hilken, *Engineering at Cambridge University 1783–1965* [Cambridge, 1967], 25.)

the point where the better known of them were simultaneously engaged in a dozen or more projects. The first engineering chairs were established in the 1840's. The greatest pioneer was Glasgow University; close behind was University College, London;[10] at Cambridge, the first proposal to establish an engineering chair was made in 1845, but the plan was dropped for economic reasons, and later the dons thought that the chair was no longer necessary with the collapse of the railroad construction boom in 1845.

Perhaps Cambridge dons considered that their University already had an engineering chair in the recently (1783) established Jacksonian professorship of experimental philosophy. Although originally intended for subjects on the borderline between chemistry and medicine, the Jacksonian professors had deviated considerably from this intention. Robert Willis, elected to the chair in 1837, converted it into a chair of mechanism, illustrated by demonstrations of models of all types of machinery.[11] A Royal Commission, appointed in 1850, reviewed his work and praised it, recommending that an engineering examination and a chair of engineering should be established. The recommendation fell through, victim of committees and lack of funds. In 1859 a fellow of Trinity Hall, Henry Latham, put forward a complete plan for an engineering school at Cambridge, but this also was abandoned after the Colleges refused to contribute money. No other attempts were made to establish an engineering chair until the death of Willis, sixteen years later; even at this late stage, his death provided an opportunity for a group of conservative tutors to launch an attack on engineering.[12] They drew the attention of the Council of the Senate to the fact that the teaching of courses of mechanisms was contrary to the original terms of the endowment

10. Glasgow University established a chair of engineering in 1840 and attracted to it a series of eminent men. The first was Lewis Gordon, who taught for fifteen years and withdrew to work in the cable manufacturing partnership that he had formed. His successor was William J. M. Rankine, who developed the first really systematic course of instruction, and was succeeded upon his death in 1872 by James Thomson. At University College, the first chair of engineering was established in 1841, with Charles B. Vignoles, who had been involved in railway engineering, and who resigned two years later to resume his professional practice on the Continent. (Olinthus John Vignoles, *Life of Charles Blacker Vignoles* [London, 1889], 262–264.)

11. Attendance at Willis' courses was great, although there were relatively few registered students later in his teaching career. (Hilken, 50–57.)

12. Hilken, 25–29.

of the Jacksonian professorship. Their scheme, however, backfired. The increasingly more powerful professorial body went along with the suggestion and asked to revert the chair to chemistry, because they were confident that they had sufficient votes to establish a chair devoted to engineering. By reverting the Jacksonian chair to chemistry, they hoped to ease the heavy workload of Liveing. Thus, in 1875, Cambridge received a second chair in chemistry and established its own chair of engineering, some thirty years behind the other institutions.

At the other institutions, renewed support for engineering education came from industrialists and practicing engineers, as a result of Continental developments in technological education. The Continental system of training, especially the polytechnic schools, were thought to be far in advance of Britain's. Moreover, the seven-week war in the summer of 1866, in which Prussia soundly defeated Austria, further worried many industrialists who had already observed Prussia's bid for technological leadership. The turning point came at the Paris International Exhibition of 1867, which caused shock and astonishment among the British who witnessed there a decline of their industrial might. Several eminent British industrialists who had served as judges at the Exhibition declared that the rapid progress of Continental industries was the result of the superior scientific and technical training of managers and engineers. Response in Britain was immediate. In Manchester, leading engineers raised almost ten thousand pounds in 1867 to endow a chair of civil and mechanical engineering, electing Osborne Reynolds to the chair the following year.[13] Another chair was established in 1868 at the University of Edinburgh, endowed by an industrialist, David Baxter.[14] Even the British government was moved by the decline of engineering to establish the Royal Indian Engineering College at Cooper Hill. But again, it was one of the smaller and poorer institutions

13. The Manchester engineers held a meeting at the Town Hall on 11 December 1866. A subcommittee composed of Joseph Whitworth, William Fairbairn, Charles F. Beyer, and John Robinson raised the money. See Joseph Thompson, *The Owens College: Its Foundation and Growth; and its Connection with the Victorian University, Manchester* (Manchester, 1886), 295–297.
14. David Baxter (1793–1872) was one of the largest linen manufacturers in Britain; in 1836 he introduced power-loom weaving. He gave six thousand pounds to endow the chair of engineering, leaving an additional forty thousand for the University of Edinburgh in his will.

that produced the greatest innovation in academic engineering education. In 1874, University College, London, appointed to the chair of engineering Alexander Kennedy,[15] who four years later established there the first engineering teaching and research laboratory; all subsequent laboratories were modeled after his. By 1884 there were seven engineering laboratories[16] in Britain and several more in the process of construction.

The Cambridge professor of engineering, James Stuart, received no support from the University for a laboratory. He established a workshop in which he constructed scientific instruments for the other departments of Cambridge, and with the income from the instruments he paid for his own assistants and bought his own special equipment. The success of his department was evident; students flocked to him even though no degree in engineering existed yet. Gradually, the University came round to his support and paid, in 1881, for the addition of a provisional wing to his workshop. Stuart was elated over his success, and contemplated the establishment of an engineering degree and of a laboratory competitive with the other institutions.

The increasing financial support that Stuart received from the University was suddenly cut short, and his workshop did not develop into a research laboratory. The reason was that Stuart had the misfortune to annoy the Cambridge dons by his radical political views and his public political campaign.[17] They proceeded to block several

15. Alexander W. B. Kennedy (1847–1928) was educated at the City of London School and the Royal School of Mines. He entered a firm of marine engineers and set up practice as a marine engineer in Edinburgh until his election to the chair. He retired in 1889 to devote himself to the construction of electrical power stations. In connection with his engineering laboratory, see his "The Use and Equipment of Engineering Laboratories," *Proceedings of the Institution of Civil Engineers, 88* (1886–1887), part II, 1–80.

16. University College (1878), Finsbury Technical College (1881), Mason Science College, Birmingham (1882), University College, Dundee (1884), Royal Indian Engineering College (1883), University College, Bristol (1883), City and Guilds Central Institute, London (1884), followed at a later date by Firth College, Sheffield (1885), Yorkshire College, Leeds (1886), Owens College (1886), University College, Liverpool (1887), and Edinburgh University (1890).

17. See Hilken, 63–74, for the development of Stuart's workshop and teaching activities. Stuart's entry into politics was commented on in the following terms by a former friend: "Professor Stuart's address and speeches at Hackney are a terrible blow to the reputation of those who pressed him upon the University constituency two years ago. . . . Their academic and cultured candidate . . . this chosen friend of Bishops and Deans and Masters of Colleges, goes to Hackney and pours forth

of his proposals in retaliation. Stuart's major defeat came in 1887 when his plan to institute an engineering tripos examination was put down, and opposition from all sources forced his resignation two years later. By seeking his resignation, the conservative dons were trying to smother engineering, because, according to the terms of his appointment, the chair was to lapse after his death or resignation. Only the Chancellor of the University, to whom this matter was referred for a decision, saved the chair from extinction by pronouncing that it really belonged to the permanent staff of the University. Thus, the Chancellor, himself an industrialist, permitted Stuart's successor[18] to establish an engineering laboratory, after obtaining funds from various friends. In this campaign to raise money for the laboratory, the opposition claimed that such a large expenditure, a total of thirty-two thousand pounds for the site and building, amounted to "science first and the rest nowhere"[19] policy. Only the infusion of donations permitted Cambridge to have an engineering laboratory just a decade after the innovation had occurred at other British institutions. Without the tapping of extra-University funds, Cambridge would probably have paralleled Oxford, where the engineering chair was established in 1907 and a laboratory constructed in 1913, roughly thirty years after it became part of the other institutions.

Engineering and chemistry at Cambridge largely conformed to the general pattern of a thirty-year lag between the adoption of an innovation by the pioneering educational institutions and the time of adoption at Cambridge. This lag was not just the result of opposition to the introduction of science. It occurred, for example, in the introduction of such reforms as the creation of a research degree, the admission of dissenters, and the reorganization of finances at Cambridge. The latter reform, first suggested in 1850—namely, that

---

a flood of ultra-Radicalism, of crazy fads, of coarse misrepresentation and vulgar abuse, which must make his Cambridge friends shudder" (Hilken, 75). Stuart presents in his autobiography his own view of the opposition; he thought that the dons were opposing his workshop on the principle that anything that was not purely abstract was unsuitable for a university. Yet he does not comment on the lack of opposition during the first ten years and on the fact that it began only after he entered politics. See James Stuart, *Reminiscences* (London, 1911), 183.

18. For details on the election of the successor, James Alfred Ewing, and on his career, see Hilken, 107–112.

19. Hilken, 117.

the Colleges contribute some small fraction of their annual income to the needs of the University—was rejected outright. Royal Commissioners suggested in 1856 that each College contribute five percent of its annual distributable income to the University. Thirteen Colleges rejected the plan by a vote of two-thirds or more of their fellows. The same plan was rejected once more in 1869, when a committee appointed by the Council of the Senate revived the plan in their search for funds to establish a chair of experimental physics. And yet, in the 1870's, Colleges began to establish a system of intercollegiate courses, lectureships, and even chemical laboratories, sometimes costing far more than the five percent outlay that would have provided these services at the University level. The only difference was that the intercollegiate lectures and courses were controlled by fellows, whereas anything that went to the University could come under the control of the increasingly more powerful professorial body. Only in 1877, and with the appointment by the government of a Royal Commission to revise the statutes of the University, was the annual income of Colleges taxed; this began in 1883. The same situation existed in the establishment of Cambridge's research degree in 1895, thereby ending the illusion that Cambridge and Oxford were the only universities in Britain and that the degrees given by the Scottish universities or London University (since 1858) were not awarded for academic work.

Institutions change when men's opinions change or when the incumbents are replaced. The innovative lag at Cambridge is the time it took young Cambridge fellows to reach positions of power within their Colleges by strict seniority. The introduction of changes depended to a large extent on the masters and senior fellows who governed the Colleges, and at any time during the nineteenth century only a handful of them were supporters of science or, even rarer, scientists. By the time they and other potential innovators reached power-wielding positions, the reforms they wanted to introduce were some twenty to thirty years behind the times. A typical example of a thwarted reformer is Henry Latham. Graduating in 1845, he was, we may recall, one of those who urged Willis to seek the establishment of a school of engineering at Cambridge that year. On various other occasions, in 1859 and 1875, he was a firm supporter of the chair of engineering. But upon Stuart's election, he increasingly opposed him and his political views. As to engineering laboratories, which Stuart tried to promote, Latham considered the idea an unnecessary luxury.

The mechanism of change depended, thus, on the particular organization of Cambridge as an aggregation of autonomous Colleges. Power was diffused among masters, tutors, and Cambridge fellows, and consequently there was no central body that the scientists or innovators could attempt to infiltrate or dominate. Cambridge had its fair share of innovators, but their reforms had to pass the full Senate, where all fellows could vote, after surviving innumerable committee meetings and then the Council of the Senate. If the innovators had held the reins of power, they would have introduced scientific changes at Cambridge at about the same time as they were introduced elsewhere.[20] But given the social organization of Cambridge, lack of will and interest by those in power stopped them. The most common excuse, "lack of funds," was invalid and irrelevant because science was just as expensive at Glasgow or Manchester, economically weaker institutions. That this was the case is confirmed by those Cambridge scientists who, when they had independent economic resources, were able to bypass the regular financial channels and introduce changes on their own.

Until about the mid-1870's, the strategy of those in power was to preserve Cambridge as the training ground of the British social elite. Internal reform at Cambridge was reform that should have occurred a generation earlier. The supporters of science were outmaneuvered time and again by the dons, who, when they had to, would sweep the floor from under the reformers' camp by adopting some of their banners. Outside critics correctly viewed Cambridge reforms as compromises designed to split and appease the reformers, and they characterized such reforms as too little and too late. The conservative dons' strategy was complemented with a few simple

20. The example that comes to mind is the laboratory of physiology. Although Cambridge University lagged more than thirty years in establishing a chair of physiology, it was not too far behind with the establishment of a laboratory. The master of Trinity College and some Trinity tutors and lecturers such as Henry Sidgwick and Coutts Trotter gave Michael Foster all the help he needed for his work in physiology. In 1870 he was elected to the post of Trinity praelector in physiology; and his laboratory came a year or so later. See Henry Dale, "Sir Michael Foster, K.C.B., F.R.S. A Secretary of the Royal Society," *Notes and Records of the Royal Society of London, 19* (1964), 10–32, esp. 18–25.

tactics. First, the reforming party was admitted into the ranks of the majority, and the reform impulse was blunted during the extended debates so well suited to the leisurely life of the dons. Second, those holding power preserved the initiative in the committees that they appointed to discuss the proposed changes, thus forcing the reformers to push proposals through the existing mechanisms of change. Many reforms, if they survived the gauntlet, emerged from committees in milder versions. And until 1883, and often afterwards, the reforms, even if adopted, were often inoperative because of lack of funds, since College authorities were unwilling to give up their own financial resources to implement changes that weakened them and the mode of life of their fellows. For most of them, the accepted meaning of reform was imperceptible evolution, something that should not affect things in their lifetime. Nevertheless, at least in one case, Cambridge fellows were caught off guard by a reformer who understood their tactics: Maxwell's introduction of the Cavendish physics laboratory constitutes a striking exception to the general process of institutional change at Victorian Cambridge.

The physics laboratory, devoted to research and teaching, was an institution that came to the forefront in British scientific education during the years 1866–1873, when at least eight such laboratories were established.[21] According to the mechanism of change at Cambridge in the nineteenth century, the chances of establishing a laboratory there before 1890 or 1900 would seem slight. As it hap-

21. The oldest physics laboratory at an institution of higher education was established by William Thomson in 1850 in a disused cellar. Only after the success of the 1866 Atlantic submarine cable, in which Thomson was involved, did officials of the University recognize the laboratory. They made provision for a laboratory in the new building to be erected in Gilmore Hill, to which he moved his laboratory during the summer of 1870. See *Nature, 6* (1872), 29–32 for an account. University College established one in 1867, under George C. Foster, while King's College followed suit in 1868. Peter G. Tait established a physical laboratory at Edinburgh University in 1868. (C. G. Knott, *Life and Scientific Work of Peter Guthrie Tait* [Cambridge, 1911], 70–72.) In 1868 construction also began on a physical laboratory at Oxford University, under Clifton. In 1870 Balfour Stewart founded a laboratory at Owens College. Another was established by Frederick Guthrie in London at the Royal School of Mines in 1872, upon its transfer to South Kensington. The following year, William Barrett established a laboratory at the Dublin Royal School of Science. (*Royal Commission on Scientific Instruction and the Advancement of Science* [London, 1872–1875], *1,* 160–171 [William Thomson], 186–194 [Robert B. Clifton], 534–538 [George C. Foster], and *2,* 12–16 [Peter G. Tait].)

pened, however, the laboratory was founded much sooner than that.

In 1866, a special building was completed to house some new facilities and lecture rooms for several professors. Reformers and officials turned their attention to other neglected branches of science, including such subjects as heat, electricity, and magnetism. Maxwell was one of the most active reformers to recommend that these subjects be included in the mathematical tripos examinations. A committee, set up to study the question of whether or not the existing chairs in science could cover these subjects, agreed with the professors that a special physics professorship was needed. The committee recommended a laboratory at a cost of five thousand pounds and equipment at a cost of one thousand three hundred pounds, plus annual salaries of seven hundred and fifty pounds for the professor, demonstrator, and laboratory attendant. Discussions of how to obtain the financial resources stretched for over a year without results. Then, quite unexpectedly, the seventh Duke of Devonshire, who was Chancellor of Cambridge University, intervened with an offer of his own money to establish a physics laboratory on the condition that a special professorship of experimental physics be established at the same time. The offer was accepted, and after four months of haggling, a way was found to provide money for salaries.

The Chancellor of Cambridge was a rare instance of a nineteenth-century nobleman who was at the same time a mathematical tripos honors graduate and a metallurgical industrialist.[22] Beginning with the 1860's, he was president of the Barrow Haematite Steel Company which exploited high-grade iron ore on one of his estates. He was very much interested in metallurgical innovations applicable to his industry, and under his guidance Barrow Haematite adopted the Bessemer steel process and became one of the largest Bessemer steel works in the country. In his inaugural address in 1869 as first president of the Iron and Steel Institute, he stressed the importance of chemistry and physics and their applications to the iron and steel industries, favoring the "promotion of science in its practical applications."[23] As Chancellor of Cambridge, he presided over the rapid expansion of Cambridge University, stressing the importance of sci-

22. The Chancellor's letter is published in *A History of the Cavendish Laboratory 1871–1910* (London, 1910), *4*. On the Chancellor of Cambridge, see J. A. Crowther, *Statesmen of Science* (Bristol, 1966), 213–233.
23. Crowther, 219.

ence. Moreover, a few months after his inaugural speech at the Iron and Steel Institute, he was appointed chairman of the Royal Commission on Scientific Instruction and the Advancement of Science, and in this central position, he heard the testimony of two directors of physics laboratories, William Thomson from Glasgow and Robert Clifton from Oxford. Thereafter, the Commission adjourned for the summer, leaving its chairman ample time to ponder how he could improve the deplorable state of physical science at Cambridge. When he made his offer in October 1870, he acted with a genuine interest in the advancement of science at Cambridge, although, as a Cambridge graduate, he may also have been motivated by a desire not to see his alma mater left behind by its rival, Oxford. No doubt he also appreciated the industrial value of research done in a well-equipped laboratory.

James Clerk Maxwell or one of his scientific friends may also have had some influence in prodding the Chancellor to endow a physics laboratory. Maxwell, an eminent graduate from the same College as the Chancellor, had become, by 1870, one of the most effective persons in the promotion of physical science at Cambridge, and one who was greatly responsible for changing the mathematical tripos from a sterile competition into an examination that included exciting problems drawn from physics. In 1865 he left the chair of natural philosophy that he had held for five years at King's College, London, having become increasingly concerned with the unsatisfactory state of the Cambridge mathematical tripos examination, which excluded most modern developments in physics since 1848. Cambridge's private mathematical coaches were not teaching the modern developments of these subjects because they were not placed on the tripos examinations. The responsibility of introducing innovations rested with the tripos examiners, but for almost two decades they had failed to use this privilege. Maxwell was first elected to the office of examiner in 1866, and he created a stir at Cambridge as a result of the interesting and challenging questions that he set. He proselytized in favor of changes, especially of including the new physical subjects that were on the research frontier, and he played an important role at the Board of Mathematical Studies, where the new subjects were discussed. On 2 June 1868, the Senate approved the new subjects and included them for the first time in the examination of January

1873. This innocuous approval was a Trojan horse that led to unexpected developments; by legally binding the University to provide instruction in the new field of electricity, it forced the creation of a special chair of experimental physics and the provision of facilities, including a physical laboratory. The added burden of new subjects was too great for the private coaches, gradually leading to the transfer of these subjects to specialized University lecturers, and greatly strengthening the professorial body.

The Cavendish Laboratory was ready to admit its first student in the spring of 1874, but the dons by and large were not ready to accept an innovation foreign to the traditions of the University. The Laboratory had mushroomed into existence too quickly, without allowing them time to adjust to the changed situation and to consider properly its role within the structure of the University. The dons did not consider the Laboratory a fit place for the instruction of gentlemen. There were no University regulations to force students to come to it and take practical laboratory instruction. Maxwell concentrated, thus, on research and directed from the beginning a research rather than a teaching laboratory. This innovation was hardly possible at the other institutions, where elementary lectures and laboratory practice absorbed professors' energies. Maxwell, unlike other directors of laboratories, directed research for a total of at least forty-eight students during his five-year tenure. Students came to his laboratory to work on challenging research problems, usually drawn from his electromagnetic field theory, although there was the factor of a recent requirement of a research dissertation, at least at Trinity College, for the award of fellowships; this brought to Maxwell's laboratory graduates interested in doing research for a year or two. The Cavendish became under Maxwell the first graduate student laboratory, and it had a large enough number of researchers working full time to create an effect analogous to critical mass. Productivity of the Cavendish workers leaped to new levels by the constant interaction; one man's work was the stimulus for another to write a paper discussing his methods and sources of error, and extending his results or applying the same experimental technique to different phenomena. Cambridge University was the only institution that could provide a group of students, well trained in mathematics, who were able to handle the laborious experiments that

141

Maxwell assigned them. Maxwell exploited very well the only asset that his institution offered, namely, people with leisure.

Research on electricity and on electrical standards became the hallmark of the laboratory. Maxwell transferred to it the equipment and the standard copies produced by a committee on electrical standards of the British Association, which in the course of the previous decade had produced a coherent system of units in electricity and magnetism. Maxwell had served on the committee, and he had led the teamwork research of several of its members on determining the value of the ohm. He continued to stress research on electrical standards, providing a link with the community of practical electricians and with the electrical industry. Maxwell's laboratory was in effect an incipient electrical standards testing laboratory, and, very appropriately, Maxwell accurately tested Ohm's law, which at the time was essential to the basis of the electrical industry. Under his successors, other standards were added, turning the Cavendish into the home of British standards until 1900, at which time the National Physical Laboratory took over from it all work and custody of standards.

Maxwell's record at the Cavendish was impressive enough to provide a new lease on life for the professorship of experimental physics, which, like the chair of engineering, was to expire at the end of the first occupant's tenure. But when Maxwell died in 1879, the Duke of Devonshire personally asked Lord Rayleigh to replace Maxwell.[24] The chair was made permanent, and with Rayleigh's election the reformers sought a greater social respectability for the laboratory and for those who worked there. Rayleigh had no difficulties raising twenty hundred pounds to provide new scientific instruments; he obtained the appointment of two demonstrators, splitting the salary and duties of the office to permit them to devote half of their time to research. Most important, he wanted to identify the Cavendish with a concrete area of research, uniting the laboratory workers around a common research theme, and he selected the redetermination of electrical standards for this purpose. In 1883, the Cavendish Laboratory became an official electrical standards

24. The Chancellor's letter, dated 15 November 1879, has been published in Robert John Strutt, *John William Strutt, Third Baron Rayleigh* (London, 1924), 100–101.

laboratory charging a fee for the testing and comparison of electrical standards.

The industrial tone that the Cavendish took on under Rayleigh's directorship was not to the liking of every don at Cambridge. The recent political activities of Stuart had made engineering and science suspicious. The candidate whom Rayleigh warmly recommended as his successor, the man in charge of testing electrical standards, was turned down as were five other men.[25] The choice of the electing board fell upon a young mathematician who was known to be quite clumsy with his hands. The election of Joseph John Thomson[26] alarmed the community of eletricians and industrialists who feared that the testing of electrical standards would be relegated to a subordinate position. One of the electrical engineers, a graduate of Cavendish, suggested in a paper to the Society of Telegraph Engineers and Electricians that a national standardizing laboratory was necessary to test standards and electrical measuring equipment. The proposal did not fall on deaf ears, and in 1887 the Board of Trade established such a laboratory.

Thomson turned out to be as good a director as his two predecessors. The development of the Cavendish during the first decade of his tenure mainly took the form of a great expansion of undergraduate teaching and laboratory work. A much needed wing was added in 1894, financed entirely by student laboratory fees collected for a decade. However, the number of graduate students remained stationary because there was intense competition from other physics laboratories, at least half of which were directed by Cavendish graduates. Lack of a University degree rewarding research at Cambridge prevented many men from coming. The research degree that was instituted in 1895 marked the second and more fruitful period of Thomson's directorship. This degree strengthened the Cavendish, because the best graduate students flocked to it, giving it a decade of unprecedented growth.

The early emergence of a research physics laboratory at Victorian Cambridge, which was otherwise so resistant to early innovations,

25. The other candidates were Osborne Reynolds, Arthur Schuster, Richard T. Glazebrook (the candidate supported by Lord Rayleigh), G. F. FitzGerald, and A. W. Rucker.
26. On Thomson's election, see R. J. Strutt [Lord Rayleigh], *The Life of Sir J. J. Thomson* (Cambridge, 1943), 20–23.

143

suggests that at least part of the Cavendish's vitality had its roots in its financial independence. The opponents of science could not retard or control the growth of the institution by controlling the allocation of finances for the laboratory building and for its maintenance and expansion. The able guidance of Maxwell, whose tactics were to undermine the opposition from within rather than to launch frontal attacks, was placed at the service of his students, who began producing high-quality research that put the Cavendish in a position of leadership almost from its inception. To the extent that other physical laboratories gradually added research to their main function of teaching, the Cavendish provided a model that other British institutions adopted. And far more important than pioneering a new development, the Cavendish altered Cambridge University. It was the first laboratory to stress research. Women were admitted on equal terms long before women's education became an issue at Cambridge. The ideals of liberal education gave way to what amounted to specialized professional training, for no less than twenty of Maxwell's Cavendish students became professors and lecturers of physics and of other scientific fields. The system of private coaching was replaced by a system of University professors and lecturers whose fees were a fraction of the fees of private coaches and whose lectures were open to all students of the University.

The growth of the Cavendish Laboratory and the introduction of physics illustrate the mechanism of change that had operated at Cambridge for so long. In the wake of the example set by Maxwell and the Cavendish Laboratory, other reformers used the same techniques that he had. Physiology became another field that rose to preeminence at Cambridge at the hand of Michael Foster and a group of reformers at Trinity College. We have seen how, later on, the successor to Stuart did not depend on the University for funds, but actively sought them from other sources. The success of the professors was so complete that they became associated with many of the ideals of the dons. They gained in prestige, responsibilities, and privileges, and in turn they made the lot of the younger innovators unhappy. In the last years of his long directorship, J. J. Thomson was cool toward most of the developments of the new quantum physics. Bohr, who came to his laboratory, left after a brief stay for Manchester, where he found a more sympathetic environment.

144

To the critics of Cambridge, most changes introduced at Cambridge in the nineteenth century seemed to be restricted to things that had successfully proven themselves at other institutions. Thus, when Cambridge finally accepted innovations such as the admission of dissenters, the engineering research laboratory, or the reserach degree, it weakened the other British institutions by siphoning off the superior talent which flocked to it because of its immense prestige. This talent it then transformed into the social and, later, the scientific elite that further increased its hold on higher education in Britain.

Commentary BY ARNOLD THACKRAY

A certain familiarity with what might be called the timeless aspects of the Cambridge scene makes me peculiarly receptive to the theses of Professor Sviedrys' original and important paper. His account reveals just how accurately the opponents of science applied the principles F. M. Cornford was soon after to enshrine in his *Microcosmographia Academica*. Those familiar with that delightful work will have no difficulty in recognizing the application of the Principles of Unripe Time, the Washing of Linen, and the Dangerous Precedent, to say nothing of the Principles of the Wedge (which holds that "you should not act justly now, for fear of raising expectations that you may act still more justly in the future").

The particular importance of Professor Sviedrys' paper lies in the way it takes its subject out of the realm of the anecdotal into that of systematic investigation. Instead of regaling us with stories about one of the most colorful epochs in British science, a start has been made on the serious analysis of the broader social context and the particular causative factors at work in the emergence of the Cavendish Laboratory. Such an analysis is especially important for two reasons. The first is that we know so very little about late Victorian science. At present it lies in uneasy limbo—just beyond the range of any living memory, but not far enough away to have yet been subject to rigorous historical enquiry. The second reason is that we know

145

even less about the growth of the characteristic social institutions of modern science. The teaching laboratory, the research team, the center of excellence, and the invisible college are the embarrassed subjects of historical ignorance—and we do not even begin to understand the way ideas and institutions interact to create the very intellectual texture of science itself. For all these reasons, I wholeheartedly welcome this paper, and wish to congratulate its author on the excellence of his pioneering work.

I take it the presentation has two main theses. The first concerns what I will dub the "generation gap" between Cambridge and other British Universities: the thirty-year time lag apparent in scientific innovation between the rest of Britain and Cambridge (and in true English fashion, when I say Cambridge I mean Oxford). The second thesis is to do with the importance of outside funding in bridging that gap. The influence of such funding is seen in Professor Sviedrys' twin illustrations of the relatively rapid creation of engineering and physics laboratories at Cambridge. I would like to comment briefly on each of these theses, and finally to raise the question of the Cavendish students.

Consider first the generation gap. To quote: "A lag of about thirty years is clearly apparent between the adoption of an innovation by the pioneering educational institutions elsewhere in Britain and the normal time of adoption of the same innovation at Cambridge." I am uneasy with this idea. I would therefore like to suggest that the thirty-year lag assumes a rather different aspect on closer examination. It seems to me gravely misleading to consider British education in terms of Cambridge versus the rest. Not only are the Scottish and English systems fundamentally different in history and ethos, but even within England the provincial or red-brick tradition should not be lightly equated with London.

If instead of studying the difficulties of Cambridge, we turn to the equally baffling problems the would-be reformers faced in the ancient universities of Edinburgh and Glasgow, the proposed generation gap is not so readily apparent. To quote Thomas Thomson's chemistry laboratory at Glasgow as an instance of progress is rather misleading, when the story of his thirty-four-year-long professorship is one of almost continuous difficulty and disappointment. Edinburgh in the first half of the nineteenth century presents an equally disheartening picture, while none but a joker would suggest Aberdeen or St. Andrew's as models of anything but frustration. Again, when William Thomson was elected to the old, established professorship of natural philosophy at Glasgow in 1846, his demand for a laboratory was met after only four years—but met by the allocation of a disused wine-cellar. And the apparatus he inherited dated from the 1740's! Robert Willis as Cambridge's Jacksonian professor looks positively pampered in contrast. I think all this perhaps suggests that the discussion should not be of Cambridge versus the rest, but of old, established institutions versus the new universities of the nineteenth century.

A related point may be seen in the thirty years it took the young Cambridge reformers to reach power. That men in their fifties had more power and influence than those in their twenties was not part of Cambridge's uniqueness. The point is rather that the foundation of new universities challenged the settled norms of Victorian academic life. Owens College, Manchester, provides a case in point. Within six years of its foundation this fledgling college, though well-endowed, had almost collapsed. A desperate situation called forth desperate remedies. H. E. Roscoe (then twenty-four) was appointed professor of chemistry. His radical, innovating approach, founded on his German training, was rapidly successful. He was fully alive to the technological problems and opportunities of the south Lancashire chemical industry, and within a few years had created what remained a dominant school of chemistry in nineteenth-century Britain. That such radical and research-oriented options were not easily open to older, secure institutions is revealed as much by Edinburgh's reluctance to innovate as by Cambridge's.

The case of H. E. Roscoe brings me to the second point—the importance of outside money. The new laboratories necessary to the undergraduate teaching of science and engineering were—to most observers—outrageously, almost unbelivably, expensive additions to the still pre-Gutenberg technology of education. Once again, Cornford put it succinctly: "The Adullamites [his name for those who 'inhabit a series of caves near Downing Street'—i.e., the Cambridge science laboratories]—the Adullamites are dangerous, because they know what they want; and that is, all the money there is going. . . .

147

These cave dwellers are not refined, like classical men. That is why they succeed in getting all the money there is going."

Even when provided with expensive laboratories, many scientists were less than convinced of the value of practical instruction to undergraduate education. Todhunter's cautionary remarks on pupils who doubt the word of their clergyman-instructors are well known, and as late as 1910 Arthur Schuster was still publicly decrying the utility of laboratory teaching. Again Professor Sviedrys' account reveals, though does not dwell on, the paradox that Maxwell's chair and laboratory were created to meet an undergraduate teaching need, yet few undergraduates darkened the Cavendish's door during its first director's reign. In such a context, the reluctance of all the old, established British universities to invest in laboratories is understandable.

The incentive to such investment had to be found in the lure of pure knowledge, or that of industrial yield. The latter rather than the former seems to have motivated the donors of nineteenth-century Britain. The very sequence of laboratory endowment is indicative: from chemisty through engineering to electrotechnics. I therefore wonder if such a generation gap as existed between Cambridge and other British institutions was not as much a function of its mid-Victorian estrangement from the manufacturing entrepreneurs of Manchester, Glasgow, and Birmingham as of its peculiar organizational setup. Insofar as the latter was important, it was because the prestige of the colleges hampered—and still hampers—all endeavors to attract funds to the university. The already formidable task of raising money from the alienated manufacturing classes was thus rendered still more difficult. The new Owens College at Manchester again provides a revealing contrast. Drawing on a solid basis of support among middle-class manufacturers, the college found it comparatively simple to finance the scientific and technological facilities demanded by the innovatory- and research-oriented curriculum. The very benefaction for the Cavendish itself would seem the exception that proves the rule. The seventh Duke of Devonshire was unique among Cambridge donors in the way he combined aristocratic inheritance and active technological and manufacturing involvement.

To recapitulate briefly—Professor Sviderys suggests a generation gap, or thirty-year lag, in scientific innovations at Cambridge. He

148

attributes this largely to the university's peculiar collegiate form of organization. My doubt is whether the divorce is not between Cambridge and the rest, but between the old, established and tradition-bound among Anglo-Scottish universities and the new, prestigeless, and perforce innovating institutions. Secondly, Professor Sviedrys attributes particular importance to outside money in financing Cambridge laboratories "ahead of schedule." Here my query is whether the Cavendish does not represent a phenomenon common among the new universities, if rare at Cambridge—the support of the new organs of scientific research by the new wealth of successful middle-class entrepreneurs and manufacturers.

I have commented enough, but cannot resist one last query. That is, who were the students who came to the Cavendish before 1895? What social characteristics can we detect in so un-Cambridge a group —neither old-style undergraduates, nor new-style seekers of the Ph.D.? What futures did they see in so untried a field as research physics? Where did they expect to gain employment? And what became of them in later life? Perhaps at this juncture I must modify Occam's razor, and not ask further questions without necessity, at least until we begin to have some answers to already obvious puzzles. For directing our attention to these puzzles and suggesting some persuasive and important answers, we are all indebted to Professor Sviedrys.

Reply BY ROMUALDAS SVIEDRYS

I have no criticism to make about any of the illuminating remarks that my critic has made. In this note, I wish to provide tentative answers to the several questions that he has raised with respect to the social characteristics of the pre-1895 research students who came to work at the Cavendish Laboratory. A full discussion of the subject ought to constitute a separate paper for a future occasion.

During the directorships of Maxwell and Lord Rayleigh (1874–1884), at least eighty students worked for a term or more at the

Cavendish Laboratory. Of these, twenty-four were high wranglers (among the first five), twenty-six were other wranglers, and only fifteen were either not honors graduates or took their degrees via the natural science tripos. Approximately sixty of the eighty students were oriented toward careers in science, either in research or in teaching. Their expectations were in part produced by the unprecedented expansion of the University College system with its great demand for teachers of physics, mathematics, and engineering. The high wranglers most often stayed a much longer period at the laboratory and usually became the better scientists later. They brought to the Cavendish a high level of mathematical training which, when combined with their research experience at the laboratory, allowed them to compete favorably with those who were graduates of other universities.

A finer discrimination can be made if an eye is kept on what became of these graduates in later life. A group of thirty students constitutes the eminent graduates primarily engaged in research and teaching. Twenty-eight make up the second group that devoted itself primarily to teaching, while another eighteen did not engage in science after leaving the Cavendish Laboratory. Four either discontinued physics or died soon after leaving the laboratory. The first group has the largest proportion of high wranglers, forty-three percent, as compared with twenty-two percent and seventeen percent for the second and third groups. With this in mind, I shall examine what became of these graduates in later life, beginning with those who did not become scientists.

The non-scientists distribute themselves among the following occupations: priesthood or missionary work (4), law (3), medicine (3), civil service (3), business (2), and other educational activities (1). The achievements of the second group, those primarily engaged in the teaching of science, had great impact on the development of science in Britain. Seven of them became masters or assistant masters in science at various schools and colleges, five became professors of mathematics or engineering, three taught chemistry and four physics, while the rest held lectureships in mathematics at Cambridge or acted as demonstrators at the Cavendish Laboratory some time in their careers.

The thirty graduates that make up the first group were truly a distinguished lot. Fourteen became fellows of the Royal Society. At least seven established physical laboratories at the institutions of higher education where they taught; four went into industry, to which they brought high scientific research qualifications; three became well-known experts on electrical and other standards; and several more held chairs of mathematics or engineering. Four stayed on at Cambridge to become professors; one succeeded Rayleigh as director of the Cavendish Laboratory, another became Plumian Professor of Astronomy and Experimental Philosophy, a third became the Sadlerian Professor of Pure Mathematics, and a fourth became the first Professor of Astrophysics at Cambridge. Out of this group also came the first professor of electrical engineering at University College, London, the first director of the National Physical Laboratory, and the first director of the Meteorological Office. Rayleigh became the first Englishman to win the Nobel Prize in physics for research that he began at the Cavendish Laboratory; and two others won the same prize later. Although the Cavendish had little prestige within Cambridge University, and only a small portion of the mathematical tripos graduates went to work there, the achievements of some of its workers lent it great prestige within the scientific community at large.

151

Alfred Landé and the Anomalous Zeeman Effect, 1919–1921

BY PAUL FORMAN*

1. Introduction 153
2. Alfred Landé and the Quantum Theory of the Atom, 1918–
   1920 158
3. Arnold Sommerfeld and the Anomalous Zeeman Effect, 1919–
   1920 179
4. Landé's Empirical Rules, December 1920–February 1921 . . . 195
5. Competition with Ernst Back . . . . ·. 209
6. A Quantum-Theoretical Construction, March 1921 221
7. Aftermath, April–September 1921 232
Appendix: Correspondence of February and March 1921 Bearing on
   Landé and the Anomalous Zeeman Effect 237

1. INTRODUCTION

In April 1921 Alfred Landé submitted for publication a paper "On the Anomalous Zeeman Effect."[1] In it he showed how the quantum theory of the Zeeman effect, which had had only very limited success in the five years since its construction by Arnold Sommerfeld and Peter Debye, could be modified *ad hoc* to yield the multifarious anomalous Zeeman effects—the very complicated patterns into which spectral lines are split when the emitting atom is in a magnetic field. By utilizing half integral quantum numbers, by incorporating the recently introduced "inner" (total angular momentum) quantum number, and especially by introducing for each state of an atom a

---

* Department of History, University of Rochester, Rochester, New York 14627.

1. Landé, "Über den anomalen Zeemaneffekt (Teil I)," *ZS. f. Phys.,* 5 (23 June 1921), 231–241, received 16 Apr. 1921.

special constant, g, the constant ratio of the actual additional energy of the atom in a magnetic field to the energy anticipated from the quantum theory, Landé was able to give an energy level analysis of the anomalous Zeeman patterns, and thus for the first time brought the phenomena into intimate contact with the apparatus of the quantum theory. The importance of this association during the following five-year development of atomic physics has been obscured by the physicists' discovery in 1926 that it was possible to regard the anomalous Zeeman effect as arising from an intrinsic property of the *electron,* rather than as revealing the true quantum mechanics of the atom. But in 1920, when Landé first took up the question, theorists such as Bohr and Sommerfeld already firmly believed that the anomalous Zeeman effect was intimately related to the problem of atomic structure. And in the following years, in large measure as a consequence of Landé's papers, it came to be very widely accepted that the anomalous Zeeman effect was, moreover and above all, an expression of the failure of classical mechanics and electrodynamics at the atomic level, and that, conversely, it was one of the most promising avenues by which to advance toward the new, true theory. In another essay I have tried to explain why and how such (largely mistaken) convictions arose;[2] as an introduction to the present study it is enough to emphasize that the anomalous Zeeman effect stood at the very center of theoretical atomic physics during a period of unprecedented activity culminating in the discovery of the quantum mechanics and electron spin.[3]

This paper is thus an attempt to elucidate the genesis of a relatively small, yet important, scientific innovation. And as the genesis of discoveries has long been a chief stock-in-trade of historians of science,

2. Forman, "The Doublet Riddle and Atomic Physics *circa* 1924," *Isis, 59* (1968), 156–174.

3. A preliminary account of this study was presented to the Reed College physics colloquium in April 1965. A subsequent version constituted Chapter III of my doctoral dissertation (University of California, Berkeley, 1967), *The Environment and Practice of Atomic Physics in Weimar Germany: A Study in the History of Science* (Ann Arbor: University Microfilms, 1968), *Dissertation Abstracts, 29* (1968), 200. The present publication differs substantially from the preceding through consideration of the neon spectrum in sections 2–4, reconstruction of the sequence of events in sections 4 and 5, and inclusion of the manuscript sources from February and March 1921 in the appendix.

As the notes testify, this study relies heavily upon materials collected by the project Sources for History of Quantum Physics (SHQP), 1961–1964. A detailed

perhaps no further statement of my aims is necessary. Nonetheless, I think it desirable that historians be self-conscious and explicit about their means and ends. There are, in my opinion, two goals to be achieved through a study of the genesis of a particular scientific innovation.

One goal is the identification of the *effective* principles, tacit as well as explicit, in that science at that time. The historian of science is, I assume, not primarily interested in the discoverer, or the idiosyncrasies of his thought, neither is he primarily interested in the act of discovery. The primary aim of the historian of science is to reconstruct the science—i.e., the social activity and the material and ideational artifacts associated with it—of the past. Such reconstructions must include not merely the content of the doctrines then orthodox, but also a specification of the strength with which each doctrine was held. In both these respects the close examination of scientific discoveries may be helpful, for a discovery is, by definition, a discovery of something whose existence is either explicitly precluded, or at least unforeseen, within the orthodox doctrines. Thus on the one hand essential components of the conceptual situation in the science, components which might otherwise be overlooked because they are either unanticipated by the historian or seldom made explicit by the scientists he is studying, can be identified in the course of a genetic investigation as those notions which made the discovery conceptually inaccessible to their adherents. In this study, for example, Runge's rule is found to play this role. On the other hand, because the discovery is not possible without doing some or much violence to orthodox doctrine, an exposition of the discoverer's conceptual gropings is simultaneously a test of the strength of belief in,

---

list and author catalog of these, and certain other, materials is given by T. S. Kuhn, J. L. Heilbron, P. Forman, and L. Allen, *Sources for History of Quantum Physics: An Inventory and Report, Memoirs of the American Philosophical Society, 68* (Philadelphia, Pa., Am. Phil. Soc., 1967). A symbol such as (SHQP 8, 9) locates a document in section 9 of microfilm 8 of this project. Also the microfilms of the Bohr Scientific Correspondence (BSC) are catalogued in this publication and available in the "Archive for History of Quantum Physics" at the same depositories: The American Philosophical Society; The University of California, Berkeley; The Niels Bohr Institute for Theoretical Physics, Copenhagen. The original texts of all translated quotations from unpublished materials are given in the notes or in the appendix. Mr. Murphy Smith, Manuscripts Librarian of the American Philosophical Society, has been most helpful in providing photocopies of these documents.

and the rigidity of application of, one or several components of that orthodoxy. Thus I will show that in the search for a theory of the anomalous Zeeman effect the principles of the quantum theory as expounded in contemporary treatises, or even so general a postulate as the combination principle, placed only the loosest constraints upon Landé's thought, or rather were treated by him as highly plastic categories to be remolded and reshaped at will.

There is a serious difficulty here. In order to draw general conclusions about the strength of the thought controls exercised by particular doctrines from an exposition of the discoverer's conceptual explorations, the discoverer must himself be conceptually typical—and that he never is. The discoverer always has an "extraordinary" background—either special knowledge which prepares him uniquely to "infer" his discovery, or special ignorance which allows him, oblivious to the directives of orthodox doctrine, to stumble upon his discovery. John L. Heilbron has suggested that accidental discoveries of unanticipated phenomena can best be explained by reference to this latter type of background, and on this basis has argued cogently that the tyro Cunaeus rather than the adept Musschenbroek is the probable discoverer of the Leiden jar.[4] *Conceptual* innovations in theoretical physics, on the other hand, have traditionally been "explained" by the former sort of background; i.e., by reference to insights which the innovator would have derived from his previous training or researches. My examination of Landé's discovery shows, however, that he was not merely utterly unaware of his special ability to leap directly to an energy level analysis of the anomalous Zeeman effect, but that in his manner of attacking the problem he implicitly repudiated the background and the insights by which we would wish to explain why precisely this man tackled precisely this problem, producing precisely this solution. This again strongly suggests that the usual method of reconstructing the route to a conceptual inno-

4. Heilbron, "A propos de l'invention de la bouteille de Leyde," *Revue d'Histoire des Sciences, 19* (1966), 133–142, and "G. M. Bose: The Prime Mover in the Invention of the Leyden Jar?" *Isis, 57* (1966), 264–267. My analysis originates in Heilbron's demonstration that the historian of science can face up to, and turn to his own advantage, the awkward fact that the most fundamental *definiens* of an unanticipated discovery is its nondeducibility from the scientific doctrines of the period. Thus, for example, Heilbron has taken the conceptual and manipulative inaccessibility of the discovery of the Leiden jar as a criterion for the verisimilitude of a reconstruction of electrical theory in the early 1740's.

vation—by drawing straight, logical lines from one research problem or result to another—should be regarded as a conventional stylization, not as a reproduction, of historical reality. At the same time, the fact that Landé's scientific education was most orthodox, while his special insights were long inoperative, suggests a general ineffectiveness of the principles of the quantum theory in channeling the thought of theoretical atomic physicists circa 1920.

The second goal to be achieved through a study of the genesis of a particular scientific innovation is the excision of a slice of the scientific life of that place and period. In addition to its unique potential for displaying the centrality of the notion of priority, and the elaborate system of mores built around it, this mode of sectioning the scientific life shares with several other modes a potential for displaying the interlocking of conceptual and social factors. Or, expressed in the jargon of our discipline, geneses too are well suited to demonstrate the untenability of the internalist–externalist distinction— a distinction so foreign to contemporary historical scholarship that we must regard its persistence among historians of science as one of the more blatant ideological atavisms testifying to our phylogenetic (and frequently ontogenetic) connection with the sciences. For if one asks how this particular physicist came to this particular problem at this particular time and why he communicated these particular results in these particular forms, how indeed can one avoid regarding an innovation as the outcome of the motivated acts of an acute man, working in a particular social environment for his own advancement as well as for the progress of science? The fiction of an autonomous development in the world of scientific ideas has been maintained by systematically mistaking description for explanation, and by systematically refusing to look for or at contrary evidence. The internalist, unaware, for example, of the conventions surrounding the German academic institution of *Habilitation,* would never feel the utter inadequacy of his account of the marked shift in the direction of Landé's work in the spring of 1919. By examining the prehistory of Landé's first paper on the anomalous Zeeman effect, that is, the genesis of an artifact of the community of physicists, I wish to present a concrete picture of the reciprocal interaction of self-interest and the discipline's interest as they are channeled and constrained by the political environment, by economic circumstances, by the structure

157

of academic life and careers, by the organization and mores of the community of physicists, by personal relationships, by accepted physical theories, and by experimental facts.

In an appendix to this study I am publishing the most important manuscript sources. By choosing a sufficiently short time interval, I have found it feasible to include every document known to me bearing upon Landé's work on the anomalous Zeeman effect. Publication of documents, along with bibliographies and compilations of data, is a form of historical publication which needs no apology, for it is a greater service to the community of historians than to the historian-editor. Moreover, the documents will provide the reader with ample opportunity to assess my readings and reconstructions.

I am greatly indebted to Professor Landé for his appreciation as well as toleration of my efforts, and for his consent to the publication of this paper and the appended correspondence. I am also grateful to Professor Aage Bohr and to Dr. Ernst Sommerfeld for permission to publish letters by their fathers. The paper has profited greatly from the discussion and criticism of earlier drafts by A. Hunter Dupree, Charles S. Fisher, John L. Heilbron, Karl Hufbauer, Thomas S. Kuhn, Charles B. Paul, F. Reif, and, most recently, by the editor of this journal, Russell McCormmach.

2. ALFRED LANDÉ AND THE QUANTUM THEORY OF THE ATOM, 1918–1920

In the winter of 1920/21 Alfred Landé, almost thirty-two years old, was in his third semester as a *Privatdozent* for theoretical physics at the University of Frankfurt a. M. The scientific company could hardly have been more lively—including Max Born, *Ordinarius* for theoretical physics, whose departure, however, for Göttingen had already been arranged;[5] Otto Stern, *Privatdozent*

5. Already on 1 May 1920 the *Personalien* column of the *Physikalische Zeitschrift* (*21*, p. 248) announced that Born had been appointed Debye's successor as the second professor of *Experimentalphysik* at Göttingen. It was to be a year, however, before Born assumed his duties there, and in the meantime, exploiting a technical error in the budget, Born managed to evade the pressures forcing ambitious theoreticians into experimental chairs by having his chair divided, the experimental half going to James Franck. R. Pohl, interview with SHQP, 25 June 1963, p. 5. Cf. Born's recollections in his commentary to Albert Einstein, Hedwig and Max Born, *Briefwechsel 1916–1955* (Munich, 1969), pp. 47–50.

with *Lehrauftrag;*[6] and Walther Gerlach, *Privatdozent* and senior assistant in experimental physics. Landé's career prospects, on the other hand, were precarious.

Landé, the son of a lawyer resident in Elberfeld in the northern Rhineland, had studied mathematics and physics at Göttingen and Munich, obtaining his D.Phil. in the spring of 1914. Anticipating that he might have to make *Gymnasium* teaching his livelihood, he, like most other aspiring physics students, had taken the requisite Prussian State examinations.[7] But in the spring of 1913, while working on an unexciting Munich doctoral dissertation under Arnold Sommerfeld, *Ordinarius* for theoretical physics, he received a valuable boost up onto the academic ladder. Through the good offices of Max Born, then *Privatdozent* at Göttingen, Landé spent the year before the outbreak of the war as a research assistant to the luminous mathematician David Hilbert. This position, which was jokingly referred to as Hilbert's *Hauslehrer,* Hilbert's private tutor in physics, "was very decisive, because as Hilbert's *Hauslehrer* in Göttingen I certainly had official status in the scientific community."[8]

Suddenly, however, the war broke off this promising start. Membership in the Mathematisches Lesezimmer, the headquarters of the circle of young physicists and mathematicians at Göttingen, dropped from 350 in the summer semester 1914 (Easter through the end of July) to 98 in the winter semester 1914/15 (November through February).[9] Landé enrolled in the Red Cross and served for some time as a medical corpsman on the eastern front. Subsequently he was drafted into the army, "but never fired a shot."[10] Through Born's intervention, in 1917 he was recalled to one of the few openings for scientists in military research. At the Artillerie Prüfungs-Kommission

6. Born to Landé, 27 Jan. 1919 (SHQP 4, 2). An exposition of German academic titles, as well as of the social and economic circumstances of the academic physicist, is given in my dissertation (note 1), Chapters I and II. I intend that this material, revised and amplified, will appear in the near future in book form.

7. Adolf Beier, *Die höheren Schulen in Preussen . . . und ihre Lehrer* (3rd ed.; Halle, 1909), pp. 517–540. Direktion des math.-phys. Seminars, *Ratschläge und Erläuterungen für die Studierenden der Mathematik und Physik an der Universität Göttingen* (Neue Auflage; Leipzig, 1913).

8. Landé, interview with SHQP, 5 Mar. 1962, session 1.

9. *Chronik der . . . Universität zu Göttingen für . . . 1914* (Göttingen, 1915), pp. 63–64.

10. Landé, autobiographical statement, 3 pp., Apr. 1962. Center for History and Philosophy of Physics, American Institute of Physics, New York.

in Berlin Born and Landé and a few other physicists and experimental psychologists worked in determining the range of enemy artillery from the sound of the discharge—a problem upon which their British, French, and American counterparts were likewise engaged.[10a]

But in Germany, after the first rush to the colors, the physicists, whatever their involvement in the military, were by and large anxious to forget the war and to devote any spare time to their own research. Papers began to appear in the physical journals bearing datelines from obscure villages on the eastern front. And when Landé arrived in Berlin—where institutions such as the Prussian Academy and the Deutsche Physikalische Gesellschaft continued to function— he found Born and the other physicists at the A.P.K. with drawers full of their own books and papers. Landé had only to choose a problem.

In his fundamental trilogy of 1913 Niels Bohr had not merely given a model and a theory of the spectrum of the hydrogen atom, but—and this had been his original and principal goal—had specified "the Constitution of Atoms and Molecules" in their normal, unexcited, nonradiating, ground state.[11] Confining the electrons in Rutherford's nuclear atom to coplanar circular rings, Bohr allowed each electron to have only one "quantum" of angular momentum, $h/2\pi$ (where h is Planck's constant), or, more generally, one quantum of kinetic energy, $h\nu_{classical}/2$ (where $\nu_{classical}$ is the frequency of the motion). Bohr then used energy and stability criteria to fix the number, size, and population of the electronic rings in the normal states of the lighter elements and simpler molecules. (Higher quantum states—i.e., those with quantum numbers greater than one—were to occur only in excited atoms.) John L. Heilbron has shown that during the war the burden of advancing these same questions fell in large measure upon the German theorists concerned with the emission of X-ray spectra by atoms so constituted.[12] And although in these

10a. Daniel J. Kevles, "Flash and Sound in the AEF: The History of a Technical Service," *Military Affairs, 33* (1969), 374–384.

11. Niels Bohr, *On the Constitution of Atoms and Molecules. Papers of 1913 reprinted from the Philosophical Magazine with an Introduction by L. Rosenfeld* (New York, 1963). John L. Heilbron and Thomas S. Kuhn, "The Genesis of Bohr's Atom," *Historical Studies in the Physical Sciences, 1* (1969), 211–290.

12. See Heilbron's pioneering survey of the development of atomic models (Ph.D. diss., University of California, Berkeley, 1964), *A History of the Problem of Atomic Structure from the Discovery of the Electron to the Beginning of Quan-*

five years, until the spring of 1918, the picture of circular rings was modified, and the restriction to "one quantum" orbits was relaxed, still the coplanar "pancake" atom remained the basis of all theoretical work.

A technique, the "correct" technique, for treating spacial atomic structures, in (and only in) the form of inclined electron rings, had been developed by Sommerfeld in 1916, but had never actually been employed for that purpose. Sommerfeld had been much impressed by Bohr's papers, and in September 1913, displaying his rather old-fashioned notions of the "Game laws and Patent laws of science," asked Bohr's leave to stake a claim to the problem of applying the new model to the Zeeman effect.[13] In order to handle this problem, Sommerfeld found it necessary to find a more general prescription for quantizing mechanical systems, one which could be applied to systems with more than one degree of freedom. His work on this problem in 1914 and 1915 produced fortuitously a marvelously successful relativistic theory of the fine structure of the hydrogen lines,[14] but it was only in 1916 that Sommerfeld found his way to a quantum theory of the Zeeman effect. The key was contained in his final exposition of his quantization rules, $\oint p_i dq_i = n_i h,$ in the *Annalen* in 1916.[15] (The q_i, $i = 1, 2, \ldots, f$, are the coordinates, and p_i the conjugate momenta, of an atomic system with f degrees of freedom. The integral, which has the dimensions of action or angular momentum, is carried over one cycle of the coordinate in question,

tum Mechanics (Ann Arbor: University Microfilms, 1965), *Dissertation Abstracts*, *25* (1965), 7216, and his articles on "The Work of H. G. J. Moseley," *Isis, 57* (1966), 336–364, and especially, "The Kossel-Sommerfeld Theory and the Ring Atom," *Isis, 58* (1967), 451–485.

13. Sommerfeld to Bohr, 4 Sept. 1913: "Will you apply your atomic model to the Zeeman effect? I would like to concern myself with that problem." Published by Rosenfeld, *op. cit.* (note 11), in his introduction to Bohr, *On the Constitution* . . . (1963), p. lii. Cf., Maxwell to William Thomson, 13 Sept. 1855: "I do not know the Game laws and Patent laws of science. Perhaps the [British] Association may do something to fix them, but I certainly intend to poach among your electrical images." (Joseph Larmor, ed., *Origins of Clerk Maxwell's Electric Ideas as described in familiar letters to William Thomson* [Cambridge, 1937], pp. 18–19.)

14. Sommerfeld, "Zur Theorie der Balmerschen Serie," *Bayer. Akad. d. Wiss., München, Sitzungsber. math.-phys. Kl.* (1915), pp. 425–458, read 6 Dec. 1915; "Die Feinstruktur der Wasserstoff- und der Wasserstoffähnlichen Linien," *ibid.*, pp. 459–500, read 8 Jan. 1916.

15. Sommerfeld, "Zur Quantentheorie der Spektrallinien," *Ann. d. Phys., 51* (22 Sept., 10 Oct. 1916), 1–94, 125–167, received 5 July 1916. *Gesammelte Schriften,* ed. F. Sauter (Braunschweig, 1968), *3,* 172–308.

and set equal to an integer [quantum number] times Planck's constant.) Overlooking the interaction between the external radiating electron and the inner electronic rings in the atom, Sommerfeld showed it to follow from these rules, applied to the single external electron or electron ring, that there were definite restrictions on the possible orientations of the orbital plane—implicitly assumed well defined and invariable—of this far out electron. Namely, the orbital angular momentum of the electron, $nh/2\pi$ (n a positive integer, the "azimuthal" quantum number), which Sommerfeld did not distinguish from the total angular momentum, could have a component in a given direction only of magnitude $mh/2\pi$ (m an integer, the "equatorial," or "magnetic," quantum number).[16]

With Sommerfeld's quantization conditions there were, then, clear prescriptions regarding spacial orientations. It only required that some direction in space be physically distinguished in order that it provide an axis for the coordinate system. Although it is usually assumed that these prescriptions, once discovered, guided research in the next decade, in fact what Sommerfeld set out in 1915 and 1916 he tacitly took back in 1917 and 1918, and only reaffirmed in 1919 and 1920. In 1918, when Sommerfeld himself considered crossed electron rings, the orientations were derived from minimum energy conditions, not quantum conditions. Even the basic requirement of integral angular momentum carried little weight with him at this time.[17]

16. Sommerfeld, "Zur Theorie des Zeeman-Effekts," *Phys. ZS., 17* (15 Oct. 1916), 491–507, received 7 Sept. 1916; *Ges. Schr., 3,* 309–325. The theory given there was incomplete in a number of respects. Besides the lack of a selection rule to limit the changes in the magnetic quantum number, m, Sommerfeld also refused to admit negative values of m. The priority in applying Sommerfeld's quantization rules to the Zeeman effect was gained, however, by Sommerfeld's former student, P. Debye, "Quantenhypothese und Zeeman-Effekt," *Phys. ZS., 17* (15 Oct. 1916), 507–512, received 7 Sept. 1916, "Nach einer am 3. Juni 1916 der Göttinger Ges. Wiss. vorgelegten Notiz." Understandably Sommerfeld was annoyed. We might compare the situation to Cavalieri's anticipation of Galileo in publishing the parabolic path of a projectile: "for really my first intention, that which incited me to meditate on motion, had been to find that line." Quoted in R. Dugas, *Mechanics in the Seventeenth Century* (Neuchâtel, Switzerland, 1958), p. 75.

17. Sommerfeld, "Atombau und Röntgenspektren. I. Teil," *Phys. ZS., 19* (15 July 1918), 297–307, received 6 Apr. 1918; "Die Drudesche Dispersionstheorie vom Standpunkte des Bohrschen Modelles und die Konstitution von H_2, O_2 und N_2," *Ann. d. Phys., 53* (24 Jan. 1918), 497–550, received 1 Aug. 1917. The general conclusion of this paper is that the total angular momentum of an electron ring with $2n$ electrons ought to be set equal to $2n \sqrt{n}\, h/2\pi$. (*Ges. Schr., 3,* 432–458 and 378–431, respectively.)

In these latter years of the war it was a popular—although, as was afterwards recognized, overly ambitious—program among German theorists to derive the various observed bulk properties of matter from the structure of its constituent atoms. Falling in with this program, Landé first tried to derive the optical rotary power of organic compounds, expressed in terms of the fundamental atomic constants, by endowing the tetrahedral carbon atom with a quantized electron ring in each of its four corners.[18] Then Born, for whom the theory of crystal lattices was always a pet subject, suggested that they see whether these pancake atoms of Bohr's, arranged in a lattice, would give the observed macroscopic properties of crystals.[19] In these calculations, which Born and Landé were submitting for publication in the last weeks of the war and the first days of the German revolution, they gave some attention to the tilting of the planes of the various rings, but, not surprisingly, ignored Sommerfeld's space quantization, and based themselves upon Sommerfeld's minimum energy calculations. Their first results showed that the lattice constants of alkali halide crystals fitted well with a pancake atom, but when they turned to the compressibilities they found that even *inclined* rings could not keep their crystals from being too soft. Between the 5th and 8th of November 1918, that is between the Kiel Mutiny and the proclamation of the Republic, Born and Landé also proclaimed their own small revolution: "the electrons of a single atom do not move about in planes, but are distributed uniformly over all directions in space. Thereby a new problem arises for the quantum theory; the planar electron orbits do not suffice, the atoms are obviously spacial structures."[20]

Born suggested to Landé that the relative positions of the various electrons in an atom be fixed by requiring them constantly to satisfy

18. Landé, "Über Koppelung von Elektronenringen und optische Drehvermögen . . . ," *Phys. ZS., 19* (15 Nov. 1918), 500–505, received 26 July 1918.
19. Born and Landé, "Über die absolute Berechnung der Kristalleigenschaften mit Hilfe Bohrscher Atommodelle," *Sitzungsber. Preuss. Akad. Wiss., Berlin* (14 Nov. 1918), pp. 1048–1068; submitted, through Einstein, 17 Oct. 1918. Born and Landé, "Kristallgitter und Bohrsches Atommodell," *Verhl. d. Dtsch. Phys. Ges., 20* (30 Dec. 1918), 202–209, received 5 November 1918.
20. Born and Landé, "Über die Berechnung der Kompressibilität regulärer Kristalle aus der Gittertheorie," *Verhl. d. Dtsch. Phys. Ges., 20* (30 Dec. 1918), 210–216, received 8 November 1918. The revolt against the "pancake" atom was a popular movement in late 1918 and early 1919. Compare J. L. Heilbron's article on the "Ring Atom," *op. cit.* (note 12).

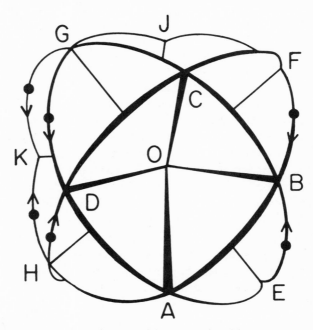

FIGURE 1. The Synchronized Polyhedric Atom of Cubic Symmetry ("Würfelver-band"). Adapted from Landé's figure, *Verhl. d. Dtsch. Phys. Ges., 21* (1919), 654.
The eight electrons travel along the edges of the four disks *E B C G K, E A D G J, H D C F, K H A B F J,* each having one of the eight spherical triangles as its orbit. The deflection of the electron traveling from *H* to *D* into the arc *DA* is effected by collision at *D* with its mirror electron on *GD*, which, reciprocally, is deflected onto *DC*.

the reflection and rotation groups of an appropriate polyhedron. The most prominent and promising case was a *Würfelatom*, eight electrons fulfilling the symmetry conditions of a cube (Fig. 1). Landé made a start on the project right away.[21]

Landé's program aroused immediate enthusiasm. Sommerfeld, the most distinguished, energetic, and influential figure in theoretical atomic physics in Germany, wrote that it "interests me extraordinar-

21. Born, "Über kubische Atommodelle," *Verhl. d. Dtsch. Phys. Ges., 20* (1918), 230–239, received 16 Dec. 1918. Landé, "Dynamik der räumlichen Atomstruktur," *Verhl. d. Dtsch. Phys. Ges., 21* (1919), 2–12, received 25 Dec. 1918; "Elektronen-bahnen im Polyederverband," *Sitzungsber. Preuss. Akad. Wiss., Berlin* (30 January 1919), pp. 101–106, communicated 9 Jan. 1919. This approach had the attractive feature of automatically reducing the intractable many-electron problem to a one-electron problem. An admirably clear account of this model is given by Heilbron, *A History of the Problem of Atomic Structure, op. cit.* (note 12), pp. 371–376.

ily,"[22] and he was soon looking to it for the "salvation" of the theory of X-ray spectra, then in considerable difficulty.[23] Debye's "greatest interest" and also Bohr's "extraordinary," indeed "highest," interest was excited.[24] But early in 1919, while the spacially symmetric atoms still seemed to be by far the most attractive problem, Landé was obliged to turn his attention more seriously to the theoretical spectroscopy of the Bohr-Sommerfeld atom.

This shift in the direction of Landé efforts was thus not dictated by the logic of scientific ideas, nor by the winds of scientific fashion, nor even by the accidents of personal influence; it was, in fact, a response to the exigencies of German academic careers. By the fall of 1918 it had been arranged—and it is characteristic that the revolution did not disrupt such arrangements—that, with the summer semester 1919, Max von Laue, *Ordinarius* for theoretical physics at Frankfurt, would receive a new *Ordinariat* in Berlin alongside his mentor, Max Planck, and that Max Born, *Extraordinarius* in Berlin, would receive von Laue's chair in Frankfurt. The plan evidently was that Born, as *Ordinarius* in Frankfurt, would sponsor Landé's *Habilitation*,[25] i.e., the process of qualification as a *Privatdozent*. In the last days of December 1918 Landé left stormy Berlin for the *Odenwaldschule*, a private progressive country school located about thirty-five miles south of Frankfurt and twenty miles north of Heidelberg in the hills along the Rhine.[26] "I had the morning free for theoretical physics, and in the afternoon earned my keep by giving music lessons, in a most stimulating atmosphere of educators, artists, nature

22. Sommerfeld to Landé, 11 Dec. 1918 (SHQP 4, 13). On 10 Jan. 1919 Sommerfeld addressed the *Münchener Physikalisches Mittwochs-Colloquium* on "Stereoatomistik (nach Landé)." (Register, pp. 69–70 [SHQP 20].)

23. Quoted by Heilbron, "The Ring Atom," *op. cit.* (note 12), p. 479. Cf., also, the preface to the first edition of Sommerfeld's *Atombau und Spektrallinien* (Braunschweig, 1919), dated 2 Sept. 1919, where he regrets not having been able to discuss "the interesting ideas of Born and Landé on spacial distributions of electrons, which perhaps presage the future of the theory of X-ray spectra."

24. Born to Landé, 27 Jan. 1919 (SHQP 4, 2); Bohr to Landé, 26 June 1919 (SHQP 4, 1).

25. Born to Landé, 27 Jan. 1919. Born's trade with Laue was announced in the *Phys. ZS.* of 15 Jan. 1919 (20, p. 49), but had been in the making for some time. See: Einstein to Hedwig Born, 8 Feb. 1918, *Briefwechsel, op. cit.* (note 5), pp. 22–23.

26. There could not have been any question of a *Habilitation* at Heidelberg, where Philipp Lenard was the one and only *Ordinarius*, although Landé actually did give a colloquium there on 17 Nov. 1920. (Baerwald to Landé, 11 Nov. 1920 [SHQP 4, 15].)

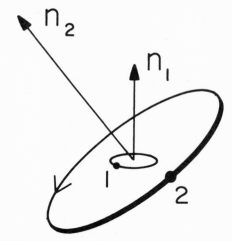

FIGURE 2. The Helium Atom, Spring 1919. Adapted from Landé's figure, *Phys. ZS., 20* (1919), 229.

Angular momentum of outer electron, n_2, space quantized with respect to direction of angular momentum of inner electron, n_1.

lovers, etc. ...."[27] Here Landé set energetically to work upon a *Habilitationsschrift,* a piece of original work substantially above the level of a doctoral dissertation. Although unquestionably he would have preferred to pursue his spacially symmetric atoms, that topic would not have been acceptable for habilitating because it was too obviously connected with work he had done jointly with Born, and the fundamental idea had originated with Born. The issue of independence was sensitive; as Born said in excusing himself for not seeing Landé while visiting Frankfurt, "perhaps it is just as well, since you were my scientific *Compagnon* so long, that your *Habilitationsschrift* arises clearly and obviously without my collaboration."[28]

Another topic had to be chosen, and Landé settled upon the problem of calculating the energy levels of a single electron moving in an orbit far beyond the nucleus and the other electrons in the atom

27. Landé, autobiog. statement, Apr. 1962. Center for History and Philosophy of Physics, American Institute of Physics, New York.

28. "Vielleicht ist das auch ganz gut; nachdem Sie so lange mein wissenschaftlicher Compagnon waren, empfiehlt es sich, dass Ihre Habilitationsschrift deutlich und sichtbar ohne meine Mitwirkung entsteht." (Born to Landé, 27 Jan. 1919 [SHQP 4, 2].) In general Born does not appear to have felt himself to be in so strong a position as we might expect. He speaks of the favorable recommendations (*Urteilen*) by Debye and Epstein as of some significance in making it easy for him to push through the *Habilitation.*

(Fig. 2). This was, thus, indifferently, a model of an alkali atom or of the excited—and, Sommerfeld thought, perhaps even the normal—helium atom.[29] Landé limited himself to the simplest case, the helium atom, which, in fact, is already such an intractable dynamical problem that it can be treated only by the approximate methods of perturbation theory.

The spectrum of helium, as it had then been resolved, consisted of two noncombining term systems, a system of single terms (parhelium) and a system of double terms (orthohelium).[30] Each system gives rise to its own principal, sharp, diffuse, and fundamental series in the spectrum. On Sommerfeld's principles one quantized the motion, and then calculated the energy of the outer electron in the field of the nucleus and of the inner electron. Because the potential was no longer that of a strictly inverse square attraction, the stationary states were no longer precisely the Bohr energy levels Rhc/n^2_B, where $R = 2\pi^2 m_0 e^4/h^3 c$ is the Rydberg constant, m_0 and e the mass and charge of the electron, c the velocity of light, and n_B is the total, or kinetic energy, quantum number—the sum of the azimuthal quantum number, n, and the radial quantum number, $n' = \oint p_r dr/h$, r the radial coordinate of the outer electron. Instead, expressing the energy as a series in ascending powers of the ratio of the radii of the orbits of the inner and outer electron, there resulted in the second approximation—the Bohr levels being the first—a formula like that which Rydberg had used to express the spectral terms, $Rhc/(n_B + f(n))^2$, and in the third approximation the further refinement of this formula due to Ritz.[31] The various sequences of spectral terms—sharp, principal, diffuse, and fundamental or bergmann—were associated with the various values of $f(n)$: $n = 1$, s terms; $n = 2$, p terms; $n = 3$,

29. The problem had been set and the methods for treating it had been developed in 1916 by Sommerfeld: "Zur Quantentheorie der Spektrallinien, Ergänzungen und Erweiterungen," *Bayer. Akad. d. Wiss., München, Sitzungsber. math.-phys. Kl.* (1916), 131–182; *Ges. Schr., 3,* 326–377, especially pp. 327–330, 347–361. The problem of the helium atom in the old quantum theory has recently been investigated by Henry Small in his University of Wisconsin doctoral dissertation (1970).

30. For the early history of the helium spectrum: H. Kayser, *Handbuch der Spectroscopie, 5* (Leipzig, 1910), 508–520. In 1927 W. V. Houston could cite a half dozen attempts since 1906 to resolve the orthohelium doublets into triplets. He, however, was the first to succeed: "The Fine Structure of the Helium Arc Spectrum," *Proc. Nat. Acad. Sci., 13* (15 Mar. 1927), 91–94.

31. For details see Heilbron, *A History of the Problem of Atomic Structure, op. cit.* (note 12), pp. 356–359.

d terms; $n = 4$, f terms. The successive terms in a given sequence were associated with successive values of $n_B = n + n'$ as n' ran through the values 0, 1, 2, 3, ... The observed spectral lines then arose as the electron jumped from one stationary state (term) to another of lower energy, $\nu_{quantum} = (E - E')/h$: a sharp series in transitions from a sequence of s terms to a given p term; a principal series in transitions from a sequence of p terms to a given s term; a diffuse series . . . d terms to a given p term, etc. The scheme included, however, no explanation of: 1) the appearance of precisely these series, namely the combination of those and only those terms whose azimuthal quantum number differed by one unit; 2) the existence of a complex structure of sublevels; i.e., more often than not the sequences of p, d, etc., terms were actually sequences of doublets or triplets; 3) the existence in one atom of more than one sequence of s, p, d, etc., terms with different complex structures—helium being unusual only in the absence of combinations between the members of the system of single terms (parhelium) and the system of double terms (orthohelium).[32]

With all of this Landé had to familiarize himself, and thus he became more aware of the explicit demands of Sommerfeld's quantization rules, as well as of the problem of accounting for the great abundance of spectral terms. Landé's solution to this latter problem—and it was an excellent idea—was to use the several, quantized orientations of the angular momentum of the exterior, radiating electron as a means of generating additional terms. His original approach was to implicitly ignore the coupling of the orbital angular momenta of the two electrons and to space quantize the orientations of the plane of the outer electron using the angular momentum of the inner electron—assumed fixed in space—as the axis of the coordinate system.[33] To this Sommerfeld did not take exception, for it was exactly how, in his original publication, he had suggested that it be done.

What discomfited Sommerfeld was that having thus generated

32. At this time only doublets and triplets were recognized; higher multiplicities were distinguished only in 1922 by M. Catalán, and by H. Gieseler working in Tübingen. V. V. Raman, "Miguel A. Catalán," *Dictionary of Scientific Biography*, 3 (New York, 1970).

33. Landé "Das Serienspektrum des Heliums," *Phys. ZS.*, 20 (15 May 1919), 228–234, received 25 Mar. 1919. Again, for details see Heilbron, *Atomic Structure, op. cit.* (note 12), pp. 376–380, and Small, *op. cit.* (note 29).

more than enough energy levels, Landé at first wished to assign them to the spectral terms merely by matching energies, unconstrained by such requirements as s terms correspond to $n = 1$, p terms to $n = 2$, etc.[34] But during the course of 1918 Sommerfeld's interpretations of the spectral terms, and also of the normal Zeeman effect, had been buttressed by the discovery of long-sought-for selection rules governing the changes in the quantum numbers in transitions between stationary states. In the arguments which Sommerfeld's student and assistant A. Rubinowicz based on the conservation of angular momentum in the process of radiation, the azimuthal quantum number, n, while yet associated with a single electron, was regarded as the total angular momentum of the atom.[35] This led to a selection rule $\Delta n = \pm 1$ *or* 0. Bohr, on the other hand, had derived a selection rule of just this form for the equatorial, or "magnetic," quantum number, m, and, albeit somewhat obliquely, extended the result to the total angular momentum of an atomic system. For the azimuthal quantum number, n, Bohr would allow only $\Delta n = \pm 1$.[36] And although these developments initiated a period of considerable confusion about which quantum numbers stood for what, they nevertheless cooperated in constraining Landé to go along with Sommerfeld's particular assignments of "n" values to the various term types.

Landé, however, soon advanced beyond his mentor and began to reconsider the problem of momentum and space quantization. He realized that Sommerfeld's neglect of the interaction between the external radiating electron and the inner electron was physically most unrealistic, that the total angular momentum of the atom, not that of the inner electron, must be taken as fixed in space. Moreover, careful study of the principles of the quantum theory, particularly as presented by Bohr in "The Quantum Theory of Line-Spectra"

34. Sommerfeld to Landé, 28 Feb. 1919, 5 Mar. 1919 (SHQP 4, 13).

35. A. Rubinowicz, "Bohrsche Frequenzbedingung und Erhaltung des Impulsmomentes. I. Teil," *Phys. ZS., 19* (15 Oct. 1918), 441–445.

36. Bohr, "On the Quantum Theory of Line-Spectra," *D. Kgl. Danske Videnskabernes Selskabs Skrifter, Naturvidensk. og math. Afd.,* ser. 8, *4*, No. 1, pp. 33–34 (Apr. 1918), pp. 59–60 (Dec. 1918). Bohr's arguments were based upon his "correspondence principle" which from the harmonic components in the Fourier series expansion of the classical mechanical motion of an electron in any given stationary state of the atom determined the physically allowed transitions from that stationary state to other stationary states. See Max Jammer, *The Conceptual Development of Quantum Mechanics* (New York, 1966), pp. 109–118.

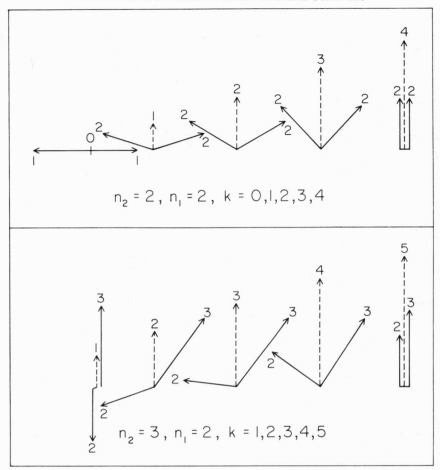

FIGURE 3. Landé's Illustration of the Addition of Quantized Angular Momenta to Yield a Quantized Resultant.

n_1 angular momentum of inner electron (or electron ring).
n_2 angular momentum of outer electron (or electron ring).
k total angular momentum, resultant of vectorial addition of n_1 and n_2.

and by J. M. Burgers in his dissertation,[37] left no doubt that the magnitude of the total angular momentum could only be an integral multiple of $h/2\pi$.

37. J. M. Burgers, *Het Atoommodel van Rutherford-Bohr* (Diss. Leiden, 1918). Also published as Part 9 of *Verhandelingen uitgegeven door Teyler's Tweede Genootschap,* new ser., 1919. See Jammer, *Quantum Mechanics,* pp. 104–105.

In the summer of 1919 Landé put these two pieces together and published a short paper giving now familiar rules regarding how, and in how many ways, one can combine two quantized angular momenta—associated with two electron rings—to give a quantized resultant[38] (Fig. 3). This neat little exercise, which in Sommerfeld's opinion was too trivial to merit separate publication,[39] made Landé quite conscious of the distinction between the azimuthal quantum number of the valence electron and the total angular momentum of the atom as a whole. It was to be an important perception, and one not always clearly grasped by Sommerfeld himself. What to do with the distinction, what to make of it in interpreting the spectral terms, was of course another question, and one to which Landé returned as he puzzled over the anomalous Zeeman effect. At this moment in the summer of 1919, when he had already given up the attempt to account for the very narrow "doublets" in orthohelium, he suggested that different values of the total angular momentum were to distinguish the two noncombining systems of terms. Very quickly, however, this proposal was quietly dropped, and in the following fifteen or eighteen months the total angular momentum of the atom was not specifically distinguished from the azimuthal quantum number, either in theoretical discussion or by attachment to any particular property of the observed spectrum.

In Landé's attack on the helium spectrum the principal methodologic advance over Sommerfeld's 1916 calculations, as also the principal source of the great difficulty of Landé's work, was the consideration of the effect of the electric field of the outer electron in distorting the orbit of the inner electron. It was this eccentric displacement —polarization—of the atom's "rump" which gave for the first time even the correct sign for the Rydberg correction, $f(n)$.[40] In his first

38. Landé, "Eine Quantenregel für die räumliche Orientierung von Elektronenringen," *Verhl. d. Dtsch. Phys. Ges.*, 21 (1919), 585–588, received 11 July 1919.
39. Sommerfeld to Landé, 18 December 1919 (SHQP 4, 13).
40. This ought to have—but in fact seems not to have—led to a recognition that in the alkali atoms among all orbits of the external electron with a given total quantum number those with low angular momentum (i.e., those passing close to the "repulsive" atomic rump) are more tightly bound than those with high angular momentum. Just the opposite was, and remained, the unchallenged preconception until the introduction of penetrating orbits in 1921. See, for example, Bohr's energy level diagram for sodium presented in his Berlin lecture, 27 Apr. 1920, "On the Series Spectra of the Elements," trans. in Bohr, *Theory of Spectra and Atomic Constitution* (Cambridge, 1922), p. 30; cf. p. 97.

skirmish Landé had deliberately neglected certain significant terms in the series expansion of the perturbing forces in order to simplify the calculations. Even so the results were surprisingly good—Sommerfeld was "frightfully," "gigantically" interested[41]—and the initial enthusiasm lasted long enough to bring Landé safely through the *Habilitation*. Now, in the summer of 1919, Landé combined his new insight regarding the addition of angular momenta with an attempt at a consistent consideration of the perturbations. Disappointingly, the results of this second calculation of the helium spectrum were not nearly so good as those of his preliminary calculation.[42] Moreover, Bohr, who, as Landé discovered in the spring of 1919, had been working for some time on the problem with his assistant, H. A. Kramers, pointed out errors.[43] Eventually, Pauli, Van Vleck, and Born did also.

Worse still was the fiasco of the normal (unexcited) helium atom. Sommerfeld's great interest in Landé's helium calculations was occasioned in good part by the fact that Landé's model of the orthohelium excited states was essentially the "double star" model of two coplanar orbits, one very near and one far from the nucleus, which Sommerfeld had proposed as the normal state of the helium atom in preference to Bohr's 1913 model of a two-electron ring. Since the orthohelium term of given total and azimuthal quantum number lay lower than the corresponding parhelium term, Landé, urged on repeatedly by Sommerfeld, extrapolated his results for the excited orthohelium *s* terms to predict the energy of the 1*s* normal state. The result was that the Sommerfeld-Landé "double star" model was slightly more stable than Bohr's.[44] On the basis of this very questionable procedure Sommerfeld stated categorically—in *Sperrdruck*—in the first edition of *Atombau und Spektrallinien* that "Bohr's helium model does not represent the most stable arrangement, but is to be replaced by our double star model."[45] But even before Sommerfeld's

41. Sommerfeld to Landé, 5 Mar. 1919, 17 Mar. 1919 (SHQP 4, 13).

42. Landé, "Störungstheorie des Heliums," *Phys. ZS., 21* (1 Mar. 1920), 114–122, received 17 Aug. 1919. J. H. Van Vleck, *Phys. Rev., 21* (Mar. 1923), 372–373, pointed out that this second paper considered only some of the terms neglected in the first, which terms, if all of them are taken into account, actually cancel.

43. Bohr-Landé correspondence, 1919 (SHQP 4, 1; BSC 4. 3).

44. Landé, "Das Serienspektrum des Heliums," *Phys. ZS., 20* (1919), 233–234. Sommerfeld to Landé, 7 Mar. 1919, 14 Apr. 1919, 2 July 1919, 18 July 1919 (SHQP 4, 13).

45. Sommerfeld, *Atombau . . .* (1919), *op. cit.* (note 23), p. 515; also p. 70 for illustration.

book could appear, James Franck's experiments showed that Bohr's model was already *too* stable, and thus *a fortiori* Sommerfeld's as well. Then, in January 1920, Franck and Fritz Reiche showed that the normal state of the helium atom was actually part of the parhelium term system. The orthohelium terms dead-ended at the 2s state; a 1s orthohelium state simply did not exist.[46] All that Sommerfeld could do was delete the rash claims in the second edition of his book; indeed he deleted every mention of Landé's work on helium.[47]

Thus, for several reasons, Landé regarded the helium calculation as merely an episode. It involved, implicitly and explicitly, several new and, ultimately, important physical ideas. But it would be some time before these, or similar, ideas would bear fruit in Landé's, or others', hands. Meanwhile the calculation was "wrong" and to be forgotten about as quickly as possible. Yet the helium calculation had served its immediate purpose; it did provide a *Habilitationsschrift*. With Born smoothing the way and soliciting his colleagues' approval, Landé received his *venia legendi*—the privilege of teaching —in the summer of 1919.[48]

Although he now had his passport to an academic career, Landé still lacked passage money. Born had felt obliged to reserve the only assistantship in the Institute for Theoretical Physics at Frankfurt for a trained experimentalist. This was one of the ironies of the theorists' effort, once established in independent chairs, to achieve an independent experimental capability. Landé had to be content with the expectancy to a second assistantship, if and when one was created.[49] There could, of course, be no question of living on *Kolleggeld*, the fees paid by the students to their instructors, certainly not now that the postwar inflation was well under way. Indeed, prices

46. J. Franck and P. Knipping, "Die Ionisierungsspannung des Heliums," *Phys. ZS., 20* (1 Nov. 1919), 481–488, received 26 June 1919. J. Franck and F. Reiche, "Über Helium und Parhelium," *ZS. f. Phys., 1* (Mar. 1920), 154–161, received 10 Jan. 1920. The absence of a 1s orthohelium state, and thus the "metastability" of the 2s orthohelium state, is a result of the Pauli exclusion principle which does not permit two *K* electrons to have their spins parallel.

47. Sommerfeld, *Atombau* . . . (2nd edition; 1921), preface dated 2 Sept. 1920. Only Landé's work on spacial arrangements of electrons is mentioned. The unusual delay in the publication of Landé's second paper, "Störungstheorie des Heliums," received 17 Aug. 1919, issued Mar. 1920, may have been due to revisions removing all assertions about the energy of the ground state and explaining why one would not expect the calculations to be applicable to it.

48. Born to Landé, 6 June 1919 (SHQP 4, 2).

49. Born to Landé, 27 Jan. 1919 (SHQP 4, 2).

were rising even faster in Frankfurt than elsewhere in Germany and by April 1920 the price indices showed it to be "by far the most expensive" of all German cities.[50] Already early in February of that year the *Privatdozenten* of the Philosophical Faculty of the university had pioneered what was soon to become a common practice—the submission of petitions to the ministry of education for the improvement of the economic circumstances of the *Privatdozenten*.[51] In addition, there was an extraordinarily severe housing shortage in Frankfurt in the immediate postwar years; apartments and even rooms were almost impossible to find.[52] Landé therefore remained at the *Odenwaldschule*. Once a week he took the train to Frankfurt and delivered his lectures.[53] It was evidently only in December 1920 or January 1921 that he settled in the city, renting a room in the home of Frau Geheimrat Freund, widow of the professor of Chemistry at the university.[54]

Even while working on helium Landé had not entirely put aside his spacially symmetric atoms. After August 1919 he gave them all his efforts, publishing nine papers in less than twice as many months. "Your productivity is really wonderful," Rudolf Ladenburg wrote him, "and all that teaching besides!"[55] Quantization of the cubic and tetrahedral atoms and calculation of their energies was followed by a survey of the periodic system attempting to assign the appropriate polyhedral symmetry to the elements.[56] Landé made a number of interesting suggestions about the structure of various atoms—in particular he argued strongly for a model of the carbon atom involving, again, a tetrahedral arrangement of the four valence electrons, each revolving in a two-quantum elliptical orbit (i.e., both the azimuthal quantum number, n, and the radial quantum number, n', were equal to unity).[57] In the summer of 1920 he had evidently even considered

50. Franz Eulenburg, "Die Preisrevolution seit dem Krieg," *Jahrbücher für Nationalökonomie, 115* (1920), 289–338; on p. 309.

51. *Mitteilungen des Verbandes der Deutschen Hochschulen, 2* (1 Oct. 1922), 263.

52. Born to Landé, 27 Jan. 1919 (SHQP 4, 2); Born-Gerlach correspondence (SHQP 19, 1), *passim;* A. Magnus to Landé, 17 Sept. 1922 (SHQP 4, 18).

53. Landé, interview with SHQP, 7 Mar. 1962, session 3.

54. Magnus to Landé, 17 Sept. 1922 (SHQP 4, 18).

55. "Ihre Produktivität ist ja bewundernswert u. daneben noch all des Unterrichts." (R. Ladenburg to Landé, 20 May 1920 [SHQP 4, 18].)

56. Landé, "Würfelatome, periodisches System und Molekülbildung," *ZS. f. Phys., 2* (Sept. 1920), 380–404, received 21 July 1920.

57. Landé, "Über Würfelatome," *Phys. ZS., 21* (Nov. 1920), 626–628. Lecture at the *Naturforscherversammlung* in Nauheim, 19–25 Sept. 1920.

seriously the problem of calculating the excited states, and thus the spectrum, of neon—the paradigm of the cubical atom with its outer shell of eight electrons. Friedrich Paschen, who had recently managed to make neon one of the best ordered spectra by analyzing almost all of its several hundred visible lines into a large number of s, p, and d terms,[58] sent Landé an offprint of this paper and wished him complete success.[59] Landé's publications, however, give no evidence that he actually undertook such a calculation.

The first national scientific congress in postwar Germany, the 86th Versammlung deutscher Naturforscher und Ärzte, was held at Bad Nauheim 19–25 September 1920. Here, after two years of development, the synchronized polyhedral atoms reached their greatest prominence. Sommerfeld addressed the Physics Section on "Unsettled Questions in Atomic Physics," and the two questions which he chose to consider were, first, the nature of the complex structure of spectral lines, and, second, "the question of the cubic arrangements of electrons."[60] To the first of these questions I will turn in the next section of this paper. The second question Sommerfeld introduced by conceding that, "The assumption of plane rings, which we made in the first stadia of our conceptions about atomic structure, was certainly only provisional. Much more appealing is the conception of cubes studded with electrons. We hope that Herr Landé will succeed or may have succeeded in proving that it is dynamically possible." This remark pointed to the virtually hopeless problem of deriving from classical dynamics the stability and synchronization of the electronic motions which the model assumed: "The slightest disturbance of an electron prevents that electron from meeting its mirror electron in a symmetry preserving collision, and consequently bursts asunder the entire system of orbits."[61] Without worrying himself further about this difficulty, Sommerfeld added an extra charge to the nucleus of

58. Paschen, "Das Spektrum des Neon," *Ann. d. Phys., 60* (20 Nov. 1919), 405–453; "Das Spektrum des Neon (Nachtrag)," *ibid., 63* (7 Oct. 1920), 201–220.

59. Paschen to Landé, 27 July 1920 (SHQP 4, 11).

60. Sommerfeld, "Schwebende Fragen der Atomphysik," *Phys. ZS., 21* (1920), 619–620; *Ges. Schr., 3,* 496–497.

61. E. Madelung and Landé, "Über ein dynamisches Würfelatommodell," *ZS. f. Phys., 2* (Aug. 1920), 230–235, received 18 June 1920. In this paper they proposed a modified model which eliminated the synchronized collisions of the electrons at the expense of the perfect cubic symmetry and which then also gave a lower potential energy than a plane ring of eight electrons.

the eminently cubical neon atom and considered the resulting structure to be the "rump" of the sodium atom $(Z = 11)$. With the electric
field in which the external electron moves arising from this nuclear
charge and the ten cubically arranged electrons, Sommerfeld calculated the functions in the Rydberg-Ritz spectral formulae and found
them inconsistent with experiment. "We do not want to conclude
therefrom that the cubic conception is false, but only that the cubes
do not remain regular cubes. Through the action of the outer electron they become shifted and deformed. . . ."[62] Thus, Sommerfeld
suggested, the spectra will rather give us information about the eccentric displacement and deformation of the cubes—whose existence is
to be postulated.

Landé knew only too well what was involved in calculating these
"eccentric displacements" in even the simplest possible case (helium),
and he must have felt the hopelessness of the future of the cubical
atoms—at least as far as he was concerned—if it lay in such calculations. But where indeed did the future of his synchronized polyhedral atoms lie? Nothing really solid and cogent had emerged in
these two years from Landé's program. Although in his own report
at Nauheim Landé maintained that the results were "encouraging,"
he had already moved on to the problem of extending his synchronized electronic motions to an entire crystal lattice. Here the idea
was that if all the corresponding electrons of all the carbon atoms
in a diamond crystal were constantly in phase, then the interaction
of the dipole moments of the atoms would be attractive and thus
explain covalent bonding.[63] Again, for a short time Landé's program

62. "Wir wollen daraus nicht schliessen, dass die Kubenvorstellung falsch sei,
sondern nur, dass die Kuben keine regelmässigen Kuben bleiben. Sie werden
durch die Wirkung des äusseren Elektrons verschoben und deformiert und wirken
dann zusammen mit dem Kern, als Dipol und Quadrupol, deren Momente vom
Abstande des äusseren Elektrons abhängt in die Ferne. . . . Bei genauerer Verfol-
gung dieses Weges scheint es mir möglich . . . Aufschlüsse nicht sowohl über die
Existenz als bei Voraussetzung der Existenz über die feineren Fragen nach der
exzentrischen Verschiebung und der Deformation der Kuben zu gewinnen. Die
Spektraltafeln würden sich dann auch hier als die eigentlichen Quellenbücher der
Atomphysik bewähren." Sommerfeld, "Schwebende Fragen . . . ," *op. cit.* (note
60).
63. Landé, "Über Würfelatome," *op. cit.* (note 57), and "Über die Kohäsionskraft im Diamanten," *ZS. f. Phys., 4* (Mar. 1921), 410–423, received 29 Dec. 1920.

and his order-of-magnitude agreement with the observed properties of diamond evoked his colleagues' interest. But when Hans Thirring achieved good results with less artificial assumptions,[64] Landé's theory ceased to be "convincing."[65]

After Nauheim Landé went to Copenhagen, where he spent the first two or three weeks of October 1920. Bohr encouraged this visit and provided 500 kroner to make it possible,[66] for he could not fail to recognize Landé as the one physicist whose research topics most completely overlapped his own. Bohr was then concerned to find some natural explanation and criterion for the existence of multiple-quantum orbits in unexcited atoms, as well as of their spacial extension and symmetry, with the intent of resuming his original program of determining the constitution of atoms across and down the periodic table. And in the few months following Landé's visit, through criticism of Landé's model of the carbon atom, Bohr developed a theory of penetrating orbits, which completely displaced Landé's program and thenceforth provided the starting point for all considerations— both quantitative and qualitative—regarding electronic motions and structure.[67] But it was mid-February of the following year before the first announcements of this new order of things appeared, and they then found Landé in the middle of quite a different problem, that of the anomalous Zeeman effect.

Just as there is some reason to trace Bohr's theory back to contact with Landé, so there is good reason to trace Landé's interest in the anomalous Zeeman effect back to contact with Bohr. Landé surely already knew Sommerfeld's opinion (1919) that those complicated splitting patterns would be explained only by "a penetrating consideration of atomic structure," and that, conversely, "one may expect

64. H. Thirring, "Über die Kohäsionskräfte des Diamanten," *ZS. f. Phys.*, *4* (Jan. 1921), 1–25, received 17 Aug. 1920.
65. Sommerfeld to Landé, 25 Feb. 1921 (appendix, document 7). Landé evidently sent the manuscript or the proofs of "Über die Kohäsionskraft . . ." to Sommerfeld, who then passed it on to P. P. Ewald, his assistant and expert in these matters.
66. Bohr to Landé, 12 Aug. 1920 (SHQP 4, 1).
67. Bohr, "Atomic Structure," *Nature, 107* (24 Mar. 1921), 104–107, dated 14 Feb. 1921. "Vielleicht regt es [the manuscript of Heisenberg's "Abänderung der formalen Regeln der Quantentheorie," *ZS. f. Phys.*, *26* (14 Aug. 1924), 291–307] ihn [Bohr] dazu an, dann selbst die richtige Theorie zu machen, wie dies einst bei Landés Kohlenstoffmodellen der Fall war." (Pauli to Kramers, 19 Dec. 1923 [SHQP 8, 9].) See Heilbron, *Atomic Structure, op. cit.* (note 12), pp. 392–404.

177

from them the deepest elucidation of atomic structure."[68] Now he found that Bohr, whose physical insight he regarded far more highly, held similar views. From subsequent allusions, one gathers not only that the topic was discussed but that Bohr made the particular suggestion that the anomaly was due to the failure of electrons in atoms to show the classically expected magnetic properties.[69] Perhaps Bohr saw this as one further instance of the suspension of the requirements of electrodynamics in the stationary states.

In any case, Landé returned home still caught up in his synchronized covalent lattice, and he was not able to find his way out of this calculation until the end of the year. By then it was probably quite clear to Landé that this was no longer the crest of the wave, indeed, that this entire *a priori* approach to the problem of atomic structure had not achieved that close contact with experiment on which solid reputations were built. In December 1920 Landé began to study carefully the most recent and extensive analysis of the anomalous Zeeman effect, a long paper by Sommerfeld published in the *Annalen der Physik* in October—"General Spectroscopic Laws, Especially a Magneto-Optic Splitting Rule."[70]

68. Sommerfeld, *Atombau* . . . (1919), *op. cit.* (note 23), pp. 427, 438–439. "Die Beschaffenheit des Zeemaneffektes . . . zeigt überdies an, dass es mit der einfachen Lorentzschen Theorie nicht getan ist, sondern dass an deren Stelle eine eindringende Berücksichtigung der Atomstruktur treten muss. Die komplizierten Zeemanzerlegungen sind so mannigfach und dabei so gesetzmässig und charakteristisch, dass man von ihnen die tiefsten Aufschlüsse über den Atombau erwarten darf. In dem Auftreten der Rungeschen Zahl, wie in allen ganzzahligen Verhältnissen der Spektroskopie, hat man ohne allen Zweifel das verborgene Wirken von Quantengesetzen zu erblicken. Es fehlt nicht viel daran, dass wir diese Gesetze entziffern können." However after writing another dozen pages on the Zeeman effect, Sommerfeld's mood was less optimistic: "Eine wirkliche Theorie des Zeemaneffektes bei wasserstoffunähnlichen Atomen ist so lange aussichtslos, als nicht der Grund für die Vielfachheit der Spektrallinien geklärt ist. Hiervon sind wir . . . noch weit entfernt." These same sentences appeared in the second edition (1921) on pp. 421 and 433.
69. Landé to Bohr, 29 June 1921 (BSC 4, 3); Bohr to Landé, 4 July 1921 (SHQP 4, 1). In the first of these letters Landé proposes this magnetic "Unwirksamkeit," as the physical cause of his anomalous splitting factor, *g*. Bohr replies that, "wie Sie wissen," this explanation completely agrees with views he has cherished for a long time. At Nauheim, Sept. 1920, discussing the first of his two "Schwebende Fragen," Sommerfeld had stated his expectation that from Bohr would come the clue to the interpretation of the complex structure and the anomalous Zeeman effect in terms of a model (*Ges. Schr., 3*, 497).
70. Sommerfeld, "Allgemeine spektroskopische Gesetze, insbesondere ein magnetooptischer Zerlegungssatz," *Ann. d. Phys., 63* (7 Oct. 1920), 221–263, received 23 Mar. 1920; *Ges. Schr., 3*, 523–566. A translation of the introduction and sections I and II, pp. 221–234 (*Ges. Schr., 3*, 523–536), is to be found in W. R. Hindmarsh, *Atomic Spectra* (Oxford, 1967), pp. 145–159.

3. ARNOLD SOMMERFELD AND THE ANOMALOUS ZEEMAN EFFECT, 1919–1920

Having introduced Landé, and followed him to the point, in late 1920, when he took up the problem of the anomalous Zeeman effect, I must now go back to describe the origin and content of Sommerfeld's paper. The importance, for understanding Landé's subsequent work, of an analysis of the *content* of Sommerfeld's paper is immediately clear; of equal importance, however, are the peculiar circumstances leading up to Sommerfeld's publication.

In the fall of 1896 Pieter Zeeman obtained positive evidence of the influence of a magnetic field upon the emission of spectral lines.[71] Serving as midwife at this difficult labor was H. A. Lorentz, and Zeeman believed he observed exactly what Lorentz' electron theory predicted. Namely, viewed in a direction perpendicular to that of the magnetic field, H, the spectral line was to split into three: one undeviated component polarized parallel to the magnetic field, with components on either side, displaced in frequency by an amount $\Delta\nu_{classical} = \pm\, eH/4\pi m_0 c$, and polarized perpendicular to the magnetic field. It soon turned out, however, that this "normal triplet" was the exception rather than the rule. Indeed, Zeeman's own observations, made on the sodium D lines (NaD_1 and D_2 in Fig. 4), despite his valiant attempt to construe them as a complete confirmation of Lorentz' theory, were actually of one of the most common "anomalous" patterns.[72]

The process of regularizing the anomalous effects was considerably advanced by Carl Runge and Friedrich Paschen in 1900–1902.[73] Most important was their clear formulation and support of what is known as Preston's rule: namely, that all the lines of a given series, and of corresponding series in elements with analogous spectra, show the

71. A detailed account of the discovery of, and early experimental work on, the Zeeman effect is given by James Brookes Spencer (Ph.D. diss., University of Wisconsin, Madison, 1964), *An Historical Investigation of the Zeeman Effect, 1896–1913* (Ann Arbor: University Microfilms, 1964), *Dissertation Abstracts 25* (1964), 3547.
72. The "irony" is pointed out by S. Endo and S. Saito, "Zeeman Effect and the Theory of Electron[s] of H. A. Lorentz," *Japanese Studies in the History of Science*, 6 (1967), 1–18.
73. Runge and Paschen, *Astrophys. J., 15* (1902), 235–251, 333–339; *16* (1902), 123–134. Spencer, *Zeeman Effect, 1896–1913, op. cit.* (note 71), pp. 49–51, 61, 92–93.

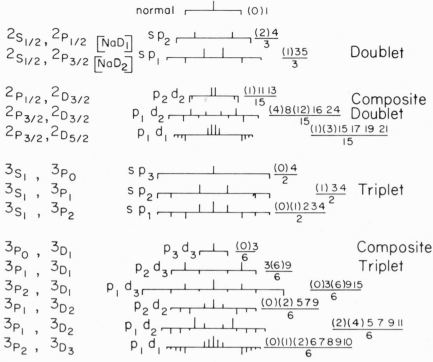

FIGURE 4. Zeeman Splitting Patterns of the Doublets and Triplets.
(Adapted from Back-Landé, *Zeemaneffekt und Multiplettstruktur der Spektrallinien* [Berlin, 1925], p. 19.)

Normal (Lorentz) triplet at top for comparison. Components polarized parallel to the magnetic field are drawn up, perpendicular down, with lengths proportional to their intensities. On the right the Runge numerators of the components (parallel components in parentheses) are set over the Runge denominator of the line group. On the left is the contemporary sp_i, p_id_j notation for the combining terms (*sharp*, *principle*, *diffuse*) and their complex structures (subscripts). At the far left is standard Russell-Saunders (spin-orbit) spectroscopic notation for the combining terms; *S*, *P*, *D* indicate the total *orbital* angular momentum of the electrons in that stationary state, the subscript gives the *total* angular momentum of the atom in that state (vectorial sum of orbital and electron spin angular momentum), and the preceding superscript gives the maximum multiplicity of that system of terms.

same splitting pattern.[74] Thus, for example, the lines of the doublet sharp series of sodium showed the same pattern as those of the doublet sharp series of potassium, and indeed this was the same pattern as

74. Sommerfeld, "Allgemeine spektroskopische Gesetze . . . ," *op. cit.* (note 70), p. 242; E. Back, *Zur Prestonschen Regel* (Diss., Tübingen, 1921), p. 6. The dissertation was accepted Feb. 1913.

that of the doublet principal series of both elements.[75] Thus the pattern depended only on the multiplicity of the lines and the nature of the combining terms (Fig. 4). I speak here of "the pattern of the doublet lines," for although each line had its own distinct pattern, it was regarded as a higher degree of regularity that the much more complicated splitting pattern of the "line group" as a whole was preserved both within a given series and among analogous spectra. By a "line group" the spectroscopist meant the set of lines produced by the various combinations of two complex terms. Thus the "doublet" diffuse series of sodium is in fact composed of "line groups" containing the three lines which arise from the allowed combinations of a double d-term with a double p-term. These groups of three lines, characteristic of the alkalis and of other spectra showing doublet principal and sharp series, were called "composite doublets." Similarly, in the alkaline earths, mercury, and other elements with a system of triple p, d, etc. terms, the triplet principal and sharp series were accompanied by a diffuse series of "composite triplets," a characteristic line group with six members. It was really quite astonishing that the exceedingly complicated Zeeman pattern of such a line group, containing fifty-four components, remained constant from line to line and from element to element. In the midst of these "anomalous" patterns it was the systems of single lines which alone showed normal triplets in a magnetic field, thus tying the anomaly in the Zeeman effect to the complexity of the line group.

To all this one must now add Runge's rule (1907) that the distances of the various components from the position of the undeviated line were rational fractions, q/r, of the separation of the components in the normal triplet, $\Delta\nu_{classical} = e/4\pi m_0 c \cdot H \equiv aH$, with a single denominator, r, characteristic of the entire line group.[76] Ironically, Runge had been stimulated to construct his rule by—and based it in the first instance upon—recent observations by Wilhelm Lohmann

75. This also was Runge and Paschen's contribution, and they were able to conjecture from Rydberg's series formulae that such a connection existed between the principal and the sharp series patterns. Actually, as they found, the two patterns are reversed, the splitting of the longest wavelength member of the sharp series line group corresponding to that of the shortest wavelength member of the principal series line group, etc.

76. Runge, "Über die Zerlegung von Spektrallinien im magnetischen Felde," *Phys. ZS.*, 8 (15 Apr. 1907), 232–237, received 3 Apr. 1907. Runge emphasizes that he and Paschen had already demonstrated such relationships in a good number of spectra in 1902.

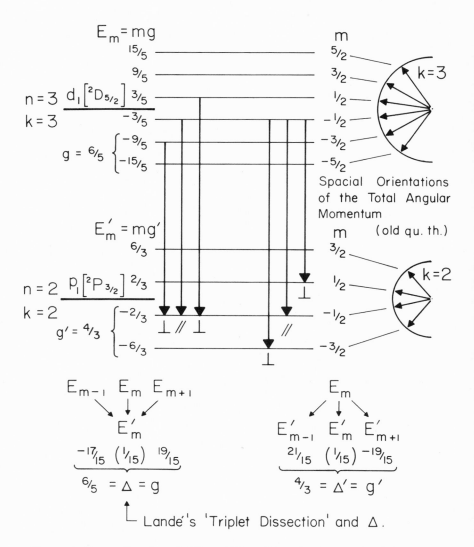

FIGURE 5. Energy Levels and Allowed Transitions in a Magnetic Field.
On the left the field-free energy levels of the doublet diffuse series line $p_1 d_1$ [$^2P_{3/2}$–$^2D_{5/2}$]; in the center the splitting of these levels in a magnetic field (the E_m are in units of $h\Delta\nu_{classical} = h \cdot eH/4\pi m_0 c \equiv h \cdot aH$, and greatly enlarged in comparison with the energy separation of the field-free levels), and five of the twelve transitions (Zeeman components) allowed by the selection rule $\Delta m = \pm 1$ or 0; on the right the interpretation of the splitting as arising from space quantization in a magnetic field, H, directed vertically upward. For the "triplet dissection" see Section 4, below.

of the Zeeman patterns of a number of *neon* lines. For eighty-five components arising from nine neon lines Runge found the distances of the components from the positions of their field-free lines to be $q/12 \cdot a H$. The next twenty years, however, saw recurring controversy, not over the concept of the Runge denominator, for the existence of such an integer was largely taken for granted, but over the particular integer appropriate to any given group of neon lines. And no wonder; for neon the Runge denominators were a will-o'-the-wisp.

[To explain this point I must anticipate the development of physics somewhat. As illustrated in Figure 5, for each stationary state of an atom there is a constant factor, g, whose value depends upon the particular configuration of angular momenta of the constituent electrons. It is g, together with m, the quantum number of the component of the total angular momentum of the atom in the direction of an external magnetic field, which determines the energy levels into which that stationary state is split by the (not too strong) magnetic field. The existence of such a constant g (different from **unity**, and varying from state to state) is essentially due to the fact that the ratio of the intrinsic magnetic moment of an electron to its intrinsic or spin angular momentum is not equal to (but very nearly twice) that ratio for the magnetic moment and angular momentum arising from the orbital motion of the electron. And the half-integral values of m are due to the fact that the spin angular momentum of the electron is only "half a quantum," $\frac{1}{2} \cdot h/2\pi$. Consequently the magnetic moment of the atom as a whole—and thus also its energy in a magnetic field—is not simply proportional to the total angular momentum of the atom, but depends upon the way in which that total is made up of spin and orbital angular momenta of the various electrons. For a wide range of spectra, including in particular the doublets and triplets, the coupling of the angular momenta is of an especially simple form ("Russell-Saunders" coupling), resulting in g factors which are rational fractions, and thus also in energy levels and separations of observed spectral lines which are rational multiples of the normal Lorentz separation. But even in these cases the various sublevels of a doublet or triplet term, arising as they do from different relative orientations of the total spin and the total orbital angular momentum of the atom, have different g factors. Thus a single Runge denominator for all the members of a line group could be nothing

183

more than a least common multiple of the denominators of the several g factors, and could only lead the theoretical spectroscopist astray. In the case of neon, however, Runge's rule was simply an illusion. The complicated coupling of the angular momenta in the excited states of the neon atom results in g factors which are not rational numbers, and thus the separations of the components in the Zeeman pattern of even a single neon line are not rational multiples of the normal splitting.[77]]

Thus, although we would wish to ignore the line group and consider each line separately, to do so would have appeared, both to the experimental and the theoretical spectroscopist of the period, as a senseless sacrifice of the most striking of the spectral regularities; and these regularities, it was generally agreed, were to provide the clues for a theoretical interpretation. The Paschen-Back effect (1912) then seemed emphatically to confirm this point of view.[78] Who could doubt that in the Zeeman effect the line group ought to be treated as a unit once it was shown how, in magnetic fields so strong that the components of the splitting patterns of the various members of a line group overlapped, the entire structure was transformed into a single normal triplet?

In the years following the enunciation of Runge's rule, there was often considerable disagreement about the Runge denominator to be assigned to the various line groups—1/7 and 2/13 only differ by seven percent. Ernst Back of the Paschen-Back effect was a law student turned experimental spectroscopist with a passion for meticulous detail and for leaving no loose ends. Back now undertook much more accurate measurements of the patterns of the doublets and triplets. The completion of this work, which he had begun under Friedrich Paschen's direction for his doctorate at Tübingen, was interrupted by the war. Paschen, thinking Back might not take it up again, communicated Back's results to the two leading theoretical spectrosco-

77. While the states with the excited electron in an s orbit have g factors within a few percent of those to be anticipated from normal LS (Russell-Saunders) coupling, the states with the excited electron in a p orbit have g factors which are "very irregular. They do not agree at all with those calculated from any possible coupling, showing clearly that the levels under consideration form an intermediate case." (L. Pauling and S. Goudsmit, *The Structure of Line Spectra* [New York, 1930], p. 254.)

78. Paschen and Back, "Normale und anomale Zeemaneffekte," *Ann. d. Phys.,* *39* (1912), 897–932; "Nachtrag," *ibid., 40* (1913), 960–970.

pists, Woldemar Voigt and Sommerfeld.[79] And for some time the subject rested.

Early in 1916 Sommerfeld had introduced space quantization and he (and also Debye) had then made use of it to attempt a quantum theory of the Zeeman effect. The quantized components, $m \cdot h/2\pi$, of the total angular momentum (mistakenly identified with the orbital angular momentum of the radiating electron, p, and represented by the azimuthal quantum number, n) in the direction of the magnetic field would give rise to magnetic interaction energies $E_m = p \cdot w_{Larmor} = m \cdot h/2\pi \cdot eH/2m_0c = m \cdot h \cdot \Delta\nu_{classical}$, where $|m| \leq n$. Thus the change in energy of an atom in a magnetic field, E_m, had the same series of values for all atoms and all stationary states. If in the absence of a magnetic field an atom emitted a spectral line of frequency $\nu_{quantum} = (E_1 - E_2)/h$, that line would be split by a magnetic field into $\nu_{quantum} + \Delta\nu_{quantum} = ([E_1 + E_{m'}] - [E_2 + E_{m''}])/h$, or $\Delta\nu_{quantum} = (E_{m'} - E_{m''})/h = (m' - m'') \cdot \Delta\nu_{classical}$, where the primes on m indicate some one of its several allowed values. In order to obtain the three, and only the three, lines of the Lorentz triplet, one had, *ad hoc*, to restrict $m' - m''$ to the values ± 1 and 0. Without any theoretically derived selection rules one could not explain why the additional lines at 2, 3, etc., times the normal separation (which, to be sure, corresponded to no known *anomalous* Zeeman effect) were to be excluded, or why the allowed lines were polarized as they were.

These grounds alone, added to Sommerfeld's general loss of confidence in his principles in the period 1917–1918, make it understandable that Sommerfeld did not feel encouraged to attack the anomalous Zeeman effect.[80] But there were still other circumstances which made the problem seem quite unapproachable. First of all, the anomalous Zeeman effect was intimately connected with the fact of complex terms, and no one really knew to what physical phenomenon to attribute the presence of complex structure. In the second place, the existence of the Paschen-Back effect made the entire prob-

79. Paschen to Landé, 8 Mar. 1921 (appendix, doc. 11).
80. Sommerfeld, "Zur Theorie des Zeeman-Effekts," *op. cit.* (note 16), had hoped that from a consideration of the simultaneous effects of the relativistic change of mass of the electron and a magnetic field upon the hydrogen spectrum some insight into the anomalous Zeeman effect would be achieved. "The success is, however, not the expected one: namely, there is also here only the normal Lorentz triplet."

185

lem even more obscure, while also arguing that the entire line group had to be treated as a unit. Finally the excessive unification provided by Runge's rule made the problem of term analysis appear much more complicated than in fact it was; indeed it set the theorist a problem with no solution.

By 1919, however, Sommerfeld was becoming discouraged with the lack of progress he and his students were making, using what we may call the *a priori* approach to accounting for the spectra of atoms and molecules. In this approach one began with a definite model of the atomic system, then, applying classical mechanics and the quantization rules, one calculated the stationary states. Only at the end did one attempt to identify the energy differences and allowed transitions between these states with the observed spectral lines. This was, of course, precisely the approach which Landé had applied to helium without results of very much material use to the experimental spectroscopist. Sommerfeld evidently became convinced that if the theorist was to get back on the track, as well as reestablish close contact with experiment, he would have to adopt a different approach. This new approach, which Sommerfeld and his students began applying to X-ray and optical spectra in 1919, 1920, and 1921, we may call *a posteriori*. It began with the observed spectral lines, and worked back to the energy levels. These levels were then characterized by quantum numbers and selection rules—invented *ad hoc* if necessary.

In the course of 1919 Sommerfeld began to apply this *a posteriori* approach to the previously intractable problem of the complex structure of spectral lines, and, in particular, to the anomalous Zeeman effect. The basic idea was very simple: if all spectral lines are really energy differences of quantized stationary states, then the combination principle must apply to the anomalous Zeeman effect also. Sommerfeld maintained that this proposition, which we might even regard as implicit in Preston's rule, had never yet been mentioned in the literature.[81] Indeed, it seems to have enjoyed at the time the hazy status of one of those tacit scientific principles which, as Berzelius said, "although never positively denied, have also never been

81. This is stated by Sommerfeld ("Allgemeine spektroskopische Gesetze . . . ," *op. cit.* [note 70], p. 241; *Ges. Schr., 3,* 543) in assigning van Lohuizen priority on this point. See notes 87 and 88 below.

generally admitted."[82] Once explicitly enunciated, the consequence, given the line group approach, was to derive from the observed pattern the "frequency splitting" of the complex terms. I.e., one believes one knows that $\Delta \nu = q/r \cdot aH$, but the combination principle requires $\Delta \nu = \Delta \nu_1 - \Delta \nu_2$, which suggests $\Delta \nu_1 = q_1/r_1 \cdot aH$, $\Delta \nu_2 = q_2/r_2 \cdot aH$. Then $q/r = q_1/r_1 - q_2/r_2 = (q_1 r_2 - q_2 r_1)/r_1 r_2$, and $r = r_1 r_2$. Thus each complex term was imagined to have its own "term denominator," and the Runge denominator was the product of the two. Sommerfeld set one of his doctoral students to figuring out, from the observed patterns, the term denominators, and these were then checked to make sure that they were characteristic of the terms and not merely of the combinations.[83]

As one of the early efforts by the neutrals to reestablish scientific contacts, Manne Siegbahn, who had been Sommerfeld's principal source of X-ray spectra during the war, brought the German theorist to Lund in September 1919. Sommerfeld's lectures there between the 10th and the 20th drew Bohr and his young co-workers across the sound. Then on the invitation of the semiofficial Danmarks Naturvidenskabelige Samfund, elicited of course by Bohr, Sommerfeld returned with the "Copenhageners" to give two lectures.[84] Sommerfeld devoted at least one of his Lund lectures to his "splitting rule," as he called it, and to a "Zahlenmysterium" among the term denom-

---

82. *Taylor's Scientific Memoirs, 4* (1846), 663, trans. from *Ann. d. Phys., 68* (1846), 161. Heilbron, ". . . the Ring Atom," *op. cit.* (note 12), p. 475, points to the controversy at just this moment over the first enunciation of the principle of multiple quantum orbits in normal atoms as an instance of "the well-known but often neglected truth that some of the most significant aspects of a physical theory often remain tacit."

83. J. Krönert, "Gesetzmässigkeiten beim anomalen Zeemaneffekt" (Diss. Munich, 1920), cited by Sommerfeld, "Allgemeine spektroskopische Gesetze . . . ," *op. cit.* (note 70), p. 257, and Landé, "Über den anomalen Zeemaneffekt (Teil I)," *op. cit.* (note 1), p. 234. Krönert's dissertation remained unpublished, probably because of the economic conditions. (The high cost of printing during the inflation led to the suspension of the requirement that doctoral dissertations be printed.)

84. Siegbahn-Sommerfeld correspondence (SHQP 34, 3); Bohr-Sommerfeld correspondence (BSC 7, 3). On Monday 22 September Sommerfeld addressed the Fysisk Forening on molecular, atomic, and nuclear models, and following his talk there was a reception at the pavilion on Langelinie. On Tuesday he spoke less popularly on band spectra. (Record book of the Fysisk Forening, SHQP 35.) Sommerfeld then returned to Sweden, lecturing in Uppsala and Stockholm at the end of the month. From Stockholm Sommerfeld wrote Rubinowicz on 1 October that in Uppsala he had given "einen ganzen Vortrag über Wellenth. u. Quantenth., d.h. über Ihre Arbeit. . . . Es war sehr schön in Schweden, habe viele warme Herzlichkeit gefunden, und viel Interesse für meine Vorträge." (SHQP 17.)

inators which emerged from the application of his rule to Back's excellent observations.[85] Displaying to his audience the Runge denominators of the s, p, and d terms of the doublet and triplet systems, Sommerfeld induced the mysterious law they obeyed, and produced the following striking table by deducing the denominators of the as yet unobserved f terms:

	term type:	s	p	d	f
	singlet lines:	1	1	1	1
term denominators for	triplet lines:	1	2	3	(4)
	doublet lines:	1	3	5	(7)

As Sommerfeld was confidently affirming that "no one will doubt" the extrapolation, Oskar Klein, who was sitting next to Bohr, turned to him and asked what he thought. Bohr, with an amused smile, replied that "he didn't believe that."[86]

Was it the extrapolation, or the attempt to characterize all the levels of a complex term by a single denominator, that awakened Bohr's scepticism? Klein implies that it was the former—and indeed that was not Bohr's way of doing theoretical physics. Actually the extrapolations were rather good. We know that to introduce the g factor and solve the problem one must *eventually* abandon the "line group" approach and treat each level of the complex term separately. But indeed the g factors of the doublet terms do all have a denominator $2n - 1$, and though the g factors of a triplet term do not all have the same denominator, Sommerfeld's prediction of a Runge de-

85. Sommerfeld, "Ein Zahlenmysterium in der Theorie des Zeemaneffektes," *Die Naturwissenschaften, 8* (23 Jan. 1920), 61–64; *Ges. Schr., 3,* 511–514. Sommerfeld does not say that this was a lecture, let alone where it was delivered, but the subject, content, and style all argue that it was his Lund lecture. Writing to Bohr, 11 Nov. 1920 (BSC 7, 3), Sommerfeld referred to "Meinen 'Zerlegungssatz,' über den ich schon in Lund in Ihrem Beisein vorgetragen habe. . . ." The characterization of the results as a "Zahlenmysterium" originated with A. Rubinowicz: "Ich habe noch verschiedene Controllen dieses 'Zahlenmysteriums'. Den Ihrem Briefe entnommenen Namen habe ich als Überschrift über eine vorläufige Notiz für die Naturwissenschaften gesetzt." (Sommerfeld to Rubinowicz, 26 Dec. 1919 [SHQP 17].)

86. Sommerfeld, "Ein Zahlenmysterium . . . ," *ibid.,* p. 64. Sources for History of Quantum Physics, transcripts of interviews with O. Klein, 1962–1963. Session 1, p. 10; Session 2, p. 3; Session 5, p. 14. The implicit, but unstated, generalization underlying the extrapolation is that for "triplet lines" $r = n$, for "doublet lines" $r = 2n - 1$. These expressions in terms of quantum numbers are given explicitly in "General Spectroscopic Laws."

nominator of $3 \times 4 = 12$ for the df composite triplets was exactly right.

On the other hand, it is entirely possible that Bohr thought that dealing with the line group as a whole was an untenable halfway house. And in this connection Bohr, I imagine, had something else to tell Sommerfeld: the proposition that the combination principle applies to the anomalous Zeeman effect had already been published. Bohr had been in Holland in the spring of 1919 and there met Dr. Teunis van Lohuizen, a secondary school teacher in Den Haag, who had received his doctorate with Zeeman a few years before. Van Lohuizen was then working on a paper in which he attempted to test the proposition on the available published observations. Although lacking Back's measurements, and thus relying on incomplete and sometimes faulty data, van Lohuizen convinced himself of the applicability of the combination principle and published his results shortly after in *The Proceedings of the Amsterdam Academy*.[87] And this paper is all the more interesting because van Lohuizen, probably through oversight rather than through insight, entirely ignored the Runge denominator and the line group and considered separately the magnetic splitting of each of the energy levels of a complex term.

Sommerfeld did not see the significance of van Lohuizen's approach; his reaction was rather relief that "our splitting rule remained hidden from Herr Lohuizen."[88] Up to this point, although he had taken the liberty of giving public lectures on results abstracted from Back's observations, Sommerfeld had not intended to publish *in print* until Back had himself published his complete results. But now, nervous about securing priority for his "splitting rule," Sommerfeld obtained from Back permission to go ahead.[89] The first publication was essentially the Lund lecture. While it contained the rule it did not give the data, Back's data, on which it was based. The second and complete publication—data and all—was "General

87. T. van Lohuizen, "The Anomalous Zeeman-Effect," *Koninklijke Akad. v. Wetensch., Amsterdam, Proc.*, 22 (1919), 190–199, dated 7 May 1919.

88. "Unser Zerlegungssatz musste Hn. Lohuizen verborgen bleiben, da ihm die Backschen Präzisionsmessungen nicht zur Verfügung standen. Während ich ursprünglich die Veröffentlichung der vollständigen Resultate des Hn. Back abwarten wollte, veranlasst mich die Arbeit des Hn. Lohuizen zu dieser vorläufigen Note." (Sommerfeld, "Ein Zahlenmysterium . . . ," *op. cit.* [note 85], p. 64.)

89. This may be inferred from *loc. cit.*, and is explicitly stated in Sommerfeld to Landé, 3 Mar. 1921 (appendix, doc. 9).

Spectroscopic Laws . . . ," submitted to the *Annalen der Physik* in March 1920 and published in October.

The long delay seems to have been due only to the notorious slowness of the *Annalen*. It would appear that after he had submitted his paper, Sommerfeld had second thoughts. Aware that Back, recently returned to Tübingen, was working up his data and methods for publication, Sommerfeld addressed a note to the editorial office of the *Annalen* requesting that publication of his paper be delayed until it could appear together with Back's. He then sent this note to Back, evidently suggesting that if Back approved he should send it off to the editorial office. But Back, never able to convince himself that his results were quite perfect, and unwilling to be satisfied with anything less, wrote Sommerfeld a few weeks later that there were new difficulties, that his manuscript was not ready, and that he would not make use of Sommerfeld's note. Indeed, it was to be another two and a half years before Back submitted his own paper to the *Annalen*.[90]

Sommerfeld would later claim that Back gave his permission willingly because Back himself had no theoretical pretensions; and that is probably true. Sommerfeld perhaps realized only later how self-effacing Back was, that he would have granted the request even had he been reluctant. But the recriminations which Sommerfeld might have directed against himself were all, eventually, to be visited upon Landé.

In the first part of "General Spectroscopic Laws . . ." Sommerfeld admits that at present the quantum theory has "simply nothing to say" about a model that might account for the complex terms. Nonetheless, the recent successes of the selection rules argue that one should try to derive the observed line groups by assigning quantum numbers to the various levels of a complex term and then allowing combinations among the terms according to a selection rule. Sommerfeld so clearly states his line of reasoning that it is worthwhile quoting him at length:

> The structure of the complete doublets and triplets makes it natural to assume that here also a kind of selection principle is operative. To

90. Back to Sommerfeld, 29 Apr. 1920 (SHQP 29, 55). E. Back, "Zur Kenntnis des Zeemaneffektes," *Ann. d. Phys.*, 70 (1923), 333–372, received 25 Sept. 1922.

be sure, it can't be the azimuthal quantum number, for the azimuthal quantum number is assigned to the angular momentum of the entire atom, to its external rotation, so to speak. The azimuthal quantum number has, according to our interpretation of the spectral series, the value 3 for all d_i terms, the value 2 for all p_i terms. The selection principle is thus fully satisfied by *any* transition of one *d*-term into one *p*-term. The distinguishing characteristic of the various *d*- and *p*-terms must be, rather, an inner quantum number, perhaps corresponding to a hidden rotation. Of its geometric significance we are quite as ignorant as we are of those differences in the orbits which underline the multiplicity of the series terms.[91]

Sommerfeld then gives the energy level diagrams:

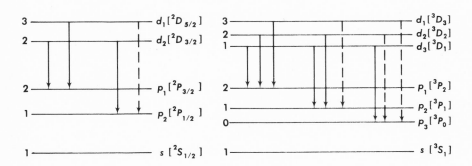

and notes further on that for each term the highest value of the inner quantum number is equal to the value of the "outer azimuthal quantum number." The transitions which Sommerfeld has drawn in are just those which are obtained "if we carry over our selection principle from the outer to the inner quantum number," although, as he concedes, "the justification for so doing can, naturally, be doubted."[92]

But having laid this foundation, Sommerfeld makes very little use of it in the second half of the paper, which is devoted to his spectroscopic splitting rule. Lauding the rule as "uncovering a harmony of integral relations of a purity which will surprise even those

91. Sommerfeld, *Ann. d. Phys.*, *63*, 231; *Ges. Schr.*, *3*, 532.
92. *Loc. cit.*

familiar with the modern quantum theory,"[93] Sommerfeld gives a detailed exposition of the observed anomalous Zeeman patterns, and illustrates the utility of his rule in working out otherwise uncertain Runge denominators. However, anticipating Landé's efforts, it is Sommerfeld's theoretical perspectives which, presumably, should command our attention. First let me extract the disclaimer—entirely analogous to the one above regarding the meaning of the inner quantum number—which Sommerfeld interjects in his first suggestions of what might be going on: the table of term denominators "is of course essentially empirical in origin and understood theoretically precisely as little as the origin of the line multiplicities generally."[94] Thus the two parts of the paper are tied together not by an inference but by a state of ignorance, conceived as a transitive relation connecting the complex structure with the anomalous Zeeman effect: we are as ignorant of the significance of the inner quantum number as we are of the cause of the complexity of the spectral terms, and that is precisely how little we understand the cause of the Runge denominators of the terms.

Unfortunately, Sommerfeld did not adhere to this impeccable logic in the analysis which follows. His dicta are here, for the sake of clarity, dissected into five propositions. I give them in the order in which they occur, but in a more directly declarative form:

1. The Runge denominator of a term has its "ultimate basis" in "hidden quantum numbers."
2. The Runge denominators are causally connected with the azimuthal quantum numbers.

93. *Ibid.*, p. 221; p. 523.
94. "In der Tat ist unsere Tafel wesentlich empirischen Ursprungs und theoretisch ebenso unverstanden, wie der Ursprung der Linienmultiplizitäten überhaupt. Nur soviel scheint sicher zu sein, dass der ganzzahlige Zusammenklang unserer Rungeschen Nenner seinen letzten Grund in dem Walten verborgener Quantenzahlen und Quantenbeziehungen hat. . . . Dass ein ursächlicher Zusammenhang zwischen Rungeschen Nennern und azimutalen Quantenzahlen besteht, ist auch theoretisch verständlich. Misst doch die azimutale Quantenzahl das gesamte Impulsmoment und daher auch das gesamte magnetische Moment des Atoms. Es ist von vornherein anzunehmen, dass die Einwirkung des äusseren Magnetfeldes auf das Atom von dem magnetischen Moment desselben bestimmt wird, wie wir es bei dem normalen Zeemaneffekt näher verfolgen können. Daher ist es begreiflich, dass der Rungesche Nenner r eine (natürlich ganzzahlige) Funktion der azimutalen Quantenzahl n sein wird: $r = g(n)$." (*Ibid.*, pp. 247–248; pp. 549–550.)

3. "The azimuthal quantum number measures the total angular momentum, and therefore the total magnetic moment, of the atom."
4. "It is to be assumed *a priori* that the effect of a magnetic field on an atom will be determined by its magnetic moment, as in the normal Zeeman effect."
5. "Therefore it is understandable that the Runge denominator of a term is a whole number function of the azimuthal quantum number, n: $r = g(n)$."
   Triplet lines: $g(n) = n$.
   Doublet lines: $g(n) = 2n - 1$.

And thus in the space of a few lines Sommerfeld had reneged on his promise to regard the *inner* quantum number as the "ultimate basis" of the Runge denominator. But this ambivalence has its counterpart in Sommerfeld's presentation of the desiderata in an explanation of the anomalous Zeeman effect:

> Our considerations regarding the Zeeman effect, based as they are on the combination principle, would only be conclusive if we were able to assign to the terms s, p_i, d_i separate definite magnetic splittings, and then, out of their combinations, deduce the observed splitting patterns and their polarizations.[95]

But before we are able to say "Amen," and consider ourselves free to take separately each level of a complex term, Sommerfeld shifts to the first person singular in order to make perfectly clear what he has in mind:

> Using the deduction of the normal Zeeman effect as a model, the following way of handling the problem would appear to me to be ideal: one assigns to the various terms the following energy levels of the magnetic splitting (suppressing the aH)

$$
\text{triplet systems}
\begin{cases}
s)\ 0,\ \pm 1 \\
p)\ 0,\ \pm 1/2,\ \pm 1 \\
d)\ 0,\ \pm 1/3,\ \pm 2/3,\ \pm 1
\end{cases}
$$

$$
\text{doublet systems}
\begin{cases}
s)\ 0,\ \pm 1 \\
p)\ 0,\ \pm 1/3,\ \pm 2/3,\ \pm 1 \\
d)\ 0,\ \pm 1/5,\ \pm 2/5,\ \pm 3/5,\ \pm 4/5,\ \pm 1
\end{cases}
$$

95. *Ibid.*, p. 253; p. 555.

193

and combines these, paying attention to a properly chosen selection principle. This selection principle must depend on the "outer" quantum number, the hypothetical "inner" quantum number, and the Runge numerator q (so to say, the quantum numbers of the magnetic energy level in question). . . .

But all the levels of a complex term are assigned the same magnetic splitting, q/r, $q = 0$, ± 1, . . . , $\pm r$, and the inner quantum number, relegated to the selection rule, has no influence upon the magnetic energy levels themselves. Sommerfeld was therefore compelled to admit that "I have not succeeded in imitating theoretically in this way the empirical splitting pattern."[96]

Surprising and significant is the context of Sommerfeld's presentation of his "ideal" treatment of the anomalous Zeeman effect, or, rather, the particular section and subsection of this long and diverse paper which he chose for this suggestion of the structural elements of a theory. It concludes the subsection (V.3) entitled "The Runge Numerators of the Doublet and Triplet Systems. Regularities in the Number and Arrangement of the Components of the Splitting Patterns," in which Sommerfeld presents solely *empirical* rules. Sommerfeld considers the cases in which the s term in the combinations sp_1, sp_2, sp_3 can be replaced by one of the d terms, e.g., p_1d_2, p_2d_2, and sets up rules for the number of additional components in the splitting patterns of such substituted lines. For example: "In both the doublet and triplet systems the d_1 term raises, without exception, the number of $\perp$ components by 4 and the number of $\parallel$ components by 2."[97] He remarks then that, "These regularities are entirely consistent with the combination principle, which indeed demands a superposition of the magnetic effects of the first and second term. . . . The additional number of components which is peculiar to the term d_j thus superimposes itself upon the original number belonging to sp_i."[98] There is of course nothing in the combination principle, nor

96. *Ibid.*, p. 254; p. 556.
97. *Ibid.*, p. 251; p. 553.
98. "Diese Gesetzmässigkeiten liegen ganz im Sinne des Kombinationsprinzipes, welches ja eine Überlagerung der magnetischen Effekte des ersten und zweiten Termes verlangt. Damit stimmt es überein, dass die Zerlegungsbilder der Terme p_i, die ursprünglich in den Kombinationen sp_i in Erscheinung traten, durch Vertauschung des s mit einem der Terme d_j in einer für jeden d-Term charakteristischen Art abgeändert werden. Die hinzukommende, dem Term d_j eigentümliche Komponentenzahl überlagert sich dabei der ursprünglichen zu sp_i gehörenden." (*Ibid.*, pp. 251–252; pp. 553–554.)

in its energy-level interpretation, which sanctions such a confusion between the terms themselves and the possible transitions between them. Moreover, Sommerfeld's suggestion of addition runs directly contrary to the combination principle which is based upon subtraction. Yet, as I shall show when I turn to Landé's work, these intuitions, which seem only vaguely connected with, or indeed contradictory to, the established principles of the quantum theory of spectra, nonetheless grasp an element of the true physical situation.

After interjecting the above justification of these empirical rules regarding the number of components Sommerfeld goes on to even more involved manipulations of the splitting patterns, showing, for example, how one can get the $p_i d_3$ splitting pattern out of that of $s p_i$. The positive $\perp$ components are obtained by doubling their distance from the position $3/2 \, a \cdot H$, then exchanging left and right, etc. He then concludes the subsection by presenting his ideal of a theory, and turns in the next subsection (V.4) to "The Splitting Rule in the Neon Spectrum." Basing himself on the dissertation of his student, J. Krönert, Sommerfeld here attempts to resolve the chronic disagreement over the Runge denominators for neon by application of his splitting rule. While we know this was hopeless, Sommerfeld was convinced that the effort was a complete success. The prerequisite for such an effort was, of course, a term analysis of the neon spectrum, and this Paschen had recently contributed.[99] Paschen's four s terms and ten p terms were assigned Runge denominators; nothing, however, was said about the number and arrangement of the components (Runge numerators) in the neon splitting patterns. These thus became obvious candidates in any search for further empirical regularities such as Sommerfeld had found for the doublets and triplets. And it was indeed these two subsections of Sommerfeld's paper— the only points at which Sommerfeld consistently considered the splitting patterns of the individual lines rather than those of the line groups—that caught the attention and excited the efforts of both the experimentalist Back and the theorist Landé.

4. LANDÉ'S EMPIRICAL RULES, DECEMBER 1920–FEBRUARY 1921

In December 1920 Landé read Sommerfeld's "General Spectroscopic Laws . . ."—a paper which certainly did not pretend to have

99. Paschen, *op. cit.* (note 58).

solved the problems it addressed. Sommerfeld hinted at a number of connections and approaches, and for the anomalous Zeeman effect he went so far as to write a prescription for an ideal solution from the point of view of the quantum theory. Having puzzled over the question of the complex structure of spectral terms in connection with the helium spectrum, Landé would certainly be interested in Sommerfeld's inner quantum number, and its selection rule. But Landé, familiar now with Bohr's correspondence principle derivations of the selection rules, would, we may suppose, immediately recognize that Sommerfeld's selection rule for the inner quantum number was not identical with that for the azimuthal quantum number, but rather with that for the equatorial quantum number, m. It would then occur to Landé, conscious as he was of the distinction between the azimuthal quantum number of each electron and the total angular momentum of the atom, that this selection rule was also the selection rule for the total angular momentum. The inner quantum number, he would speculate, was not a "hidden rotation" but the "external rotation," the total angular momentum of the atom, and the complex structure of the spectral terms was its long sought physical manifestation. Indeed, as Sommerfeld had not introduced a symbol for his hypothetical "hidden rotation," Landé did eventually label it with the same letter, k, which he had used a year and a half earlier for the total angular momentum.[100]

One might proceed from this imagined insight to an entirely plausible reconstruction of Landé's route to the solution of the puzzle. And I will do so, both in order to see more clearly where we wish, ultimately, to come out, and also in order to show how very far from the logical route, and how much longer than the logical route, is Landé's actual path to the solution.

Sommerfeld assumes that the azimuthal quantum number is the total angular momentum, and thus also the total magnetic moment, of the atom. Therefore, the inference runs, the Runge denominator of a term, r, must be a whole number function of the azimuthal quantum number, $r = g\,(n)$. Landé, we imagine, would then recognize that the premise is false and the inference illogical. Premise:

100. Landé, *op. cit.* (note 38). Just how soon Landé adopted a symbol for the inner quantum number is not clear. In the documents k does not appear until 17 Mar. 1921, and earlier statements use only verbal expressions and numerical values.

196

not the azimuthal quantum number, n, but the inner quantum number, k, is the total angular momentum (and therefore also the magnetic moment of the atom). Inference: since the magnetic perturbation energy E_m is given by $\mathbf{p} \cdot \mathbf{w}_{Larmor}$, where $\mathbf{p}$ is to be identified with the *total* angular momentum, the quantum number corresponding to this quantity must enter in the *numerator*, not the denominator, of the expression for the energy splitting. Undeniably, Sommerfeld's values for the Runge denominators of the various types of terms do work, and his expressions for r thus seemed good and necessary: $r = n$ for the triplets, $r = 2n - 1$ for the doublets. Landé would therefore keep these "term denominators" and replace the indeterminate "term numerator," q, by something proportional to the inner quantum number, k. Thus for the doublets Landé would try:

$$\Delta \nu_1 = q_1/r_1 \cdot a \cdot H = k_1/(2n_1 - 1) \cdot a \cdot H \, ,$$

$$\Delta \nu_2 = q_2/r_2 \cdot a \cdot H = k_2/(2n_2 - 1) \cdot a \cdot H \, ,$$

$$\Delta \nu = \Delta \nu_1 - \Delta \nu_2 = \left(\frac{k_1}{2n_1 - 1} - \frac{k_2}{2n_2 - 1} \right) \cdot a \cdot H \, .$$

Now because of the very remarkable accident that $k/(2n - 1)$ is one half of the true g factor for all the doublet energy levels, the $\Delta \nu$'s so formed would have given the positions of the symmetrical innermost ‖ components of all five of the doublet combinations (sp_1, sp_2, $p_1 d_1$, $p_1 d_2$, $p_2 d_2$). Using $q/r = 3k/(2n - 1)$, Landé would have gotten the positions of the remaining pair of parallel components in the $p_1 d_2$ and $p_1 d_1$ patterns. All the rest would have come very quickly—as indeed happened once Landé *had* reached this point. The interpretation of this energy level analysis of the splitting of the individual lines, using quantum numbers, selection rules, and a "g factor," would finish the story. In his first paper on the anomalous Zeeman effect Landé emphasizes that he has fulfilled Sommerfeld's ideal desiderata for an explanation based on an energy level analysis.[101] Inasmuch as the expression $g = 2k/2n - 1$ plays an obvious and important role in this paper, were the published record the only source, this reconstruction would appear cogent, though perhaps not impeccable. But the unpublished record shows that this is not at all the way that Landé approached the problem. Although some

101. Landé, "Über den anomalen Zeemaneffekt (Teil I)," *ZS. f. Phys.*, 5 (1921), 231–241, received 16 Apr. 1921.

of the constituents of our reconstruction were present at an early stage, it was some time before they were used in a consistent pursuit of the "obvious" desiderata of a quantum theoretical explanation.

It seemed reasonable to place Landé's first close examination of Sommerfeld's paper in December 1920, perhaps during his Christmas vacation.[102] By the end of January 1921 Landé had some results. On 4 February he wrote Bohr:

> I'm very anxious to know whether you have made progress in the problem of atomic structure and spectra with the help of the elliptical orbits $n = 1$, $n' = 1$, and, above all, whether results are thereby obtained which give an understanding of the Zeeman effect. With regard to the complicated types of the Zeeman effect, I have found a few empirical rules which go considerably beyond Sommerfeld's compilation in the *Ann. d. Phys.* and permit one to make predictions regarding the neon spectrum. But what these rules signify is entirely incomprehensible to me.[103]

Most striking here is the conjunction of Landé's expectation that the anomalous Zeeman effect is to be understood in terms of the geometry and dynamics of electronic orbits with his announcement that he has found a number of *empirical* rules, whose geometrical or dynamical interpretation completely eludes him. Thus our plausible reconstruction is immediately contradicted; the theoretical viewpoint had not guided him—or at least whatever may have induced Landé to take up the anomalous Zeeman effect, his first results were obtained in some way quite unconnected with any theoretical perspectives.

Landé says nothing in this letter to Bohr about the content of his rules. We can infer, however, from Paschen's letter of 5 February,[104] replying to queries Landé sent him about Zeeman patterns, that at least Landé's principal result—Rule 1—is the same as that which Landé did communicate to Bohr twelve days later.[105] The opening line of this second approach to Bohr on 16 February—"In the course of reading Sommerfeld's compilation 'General Spectroscopic Laws...' . . . I was struck by the enclosed additional empirical rules" —suggests that *all* the rules here communicated were originally en-

---

102. See section 6 below for further discussion of this point.
103. Landé to Bohr, 4 Feb. 1921 (appendix, doc. 1).
104. Paschen to Landé, 5 Feb. 1921 (appendix, doc. 2).
105. Landé to Bohr, 16 Feb. 1921 (appendix, doc. 4).

tirely empirical, and that they are the same which Landé had in hand when he wrote Bohr on 4 February and Paschen somewhat earlier. Now indeed Landé prefaced his rules with a speculation about their theoretical significance. But here I present the empirical schema first in the conviction that it was developed before the theoretic interpretation. The text is given in the appendix (document 4); for clarity of analysis the rules may be condensed as follows:

1. Every ∥ component is surrounded by two ⊥ components at distances ±Δ, where Δ is the separation of the outermost ⊥ component from the outermost ∥ component. However ⊥ components will be missing whenever two fall upon one another, as they then cancel. E.g., from Figure 4 the splitting pattern of the doublet p_1d_1 line [$^2P_{3/2}$–$^2D_{5/2}$], and its appropriate Δ, are:

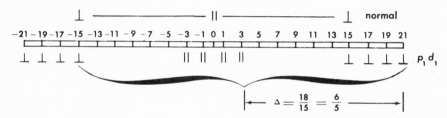

2. For all triplet lines the number of ∥ components is equal to the sum of the "inner" quantum numbers of the *aufbauend* terms. For doublet lines the rule holds only *approximately*. E.g., for doublet p_1d_1, $2 + 3 = 5 \sim 4$.

4. One gets the quantity Δ *approximately* out of the two "inner" quantum numbers of the *aufbauend* terms if one divides each of them by its particular Runge denominator and adds the two quotients.

E.g., for doublet p_1d_1, "$2/3 + 3/5 = 19/15 \sim 18/15$." Landé neglects to emphasize here (and he will emphasize it later) that Rule 1, the constancy of Δ, applies to the splitting of one line, not to the line group as a whole. Landé is probably not yet fully conscious of the extent of the novelty. The implicit neglect of the line group, and, especially, the form and procedure of Rule 1, argue strongly that Landé assaulted the problem of the anomalous Zeeman effect not by taking off from Sommerfeld's discussion of the probable features of a quantum-theoretical derivation, but rather from Sommerfeld's entirely empirical discussion of the number and positions of the components in the splitting patterns of the individual lines.

199

It would appear that in playing with the observed patterns, as given by Sommerfeld in "General Spectroscopic Laws . . . ," Landé observed that by working in from the outside of the patterns, all—or almost all—of them could be dissected into triplets of anomalous, but equal spacing. It required a little awkward straining to carry it through, but only one of the fourteen cases appeared entirely intractable—p_1d_2 of the triplets. (Landé was rooting so strongly for his triplet dissection that he drew the triplet sp_2 pattern incorrectly—in such a way as to allow this dissection. Actually this pattern presents exactly the same problem as p_1d_2, viz., the special restrictions on transitions between states of equal total angular momentum.) That this first rule was the first arrived at and the most confidently adhered to is also indicated by the queries Landé addressed to Paschen —probably at the end of January. Paschen's answer of 5 February shows that Landé was trying to alter the Zeeman patterns to facilitate his dissections, and that p_1d_2 was a particular problem. Paschen, however, was adamant: "It is out of the question that the strong component $\perp$ 9/6 lies at 13/6. The type is so often and definitely observed, already in 1900 by Runge and myself, that an error can scarcely be possible. Moreover the Zeeman types in the neon spectrum are not to be altered."

Thus far we have no inner quantum numbers, indeed no reference at all to the spectroscopic terms whose combination yields the spectral line. These appear only in Rules 2–4 where the number of parallel components in the splitting pattern of each line is given (usually) by the sum of the inner quantum numbers, $k + k'$, of the "constructive" terms, and the width of the constituent triplets is given (roughly) by the sum of the inner quantum numbers each over a Runge denominator,[106] $\Delta = k/r + k'/r'$. Here, finally, is the k/r which our plausible reconstruction cast in the leading role. It is, as we anticipated, being used as a measure of the width of the Zeeman pattern, but certainly not in the way we anticipated. It does not measure the distance from the field-free line, but the width of the purely empirical constituent triplet. More important, neither Rule 2 nor 4 is based upon the combination principle, upon term *differ-*

---

106. Landé will not give this quantity a name for another week (when it becomes the "inner quantum fraction"), nor express it symbolically for another month.

ences, but upon *sums.* Significantly, Landé does not even speak of *combining (kombinierend)* terms, but uses the neutral and ambiguous expression "constructive *(aufbauend)* terms." Thus, despite its potential theoretical significance, the quantity k/r was evidently empirical in origin, and presumably already figured in this same empirical rule at the end of January.[107]

There is, however, one striking difference between the schema which Landé sent to Bohr on 16 February and the expectations engendered by his previous letter of 4 February and by the queries he put to Paschen; namely, the restriction of his rules to the doublets and triplets and the absence of any mention of the neon spectrum. When writing on the 4th to Bohr, Landé clearly regarded his ability to make predictions of neon Zeeman patterns as a principal merit of his rules. (Their cogency was somewhat reduced by Paschen's reply of the 5th insisting that "the Zeeman types in the neon spectrum are not to be altered," and this probably led to the omission of neon on the 16th.) Thus at least as a check upon my identification of the schema of 16 February with the rules of late January, I must consider how this schema could have been applied to neon.

Paschen's term analysis of the neon spectrum had been in Landé's hands since the preceding summer.[108] Landé's attention would have been called to it again at the Nauheim meeting where Walther Grotrian, a young and able experimental spectroscopist, had given a stimulating talk on "Regularities in the Spectrum of Neon."[109] Grotrian rearranged Paschen's terms into an energy level diagram— explicitly in imitation of Bohr's representation of the sodium terms in his Berlin lecture the preceding spring[110]—and pointed out that not all of the possible term combinations actually occurred: "Here also a selection principle or a transition prohibition holds sway, and it is interesting to inquire whether any sort of regularities determine

107. The irony of the situation lies in the fact that a consistent application of the combination principle shows the number of $|\,|$ components *and* the value of Δ to depend on only one—not both—of the combining terms. Thus as Fig. 5 shows, in Landé's triplet decomposition of the pattern Δ depends only on the g factor of the *higher* total angular momentum state (indeed, $\Delta = g$). Conversely, the number of $|\,|$ components is determined by the number of orientations, in a magnetic field, of the atom in its *lower* total angular momentum state.

108. See notes 58 and 59 above.

109. W. Grotrian, "Gesetzmässigkeiten im Spektrum des Neons," *Phys. ZS., 21* (15 Nov. 1920), 638–643.

110. Bohr, *op. cit.* (note 40).

this selection." To this question Grotrian was not, however, able to give any satisfactory answer; the requisite theoretical concept—Sommerfeld's inner quantum number—was not published until a month later, and even then was not immediately seized upon by Grotrian.[111]

These circumstances, I imagine, caused Landé to be immediately and forcefully struck by an important lacuna in Sommerfeld's "General Spectroscopic Laws . . .": in the lengthy discussion of the Zeeman effect in the neon spectrum no mention was made of the inner quantum number. (This is but another example of the lack of any effective connection between the inner quantum number and the splitting rule in Sommerfeld's treatment.) Landé would then have filled this gap, using Sommerfeld's procedures and Grotrian's convenient arrangement of Paschen's terms, by assigning inner quantum numbers to the neon terms.[112] Thus with inner quantum numbers, and with Runge denominators derived from Krönert via Sommerfeld, Landé could have applied all the rules of his schema of 16 February to the neon Zeeman patterns. With what success? In principle, with about as much (or as little) as in the case of the triplet lines.

Having stressed the empirical origin of Landé's schema of 16 February, I must now consider the speculative theoretical interpretation which Landé prefaced to it.

> I believe that especially Rule 1, which allows the complicated Zeeman pattern to be broken up into individual groups $\perp-\|-\perp$, is not unimportant for a theory of the Zeeman effect. The different "inner" quantum numbers of a term will probably signify simply the total quantum numbers of the atom about its invariable axis for different spacial orientation of the valence electrons around the atomic core, and the presence of several groups $\perp-\|-\perp$ in one Zeeman pattern signifies, presumably, several orientations of this invariable axis of the atom with respect to the magnetic field.

Here then appears, as we anticipated, the identification of Sommerfeld's inner quantum number with the total angular momentum of

111. Grotrian, *op. cit.* (note 109), cited Sommerfeld's "Allgemeine spektroskopische Gesetze" in an "Anmerkung bei der Korrektur," but only for its Runge denominators for neon.
112. The first (and quite correct) assignment of inner (total angular momentum) quantum numbers to neon was indeed published by Landé: "Anomaler Zeemaneffekt und Seriensysteme bei *Ne* und *Hg*," *Phys. ZS.*, 22 (1 Aug. 1921), 417–422, received 27 June 1921. This paper is also the only one in which remnants of Landé's empirical rules appear.

the atom, for it is to this quantity that Landé refers in speaking of "the total quantum numbers of the atom about its invariable axis." And this identification is accompanied by an explicit model: different relative orientations of the orbital planes of the valence electrons produce different values of the total angular momentum of the atom, and with each value is associated one component of the complex structure of a spectral term. This essentially correct physical intuition, which ties directly onto Landé's paper on the coupling of electron rings, is not without difficulties. As soon as Landé examined the detailed consequences of the model, he would have found that with integral orbital angular momenta it is possible to obtain complex terms of odd multiplicity (singlets and triplets), but not doublets. It is perhaps for this reason that the model makes no appearance in Landé's first publications on the anomalous Zeeman effect.

It is thus fairly clear that the model which Landé suggests is indeed the result of combining his earlier investigations of the helium spectrum and coupled angular momenta with Sommerfeld's inner quantum number analysis of the complex terms. Yet the attribution of "the presence of several groups $\perp - \| - \perp$ in one Zeeman pattern" to the fact that the total angular momentum of the atom can assume "several orientations" with respect to the magnetic field is, at first, a little difficult to understand. Landé seems to be ignoring the quantum theory of the normal Zeeman effect which invokes exactly those several orientations of the atom in the applied magnetic field in order to derive the single triplet $\perp - \| - \perp$ of the Lorentz theory. But we can also see how right was Landé's intuition that the model could give his empirical triplet dissection. All that was missing, and indeed it remains absent for some weeks yet, was the combination principle. For if we turn back to Figure 5 and fix our eye on the atom in its lower total angular momentum state, p_1, we can see that each of its several orientations with respect to the magnetic field determines—when one considers the magnetic energy levels of the higher total angular momentum state with which, according to the selection rule, it can combine—a triplet, $\perp - \| - \perp$. And indeed these are precisely the triplets into which Landé decomposed the patterns.

Yet Landé's model, especially in its blurring of the distinction between the energy levels and the pattern of spectral lines resulting from transitions between energy levels, cannot help but remind us of

Sommerfeld's considerations about the number of components in the splitting patterns of the doublets and triplets. And this association seems strongest if we try to picture how Landé's model "explains" his own component number rule. If the presence of several triplets is due to the various orientations of the invariable axis of the atom in the magnetic field, then, as the number of such allowed orientations depends on the total angular momentum, the model requires that the number of ‖ components is given, in some way, by the inner quantum numbers of the combining terms. Landé, with remarkable success, lets both terms participate in fixing the number of ‖ components simply by setting that number equal to the sum of the inner quantum numbers, $k + k'$. Thus it seems clear that rather than the established concepts of the quantum theory of the atom serving as guides and controls, these concepts are themselves treated by Landé (and in a sense by Sommerfeld, too) as plastic categories which may be molded about otherwise incomprehensible empirical regularities.

Clearly, this first schema could only be a halfway house. Aside from their ambiguous theoretical character, approximate rules were of only very limited utility or cogency. Landé now concentrated on the doublet lines, where his prescription for a dissection of the pattern (Rule 1) suffered no contradictions. Convinced, as it appears, that some conjunction of the k/r of the combining terms must measure some separation in the splitting pattern, he was able, within a week, to send Bohr and Sommerfeld a new schema for the doublets.[113] The full text is given in the appendix; here I have condensed and renumbered the rules in order to display the essential improvements:

1. The ‖ components in the splitting pattern of a *doublet* line follow each other at equal separations.
2. The number of ‖ components is equal to $k + k'$, but if, for a doublet line, the sum is odd, then the number is $k + k' - 1$.[114]
4. In the splitting pattern of a *doublet* line the position of the

113. Landé to Bohr, [2]3 Feb. 1921 (appendix, doc. 6). From Sommerfeld to Landé, 25 Feb. 1921 (appendix, doc. 7) I infer that Landé sent essentially the same schema to Sommerfeld simultaneously. The form of Sommerfeld's reference there to the Zeeman patterns of the triplet lines strongly suggests that he had also received Landé's first schema, and for this reason I have hypothesized a letter of ca. 16 Feb. 1921.
114. In order to see why this rule works, we write j and j' for the "true" total angular momenta of the upper and lower terms, respectively. The selection rule

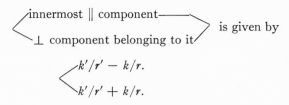

is given by

And with this new schema Landé has finally discovered what properties of the *doublet* splitting patterns are really given by conjoining the "inner quantum fractions" of the combining terms. That is, as we found in the course of our initial *a priori* reconstruction, the remarkable accident that the g factors of the doublet energy levels are expressible as $2k/(2n-1) = 2k/r$ leads in fact, upon application of the combination principle, to $k'/r' - k/r$ as the position of the innermost ‖ component, and to $k'/r' - k/r + g = k'/r' + k/r$ as the position of the positive ⊥ component associated with this ‖ component by Landé's "triplet dissection." Thus Landé can now claim that "by these rules the splitting pattern of the doublet lines is unambiguously determined from the inner quantum number and the Runge denominator,"[115] as one can see from the fact that the rules give:

1. The position of the innermost ‖ component.
2. The number of ‖ components.
3. The separations of the ‖ components.
4. The distances of the ⊥ components from the ‖ components.

---

for the total angular momentum (or the inner quantum number) then restricts the combinations to:

$$
\left.\begin{array}{l} j' = j - 1 \\[4pt] j' = j \\[4pt] j' = j + 1 \end{array}\right\} \Rightarrow \text{no. of components} = \begin{cases} 2j' + 1 = j + j' & \text{case } 1 \\[4pt] 2j' + 1 = j + j' + 1 & 2 \\[4pt] 2j + 1 = j + j' & 3 \end{cases}
$$

For triplets: $k = j$, and thus in cases 1 and 3 the number of components equals the sum of the inner quantum numbers. This holds for case 2 also, since when $j' = j$, m $= 0 \to$ m$' = 0$ is forbidden and one of the $j + j' + 1$ components is missing. For doublets: $k = j + \frac{1}{2}$, and $k + k' = j + j' + 1$. This is fine for case 2, since $j' = j$, implies $k' = k$, and as k and k' are integral, their sum is never odd. In cases 1 and 3, k and k' differ by 1, thus their sum must be odd, and $k + k' - 1$ ($= j + j'$) gives the correct number of | | components.
115. Landé to Bohr, [2]3 Feb. 1921 (appendix, doc. 6).

205

And Landé shows, using all five of the doublet lines as examples, how one is to proceed in constructing the observed pattern with his rules. Curiously, Landé does not seem to notice—perhaps because he had not written the "inner quantum fractions" symbolically—that

$$\Delta = (k'/r' + k/r) - (k'/r' - k/r) = 2k/r .$$

Instead Δ is demoted from the central position it had previously held to a quite secondary status, for Landé's perspective upon the patterns has undergone a marked shift. Where previously his attention centered upon the elementary triplets dissected from their context, now all the Zeeman components are placed with respect to a single, fixed origin, the position of the field-free line.

Although in its reproduction of the observed phenomena this second schema is a great advance over the first, in its physical conception it might well be regarded as retrograde. When the sum of the inner quantum fractions gives so peculiar a property of the splitting pattern as the position of the $\perp$ component belonging to the innermost $\parallel$ component, Landé could no longer claim (as he could with his first schema) that he had some physical picture of what was going on. Nor is there, to compensate this loss, any evident concern with combining energy levels, selection rules, and the usual apparatus of a quantum theoretical explanation. Yet by retreating we advance; by a circuitous detour we have now just about reached the end point of that specious reconstruction with which we opened our discussion of Landé's work. If, without ever introducing the combination principle, Landé has found the position of the innermost $\parallel$ component to be given by $k'/r' - k/r$, it is a sure thing that were he only to explicitly invoke the principle, the problem would break open in his lap. But that will not happen for a few weeks yet, and, wrapped up in the development of Landé's scientific ideas, I have neglected the professional environment and the collegial reactions which so largely affect their author.

Bohr was at this moment even more overwhelmed than usual by his work and responsibilities. On the 14th he sent off to *Nature* his first letter on "Atomic Structure" outlining an assignment of total quantum numbers and occupation numbers to the various electronic shells of atoms in the successive rows of the periodic table.[116] Of this

116. Bohr, *op. cit.* (note 67). Bohr to Landé, 14 Feb. 1921 (appendix, doc. 3).

manuscript Bohr appears to have distributed simultaneously just two carbon copies, one to Rutherford and one to Landé. The letter—and Bohr implied that the results it announced had been rigorously calculated from an extended version of his correspondence principle—"broke like a bomb" upon the world of atomic physics.[117] Landé was evidently responsible for spreading the first definite reports to Munich when he sent Sommerfeld his schema of 23 February. By coincidence Sommerfeld had already sent off a card on 18 February reminding Bohr how pleased he would be to get a copy of the galley proofs of anything Bohr might publish on atom models—Sommerfeld was becoming a little uneasy, for it was now two months since he had heard through Einstein that Bohr's "atom-building has again made decisive advances." He could not bring himself to admit to Landé that Bohr had not thought it necessary to send *him* the manuscript, and fortunately he did not have to feign indifference very long. Bohr sent off a copy immediately upon receiving Sommerfeld's card, and it was soon making "considerable headaches" for a number of theorists.[118] Landé responded to Bohr's revelations on the 21st—only two days before he sent Bohr the revised doublet splitting rules—that from his point of view "until your full and detailed exposition appears there is simply no sense in doing any more theoretical work in atomic theory." Without explicitly mentioning the unavailability of *Nature* in Germany due to the exorbitant price of foreign publications during the postwar inflation, or expressing pique at Bohr's preference for publishing in English journals, Landé asked permission to publish a translation of this brief communication "so that also the German physicists may be pointed as quickly as possible toward your new advance."[119]

Landé, as his first soundings at the beginning of February clearly showed, had regarded the problem of the anomalous Zeeman effect

117. The bomb simile was Landé's, echoed in Sommerfeld to Landé, 3 Mar. 1921 (appendix, doc. 9). Although Sommerfeld makes no allusion to these developments in his card of 25 Feb. (appendix doc. 7) responding to the receipt of Landé's schema of 23 February, I nonetheless assume the simile was contained in Landé's letter transmitting that schema.
118. Einstein to Sommerfeld, 20 [?] Dec. 1920 (Einstein and Sommerfeld, *Briefwechsel*, edited and annotated by Armin Hermann [Basel und Stuttgart, 1968], p. 75). Sommerfeld to Bohr, 18 Feb. 1921; Bohr to Sommerfeld, 22 Feb. 1921; Sommerfeld to Bohr, 7 Mar. 1921; Sommerfeld, Ewald, Lenz, and Mie to Bohr, 20 Mar. 1921 (BSC 7, 3).
119. Landé to Bohr, 21 Feb. 1921 (appendix, doc. 5).

as closely connected with the configuration of electron orbits in the atom and the quantum numbers assigned them. We can therefore well imagine that Bohr's letter, which enters temporally precisely between Landé's first and second schemas, was influential in relaxing Landé's reliance upon his own model for guidance, and in turning him back to empirical manipulations. However this may be, and whatever we might read in Landé's solicitous concern for the orientation of the German physicists, Bohr perceived these strong reactions of his German colleagues as a strong external pressure to work up an extensive presentation of his results. The difficulties in doing so were the more considerable inasmuch as Bohr did not possess the calculus which his German colleagues expected him to deliver.

To all this must be added the imminence of the inauguration of Bohr's Institute. The third of March had to be prepared for as an important public—indeed international—event. Thus although Landé may have regarded the silence as ominous, it is not surprising that Bohr did not respond to Landé's communications. Finally, on the last day of February, Bohr told Kramers to write Landé that he was greatly interested but did not have the time just then to get deeply into the problem or, in particular, to see how Landé's results square with the theoretical viewpoint.[120]

If Bohr sounds a little cool, and was perhaps gently hinting that attention had not been paid to the just demands of the quantum theory, Sommerfeld was enthusiastic: "Bravo, you're a regular magician. Your construction of the doublet Zeeman types is *very* beautiful."[121] Sommerfeld does not suggest that the results are less than satisfactory because theoretically ungrounded, but on the contrary congratulates himself that his inner quantum numbers, which he "hit on rather by a lucky chance," have thus proved to be so important. Entering into the spirit of Landé's work, Sommerfeld urges him now, as he had urged him two years earlier in connection with the helium spectrum, to go beyond merely accounting for the observed phenomena and to predict the splitting patterns of the fundamental series.[122] In all of this there was no hint that once Landé

120. Kramers to Landé, 28 Feb. 1921 (appendix, doc. 8).
121. Sommerfeld to Landé, 25 Feb. 1921 (appendix, doc. 7).
122. This Landé did and sent off his predictions to Paschen on 2 Mar. 1921. (Paschen to Landé, 8 Mar. 1921 [appendix, doc. 11].) However, only one doublet *f* term had then even been resolved, and the Zeeman pattern was thus beyond the reach of experiment.

had twisted the rules for the triplets out of the splitting patterns—and Sommerfeld was confident that he soon would—that he might then be expected to sit on his results and refrain from publishing.

Landé, perhaps less sure of the tractability of the triplets, was not prepared to pass up the opportunity to secure his title to the results he had already obtained for the doublets. The end of February marked the conclusion of the winter semester at Frankfurt. With the prospect of six weeks of vacation ahead, Landé wound up his lecture course, "Introduction to the Mathematical Treatment of the Natural Sciences,"[123] and then went off on the 28th to a meeting of the Gauverein Hessen of the Deutsche Physikalische Gesellschaft in Marburg. At the end of the program of this Marburg meeting Landé gave a talk entitled "Gesetzmässigkeiten im anomalen Zeemaneffekt," which we must suppose was an exposition on the schema of 23 February. But the *Verhandlungen* of the Deutsche Physikalische Gesellschaft give only the title, no summary of the contents, of this talk.

5. COMPETITION WITH ERNST BACK

In the postwar period atomic physics was burgeoning—large numbers of physicists were pressing into the field. And in German-speaking Central Europe there was a considerable degree of consensus among them about which problems were important. This circumstance, combined with the rapidity of experimental advance and conceptual turnover, led to very close and continuous informal communication among atomic physicists, even as it led to heightened competition. Such informal communication was, to be sure, intended as, and functioned as, a form of cooperation; at the same time, however, it was a novel response to the threat of competition. Where the traditional reaction had been secrecy, the new strategy for discouraging parallel efforts and avoiding being forestalled was to keep the likely competitors fully informed. Sounding out Bohr at the beginning of February about the anomalous Zeeman effect, Landé tried to mix these strategies. When no reply was forthcoming, Landé opted for full and continuous communication of his results. It is

123. "Einführung in die mathematische Behandlung der Naturwissenschaft," as given in the "Vorlesungsverzeichnis" for the winter semester 1920–1921 published by the *Phys. ZS., 21* (1920).

to his adoption of this strategy that we are indebted for the opportunity to follow the development of Landé's ideas. From Munich and Copenhagen Landé anticipated competition, and he was especially worried about Bohr. Landé probably never suspected that the most immediate threat was from Tübingen, a center of *experimental* spectroscopy.

Ernst Back had returned to Tübingen. He had had a difficult time during the war, and, as a member of a family of Alsatian emigrés, also in its immediate aftermath. For about two years he worked as a physicist at the Veifa-Werke in Frankfurt am Main. There he was occupied with X rays, and other topics far removed from the anomalous Zeeman effect. It was at this time that he gave Sommerfeld permission to publish the results he had obtained before the war. But Back had not abandoned his interest in his first love, and chose the "Zeeman-Effekt und Serienspektra" as his subject when, on 14 January 1919, he addressed the physics colloquium at the University.[124] (Here at the Frankfurt physics colloquium Back and Landé had become, somewhat distantly, acquainted.[125]) Like a number of other physicists who had gone into industry at the end of the war, Back evidently soon found that despite the economic conditions an academic career was attractive and not entirely unfeasible. Early in 1920 Back returned to Tübingen as an assistant in Paschen's institute, had his dissertation published, and started working toward his *Habilitation*.[126]

Back, like Landé, had been stimulated by Sommerfeld's "General Spectroscopic Laws . . . ," which Back had read in manuscript in the spring of 1920. Although he had none of the training and skills of a theorist, he had a strong speculative bent. This is evident in his paper with Paschen in 1912 on the transformation of anomalous Zeeman patterns into normal triplets in strong magnetic fields.[127] It is evident also in various speculations that he did not publish, for example, his suggestion to Sommerfeld that the magneto-optic splitting rule, $r = r_1 r_2$, argues "that in place of Bohr's algebraic function

124. *Physikalischer Verein zu Frankfurt a.M., Jahresbericht* (1918–1919), p. 70.
125. Back to Landé, 7 Mar. 1921 (appendix, doc. 10).
126. Forman, "Ernst Back," *Dictionary of Scientific Biography, 1* (New York, 1970).
127. Paschen and Back, "Normale und anomale Zeemaneffekte," *op. cit.* (note 78). Discussed in detail by Spencer, *Zeeman Effect, 1896–1913, op. cit.* (note 71), Ch. XII.

for singlet lines, an exponential function must enter for the multiplicities."[128] For his *Habilitation* he aspired not only to the most extensive and precise measurements, but also to discover further *Zahlenmysteria* sufficient to describe all his Zeeman patterns. With enormous perseverance, and, by way of theory, a peculiar faith that the spectroscopic notation mirrored an underlying reality, Back searched for formal numerical regularities in the splitting patterns— and indeed, like Landé, in the splitting patterns of the individual lines.

Back was not the sort of personality that thrived on competition; quite the reverse. He was a very sensitive and retiring person. Formal in manner and dress, he was among the last to drop the "Herr" when referring, in the text of his papers, to other physicists. In a way not unprecedented, he combined outward diffidence with high scientific ambition and great experimental virtuosity. This personality and virtuosity dictated his strategy of "first squeezing all of the juice out"[129] of a subject before revealing his methods and thus having to face competitors. Moreover, considered quite objectively, Back, like Landé, was simply in no position to contemplate competition with equanimity. As he himself explained, "with my advanced age —but at the same time still in the first stages of a career—I have no time and no fruits to give away."[130]

128. "Dass die Komponentenzahlen der Einzelterme sich in einer Kombination addieren, ist experimentell absolut sicher. Dass anderseits ein *multiplikatives* Gesetz im Sinne Ihres Rungeschen Zählers dennoch vorhanden sein muss, ist mir ebenfalls einleuchtend. Ich dachte schon an die Möglichkeit, dass eine logarithmische Funktion für die Typenbildung vorliegen könnte in dem Sinn, dass die massgebliche Grösse für die Komponentenzahl in den Exponenten eingeht, dann würde das multiplikative und additive Gesetz der Komponentenzahl miteinander versöhnt sein. Da eine Mehrzahl von Komponenten (Addition) nur bei den Anomalien (Multiplizitäten) auftritt, könnte man zu der Vermutung kommen, dass an Stelle der algebraischen Funktion bei Bohr für die einfachen Linien eine Exponential-Funktion für die Multiplizitäten treten müsste." (Back to Sommerfeld, 7 June 1921 [SHQP 29, 5].)
129. "Dass meine Habilitationsschrift noch nicht heraus ist, ist ein klein wenig jesuitisch von mir, denn ich möchte den Saft erst ganz auspressen, ehe ich die Methoden und Resultate veröffentliche. Doch bin ich wirklich jetzt nahe am Ende." (*Loc. cit.*)
130. Back to Landé, 7 Mar. 1921 (appendix, doc. 10). A more general, and very suggestive, statement and analysis of these issues in the context of contemporary physics is given by F. Reif, "The Competitive World of the Pure Scientist," *Science, 134* (15 Dec. 1961), 1957–1962; F. Reif and A. Strauss, "The Impact of Rapid Discovery Upon the Scientist's Career," *Social Problems, 12* (Winter 1965), 297–311. On the other hand, C. von Ferber, *Die Entwicklung des Lehrkörpers in*

Early in February Back sent off a short paper to *Die Naturwissenschaften*.[131] Justifying his intrusion upon the public with so esoteric a topic by calling attention in title and text to Sommerfeld's "Zahlenmysterium," he claimed to be able to give rules for all the splitting patterns, i.e., positions and polarizations of all the components, and promised to publish them in the *Annalen der Physik*. Clearly Back had visions of fusing experiment and "theory," all in one publication. As a foretaste he devoted this article to an exposition of his rules for the number of components. The key is an "index rule": for triplet (doublet) line groups, the sum of the indices of any two combining terms, plus the number of parallel (perpendicular) components in the Zeeman pattern of the line to which the two terms give rise, is equal to a constant.[132] This rule, which Back advertised as a splitting rule for the number of components in analogy with Sommerfeld's splitting rule for the Runge denominator, is clearly the analog of Landé's Rule 2. Purely empirical, it is already a more exact, comprehensive rule than that of Landé's schema of 16 February, although surpassed in Landé's second schema. To get the number of perpendicular (parallel) components in the case of triplet (doublet) lines, Back supplemented this index rule with a "Paschensche Regel": Paschen had pointed out to him that, excepting doublet p_2s and p_1d_2, all Zeeman types with which he was acquainted had twice as many perpendicular components as parallel components. Back made the precarious assumption that these were the only exceptions.

Back did not, however, publish his full repertoire of rules in the

deutschen Universitäten und Hochschulen, 1865–1954 (Göttingen, 1956), p. 134, found that in the late nineteenth and early twentieth centuries there was "no connection between early *Habilitation* and less than average 'Privatdozentenzeit.' In the disciplines with—relative to the total average—high ages at *Habilitation* and long periods as *Privatdozent* [e.g., medicine and experimental natural science] there even exists, above the median value of the distribution of ages at *Habilitation,* a tendency towards a shortening of the period as *Privatdozent.*"

131. Back, "Ein weiters Zahlenmysterium in der Theorie des Zeemaneffektes," *Naturwiss., 9* (25 Mar. 1921), 199–204, dated 10 Feb. 1921. In his letter of 7 Mar. 1921 to Landé Back says that the manuscript was accepted in its final form by the editorial office on 10 Feb., that it had been sent off from Tübingen on 2 Feb. and a "kleiner Schlussnachtrag," not distinguishable in the published paper, was sent off on 6 Feb.

132. Roughly, this rule works because the index on a term symbol, p_i, say, can be written $i = c - j$, j the total angular momentum. Thus for a triplet p_i term, $i = 3 - j$. Then the explanation of Landé's component number rule becomes one for Back's index sum rule.

Annalen; in June, with Landé's first paper on the anomalous Zeeman effect in press, Back felt compelled to safeguard his independence. By enlisting Sommerfeld's support, he induced Arnold Berliner, the editor of *Die Naturwissenschaften,* to accept an addendum, exceeding the original paper in length and obscurity.[133] The rules were frightfully complex, replete with recursive definitions and complicated categorizations of the lines. They strike one as a pathologic development of Sommerfeld's hints at the kinds of games one might play with the patterns, yet they also appear in many points curiously similar to Landé's rules. Back had a "Bildungsgesetz der Spannweiten" which gave the distance of the first $\perp$ component from the undeviated line in terms of Runge denominators and a regular sequence of integers. He had a "Stufenregel" which gave the distance between neighboring $\perp$ components. These, combined with his "Indexregel" and his "Paschensche Regel," were sufficient to reproduce the splitting patterns of the doublets and triplets and to predict Bergmann series patterns. Moreover, if Back's "theoretical" efforts were not generally appreciated, it was evidently not because this kind of solution was unacceptable in principle to all the quantum physicists of the day. On the contrary, Landé, a trained theoretician, was following a quite parallel, albeit somewhat more elevated, path. And, until the end of February, Landé enjoyed Sommerfeld's support.

Although Landé had contacted Paschen no later than the first few days of February, Back maintained that his first communication to *Die Naturwissenschaften* was made at a time when he "neither had nor could have had" any knowledge of Landé's work.[134] And it seems indeed that Paschen was playing a rather old-fashioned game. In contact with both Back and Landé, and giving aid and encouragement to both, throughout the month of February he told neither one that the other was working on the same problems. Rather, Paschen functioned as an agent of that older strategy for dealing with the problem of competition; namely, preventing communication and thus assuring to each the credit for an independent discovery. However, in the week between the 25th of February and the 5th of March

133. Back to Sommerfeld, 7 June 1921 (SHQP 29, 5); Back, "Zur Theorie des Zeemaneffektes: Nachtrag . . . ," *Naturwiss., 9* (22 July 1921), 569–575, dated 8 June 1921.
134. Back to Landé, 7 Mar. 1921.

Back and Landé each learned of the existence of his competitor. The character of Landé's interactions with his colleagues immediately changed. Personal and political considerations tended to displace questions of scientific merit, both in what was communicated between colleagues and in the alignment of parties.[135]

Sommerfeld had known, of course, of Back's earlier enthusiasm for his *Zahlenmysteria,* but there is no reason to suppose that he knew of Back's efforts to extend and multiply them. In his reaction to Landé's schema of 23 February there was no hint of a possible conflict, or even of Back's special claims—"I'll send your sheet [with the doublet rules] to Paschen." This he did. Back had just received the proofs of his note for *Die Naturwissenschaften* when Landé's doublet rules arrived in Tübingen on or about Saturday the 26th. Paschen then told Back of his competitor's existence, but, consistent with Paschen's strategy of protecting Back's independence, "very properly" withheld the rules from him.[136]

Back was upset. The very fact of Landé's parallel efforts disturbed him. (Had he known just how long Paschen had been aware of Landé's work he would have been more disturbed still.) Moreover, even if his independence could be established, his priority was threatened, for the article in press was intended to whet the physicists' appetite rather than prove Back could do all that he claimed. On the 1st or 2nd of March Back sent Sommerfeld the proof sheets of this first note in *Die Naturwissenschaften.* In a covering letter he evidently made it plain that he rued the day he consented to Sommerfeld's publishing his data. That Landé should be competing with him on the basis of his own measurements was hard to take.

135. An episode in many respects strikingly similar to that which we will now reconstruct, and occurring just one year later, was recounted by Norbert Wiener, *I Am a Mathematician* (Cambridge, Mass., 1966), pp. 83–85. Cf., Warren O. Hagstrom, *The Scientific Community* (New York, 1965), p. 95, *et passim,* for very suggestive, but ahistorical, discussions of many of the issues discussed in this paper, and Robert K. Merton, "Priorities in Scientific Discovery: A Chapter in the Sociology of Science," *American Sociological Review,* 22 (1957), 635–659, reprinted in Bernard Barber and Walter Hirsch, eds., *The Sociology of Science* (New York, 1962), pp. 447–485, who, under the rubric "The effort to safeguard priority" notes (p. 477) that "Numerous personal expedients have been developed: for example, letters detailing one's own ideas are sent off to a potential rival, thus disarming him."

136. The sentiment is Sommerfeld's. Sommerfeld to Landé, 3 Mar. 1921 (appendix, doc. 9).

Sommerfeld was alarmed. Remorseful for having published Back's data, and fearful of a stoppage of the flow of information from Tübingen to Munich, he was convinced that the only way to relieve Back's distress was to compel Landé to withdraw. He therefore brought all his weight to bear upon Landé, and even his maxim *pauca sed matura* was made to serve a less disinterested goal than the advancement of physics.[137] Writing to Landé on Thursday March 3rd (see appendix for text), Sommerfeld informed him that Back was working on the same problem and had a preliminary note in press. "I ask you now urgently," Sommerfeld enjoined, "to refrain from any publication of your rules until Back sends you the proofs of his full exposition which he is preparing for the *Annalen*." To Sommerfeld the key issue was the order of appearance. He quite explicitly warned Landé that when Back sends him proofs of his *Annalen* article then he, Landé, may publish what he pleases—but only in the *Annalen,* not in the *Zeitschrift für Physik,* in order that he not, thereby, get the jump on Back.

This last point was of some importance. In the archaic rules of the scientific game, dating back to the seventeenth century, priority meant precedence in discovery. Certifying this precedence had been an important function of scientific societies, and the 1904 statutes of the Deutsche Physikalische Gesellschaft still provided for the Chairman and Secretary to receive with seal and signature unpublished manuscripts read to the Society.[138] As late as 1912 Sommerfeld had protected Laue, Friedrich, and Knipping's priority in the discovery of the diffraction of X rays by crystals by depositing their statement of this discovery in a sealed envelope with the Bayerische Akademie der Wissenschaften a full month before the discovery was even announced to that body.[139] But in the postwar period such archaic practices seem to have become extinct[140]—the relevant provision was dropped from the 1919 statutes of the D.P.G.—and the conventional rules equated priority with precedence in the submission of

137. For the Latin motto, "few but ripe": Sommerfeld to Landé, 8 Dec. 1919 (SHQP 4, 13).
138. "Satzungen der Deutschen Physikalischen Gesellschaft," § 23, *Verhl. d. Dtsch. Phys. Ges., 6* (1904), 7.
139. Forman, "The Discovery of the Diffraction of X-Rays by Crystals; A Critique of the Myths," *Archive for History of Exact Sciences, 6* (1969), 38–71.
140. Except in France. See Norbert Wiener's account of his race against Lebesque and Bouligand in *Ex-Prodigy* (Cambridge, Mass., 1966), pp. 279–280.

work for publication, *in print*. Simple oral presentation, even before a scientific society, if not followed by publication of an abstract, seems not to have been regarded as establishing any formal claim to priority.

Yet at this time, as the number of physicists active in the field increased, and the pace of discovery quickened, priority tended to become, in practice, a question of the novelty or freshness of the publication. Thus the credit, so long as there is no explicit priority controversy, accrues to the physicist whose results are the first to come before the physical community. Precedence in date of appearance displaces precedence in date of submission when physicists attempt "to attract the maximum attention."[141] In the postwar period physicists were well aware of this circumstance, and even a naïve young man like Ralph Krönig knew enough to be "grateful to fate which has condemned the *Zeitschrift für Physik* to a long hibernation" during the winter 1924/25, thus allowing his article in *Die Naturwissenschaften* to appear earlier than the paper by his Munich competition.[142]

But to return to Sommerfeld's letter to Landé of March 3: Back, Sommerfeld explained, is the "experimental father" of the entire field and must be given the opportunity to pluck the fruits of his experiments himself. Sommerfeld could not bring himself to admit his sense of guilt and responsibility here, but instead tried to justify his own use of Back's results—while omitting to mention his publication of them—with the argument that "also before publishing my splitting rule I first got his consent, which he gave without reservation because he then had no theoretical perspectives whatsoever; rather, these have come to him only through that rule." He appealed to Landé's sympathy for the "expelled" Alsatian who has been physically and emotionally worn out by the war, maintaining that Back "needs quiet to work," and—what was certainly true—that Back was upset by Landé's parallel efforts.

141. Reif, "The Competitive World of the Pure Scientist," *op. cit.* (note 130), p. 1959.
142. Krönig to Goudsmit, 9 Feb. 1925 (SHQP 60, 4). Krönig's paper, jointly with Goudsmit: "Die Intensität der Zeemankomponenten," *Naturwiss., 13* (30 Jan. 1925), p. 90, dated 17 Dec. 1924. The competing paper which Krönig had in mind was, evidently, H. Hönl, "Die Intensitäten der Zeemankomponenten," *ZS. f. Phys., 31* (11 Feb. 1925), 340–354, received 26 Nov. 1924. Because of the reorganization of the *ZS. f. Phys.* no issues were published during Jan. 1925.

Then Sommerfeld took another tack, shifting from an appeal for courtesy toward Back to a depreciation of Landé's results. "Back has, in any case, already preceded you in publishing, and though his communication to *Die Naturwissenschaften* deals only with the number of components, everything else is given to be understood; on top of that Back seems to have better triplet rules than you; yours can't be regarded as final as they are too different from your doublet rules." Then, squeezed between the lines, "Back can also predict in advance everything regarding the triplets."[143] Finally, at the close of the letter Sommerfeld's tone softened in what was ostensibly an appeal to the higher interests of theoretical physics, but was equally a self-serving attempt to safeguard an invaluable source of experimental data:

> Don't be angry with me if I seem to wish to influence you in this matter. However, in the interest of theoretical physics, the trusting and confident cooperation with experiment—especially that in Paschen's institute —must not be disturbed. And I'm afraid that with Paschen's touchy personality and Back's sensitive temperament, a lasting dissonance can result if we don't proceed here with all possible delicacy.

But Sommerfeld's injunction to refrain from any publication was issued too late; in Marburg four days previously, on the 28th of February, Landé had described his "Regularities in the Anomalous Zeeman Effect" to the regional section of the Deutsche Physikalische Gesellschaft.

Until Sommerfeld's letter arrived on the 4th or 5th of March, Landé had no inkling that anything was amiss. On Wednesday, the 2nd, following the course which Sommerfeld had urged on him only a few days before, Landé had sent Paschen his predictions for the Zeeman patterns of the Bergmann (fundamental) series doublets. This letter, and Landé's doublet rules of 23 February, Paschen now showed to Back, inasmuch as Back had already returned the proofs of his article to *Die Naturwissenschaften*—it would have been awkward to keep the blindfold on very much longer. Landé's predictions for the fundamental series doublets agreed substantially with

143. Sommerfeld's letter again shows that he had no objection to the *form* of Landé's doublet rules and rather considered that the ripening of the fruit would yield analogous rules for the triplets.

Back's,[144] and that only increased Back's agitation, which was heightened still further on the following Monday, the 7th, by receipt of a letter from Landé.

Sommerfeld's letter of the 3rd both hit Landé a heavy blow and placed a heavy weight upon him. He learned that an unanticipated competitor had better rules and formal priority, while he would be expected to refrain from publishing a summary of his Marburg lecture, relinquishing all claims, even to an independent discovery of rules for the doublets. He had every intention of acting decently toward Back, but *he* was not responsible for the publication of Back's data, and he found it hard to believe that Paschen or even Back could expect him to sacrifice himself completely. On Saturday the 5th, within a day or two of his receipt of Sommerfeld's letter, Landé wrote Back seeking some basis for compromise. He evidently explained that he had already presented his rules at Marburg, and, as it appears from Back's reply, asked whether it was really Back's wish that he suppress the summary of this lecture which he had intended to publish in the *Verhandlungen* of the Deutsche Physikalische Gesellschaft. At the same time Landé apparently gave as his personal attitude that viewpoint which regards priority as a social convention socially adjudicated; thus the parties involved could with all detachment and dispassion publish their respective results and let the physics community decide any questions of priority which arise thereby. On the other hand, he, as much concerned as Back about the question of independence, suggested that they give the physics community a little guidance by each disclaiming in their respective preliminary publications knowledge of the other's work. For the future, however, he proposed that they follow the new cooperative strategy, join forces, and make their detailed expositions a joint publication.

Back, even had he sincerely wished to accept Landé's proposals, was strapped. His note in *Die Naturwissenschaften* was already too far advanced in the press to be amended; the full publication was to constitute a *Habilitationsschrift* and thus it could not be shared. He

144. Or so Paschen reported to Landé on 8 Mar. (appendix, doc. 11). Landé's rules would have predicted quite definitely that the Zeeman pattern of the $^2F_{5/2}$–$^2D_{5/2}$ line contains six parallel components but only ten perpendicular components, while it is not clear that Back, with his "Paschensche Regel," would have foreseen the absence of two perpendicular components.

replied immediately to Landé's letter (Monday, 7 March; see appendix for text). With characteristic self-effacement, he thanked Landé for the proposals, put off the question of joint publication, and assured him that "of course I consider it only right and proper that you print your Marburg Lecture, and indeed exactly as if you didn't know I existed. . . . I am entirely of your opinion that a possible conflict over priority ought to be decided by the physics community, if it has occasion and need to do so. I for my part will most certainly sharpen no weapons for it."

Technically, then, Back's letter left Landé's hands free, but there was still the question of Paschen's atitude. Paschen's earlier peculiarly nonpartisan position was confirmed in a letter to Landé the following day, Tuesday the 8th (see appendix for text). Thanking Landé for his letter of the 2nd with his prediction of the doublet fundamental series Zeeman patterns, Paschen mentioned the "important" fact that the prediction agreed with Back's, and then passed on to a number of scientific topics before returning, finally, to the problem of the regulation of their competitive efforts. Emphasizing the apparent differences in the "elements" of their respective rules—despite the identity of their results—Paschen suggested that Landé and Back each pursue their own approaches independently and, while acknowledging that the experimental data were "chiefly" due to Back, he did not suggest that Landé should defer to him. Instead, there was a note of annoyance at Back's tardiness in publishing—"In any case Back has been pretty much finished for some time already"—with perhaps a hint that it would not be wise to wait for him, "since he lays stress upon completely closing off the subject before publishing" and is therefore making all the experiments presently possible. Indeed, as we mentioned above, it was to be more than a year and a half still before Back submitted a full account of his experimental work to the *Annalen*.

Thus about the 10th of March, after some days of agitation and uncertainty, Landé supposed that the obstacles to his publication had been entirely removed. But he, too, was misled by Back's proclivity to self-abnegation. On Monday the 7th, when replying to Landé's letter, Back also wrote to Sommerfeld—or rather, we must suppose that such a letter was written—thanking Sommerfeld for taking his side so fully and forcefully, and also giving Sommerfeld an account of Landé's letter. Sommerfeld thus learned that Landé had already presented his rules at Marburg, that he was not prepared to acquiesce in Sommerfeld's prohibition of publication but intended to print a summary of the Marburg talk, and that he was pressing a proposal for joint publication upon Back. To Sommerfeld Back confessed the distress and anxiety which he had hidden from Landé. It was now certain that Landé would have both formal and effective priority for a *complete* set of rules—for the doublets at least—and this, Back feared, might also wreck his *Habilitationsschrift* as then conceived.

Sommerfeld was now really indignant; Landé had disregarded his earnest remonstrances. But there was another person whose representations Landé could not disregard—Landé's friend and *Ordinarius,* Max Born. On Tuesday the 8th Sommerfeld shot off a letter to Born (see appendix for text), charging that Landé was "again in the process of publishing something totally unripe," and worse that he had plucked this fruit out of the hands of the experimenter who had cultivated the vine. Born is to use his influence—which was, however, at just this moment appreciably diminished by his departure from Frankfurt—to prevent Landé from publishing a summary of his Marburg lecture, and also from importuning Back with the proposal of a joint publication. To Born the guilt and fears could be more openly and honestly stated: "All our wisdom (mine and Landé's) is based on Back's unpublished measurements. Earlier, when I approached him, he was quite agreeable to my using them; now however he is not at all." And Sommerfeld worries that if Paschen's or Back's liberality is misused, "the natural consequence is that Paschen will never again let us know what's going on in his institute." Here, however, he leaves out of account, in reckoning the probable consequences, Paschen's unusually acute perception of the necessity of cooperating closely with theory in the interest of experiment. Finally, Sommerfeld suggests that Born make Landé see that he is not behaving decently toward Back: "He obviously has simply no idea just how much he is damaging himself." Sommerfeld closes by inviting Born to show the letter to Landé.

Born, in the throes of his transfer to Göttingen, had gone off to Berlin. It would appear that Sommerfeld's letter caught up with him a few days later. Having heard nothing previously about the Back affair, he would undoubtedly have been puzzled. He apparently did not send Landé Sommerfeld's letter, but rather wrote Landé that Sommerfeld was very upset and that he, Born, must discuss that matter with Landé, but hoped to speak with Sommerfeld in Berlin before doing so.[145] Thus Landé, perhaps also puzzled (and annoyed) that Sommerfeld should have drawn Born into the affair, would still have been able to reassure himself with Back's and Paschen's letters that the publication of his Marburg lecture could have no untoward consequences.

But the fruits were no longer so sweet as they had been at the end of February. For Landé must have seen that if Back could also predict the doublet fundamental series Zeeman patterns—as Paschen had told him was the case—and had, in addition, adequate rules for the triplets—as Sommerfeld tauntingly assured him—the market value of his own rules would be disastrously depreciated. With this perception, and perhaps also because he could not have imagined just how cumbrous Back's rules really were, Landé finally turned his attention to fulfilling the requirements of the quantum theory.

6. A QUANTUM-THEORETICAL CONSTRUCTION, MARCH 1921

From very early in his work on the anomalous Zeeman effect Landé connected Sommerfeld's inner quantum number with his own ideas about the coupling of angular momenta, and identified this quantum number as that of the total angular momentum of the atom. He then applied this model to an earlier, and remarkably successful, attempt to dissect the splitting patterns into triplets, $\perp - || - \perp$, imagining these to arise, somehow, from the various allowed orientations of the total angular momentum in a magnetic field. He tried to measure the span of these triplets with the quasi-theoretical parameter k/r, the inner quantum number divided by the Runge denominator. All this involved the combination principle only very indirectly, the apparatus of energy levels even less, and that of selection rules not at all. Whatever the quantum theoretical component of these first attempts, the

145. This is conjectural, but is suggested by the last line of Landé to Sommerfeld, 17 Mar. 1921 (appendix, doc. 14).

exact rules of the second schema (23 Feb. 1921) required revisions for which Landé probably saw no theoretical justification. At the same time, this second schema for the doublets put him on the threshold of the solution if he only chose to seek a construction of the splitting patterns along the lines which Sommerfeld had suggested, and the principles of the quantum theory perhaps required, namely, "to assign to the terms s, p_i, d_i separate definite magnetic splittings, and then out of their combinations deduce the observed splitting patterns and their polarizations."[146]

When Landé first began to explore the anomalous Zeeman effect, he may have been discouraged from attempting an energy level analysis by Sommerfeld's blunt statement that he had had no success in this direction. Van Lohuizen's significant attempt to give an energy level analysis of the splitting pattern of each line, rather than of the line group, had been lost on Sommerfeld, and there is no reason to think that Landé was led to consult van Lohuizen's paper. In fact, Landé did not at first realize that an important feature of his empirical rules was that they referred to the individual lines. By the time he recognized this (23 Feb. 1921) he had gotten so far away from Sommerfeld's paper—as well as from his own theoretical conceptions —that it could perhaps not so easily strike him as the point which Sommerfeld had missed.

At this time, in mid-March 1921, Landé was still on vacation. Staying with his parents in Elberfeld, he had the leisure to reconsider the pieces at hand and what could be done with his anomalous Zeeman effect rules. On Wednesday evening, March 16th, upon returning home from a dull movie, it occurred to him "to try the combination principle."[147]

146. Sommerfeld, "Allgemeine spektroskopische Gesetze . . . , *op. cit.* (note 70), p. 253; *Ges. Schr., 3,* 555.
147. Sources for History of Quantum Physics, Interviews with Landé, 2nd session, 6 Mar. 1962: "I was in Frankfurt, and this paper [Sommerfeld's "Allgemeine spektroskopische Gesetze . . ."] interested me very much. I remember in 1920 I was at home during my Christmas vacation, in my home town with my parents, and I went to a movie. The movie apparently didn't interest me as much as these problems; and when I came back from the movie I had the idea to try the combination principle. It was just an idea out of nothing."
The disagreement between my dating and Landé's recollection of this illumination results from the fact that Landé had entirely forgotten the three months of empirical rules, and thus pushed the crucial event back into his *previous* visit with his parents during the Christmas vacation. That there is indeed a gap in Landé's

In Section 4, using the conceptual components available to Landé in December 1920, I offered a cogent—but largely mistaken—reconstruction of Landé's route to an energy level analysis of the anomalous Zeeman effect. Here, again, it is tempting to indulge in a reconstruction of Landé's line of reasoning on the night of March 16. And I will do so with the clear understanding that the role of the reconstruction is merely heuristic. The historian can do no more than approximate the tortuous route to a scientific discovery by drawing straight, "logical" lines between the few points on that route fixed by the documents available to him. It is probable, however, that the actual route is nowhere straight, and it is possible that using this method of linear interpolation over a period of but one night I am omitting as many significant excursions as when I applied it to a period of three months.[148]

Nevertheless, let me imagine that Landé sat down and, looking over the rules of his second schema giving the splitting patterns of the doublets, considered how he might introduce the combination principle—which is almost to say, how he might introduce an energy level analysis. Landé would immediately seize upon the superior half of Rule 4. Here he had obtained the position of the innermost $\parallel$ component by subtracting the "inner quantum fractions" of the combining terms, and thus the frequency shift of the spectral line was explicitly given as the difference of two "terms." Even if he had not been conscious of it before, Landé would now recognize that he had in each case subtracted the $k/r = k/(2n-1)$ of the term with the higher azimuthal quantum number from $k'/r' = k'/(2n'-1)$ of the term with the lower. Thus $p_1 d_1$ $[{}^2P_{3/2}-{}^2D_{5/2}]$ with inner quantum

recollection can be inferred from the fact that "I had the idea to try the combination principle" does not make sense when directly contrasted with Sommerfeld's paper, but is easily understood when applied to Landé's empirical rules. The precise date is not known, but, as I argue below, should be placed very close to the *terminus ad quem* established by Landé's letters to Bohr and Sommerfeld on Mar. 17.

This may serve to show the extreme caution that ought to be exercised in making use of recollections of this sort. They can supply valuable details if, and only if, one has, through contemporary documents, secure base lines from which to judge whether particular features are likely to be accurate representations, or projections and distortions.

148. If this be the case, then one seems forced to the conclusion that the reconstruction of the *line of reasoning* culminating in a particular scientific discovery is not a very profitable enterprise for the historian of science.

numbers (2, 3) and inner quantum fractions (2/3, 3/5) gave, subtracting "d_1" from "p_1," an innermost ‖ component at $2/3 - 3/5 = 1/15$. Since however the ‖ components are symmetrically placed, there is also one at $-1/15 = -(2/3 - 3/5) = 3/5 - 2/3$. Thus one obtains the same pair of parallel components regardless of whether one subtracts the "p" term from the "d" term or vice versa, which is simply to say that it doesn't matter whether one works with energy levels or spectroscopic term values.

Nonetheless, at this moment, when Landé was working with heightened sensitivity to the physical interpretation of his rules, we might expect him to use the energy level representation and subtract the "p" term from the "d" term, especially as only the transitions $d \rightarrow p$ produce observed series lines. This issue is slightly complicated by the fact that in the prevailing interpretation of the alkali spectra, for a given value of the total quantum number the d terms lay lower than the p terms. Bohr, however, had been freeing himself from this point of view in the course of the preceding year,[149] while Landé, even earlier, had found that his helium model—which indeed was really a model of an alkali atom—put the p terms below the d terms. It may be that this insight, which unfortunately Landé had not developed, played a role here.

We might therefore imagine that Landé, seeking a schematic expression that corresponded both to the increasingly popular energy level diagram and to the actual arithmetic operation, wrote:

$$
\begin{array}{ccccl}
d_1 & & -3/5 & 3/5 & \longleftarrow\ k/r \\
\downarrow & & \downarrow & \downarrow & \\
p_1 & & -2/3 & 2/3 & \longleftarrow\ k'/r' \\
& & \overline{\left(\dfrac{1}{15}\right)} & \overline{\left(-\dfrac{1}{15}\right),} &
\end{array}
$$

i.e., the positions of the innermost ‖ components. Since the ‖ components are equally spaced (as well as symetrically placed) the next pair of ‖ components lie at $(3 \cdot 3/5) - (3 \cdot 2/3)$ and $(-3 \cdot 3/5) -$

149. Bohr, "The Structure of the Atom and the Physical and Chemical Properties of the Elements," lecture in Copenhagen 18 Oct. 1921, translated in *The Theory of Spectra and Atomic Constitution, op. cit.* (note 40), on p. 96.

$(-3 \cdot 2/3)$. The schema can thus be extended to give all four $\parallel$ components:

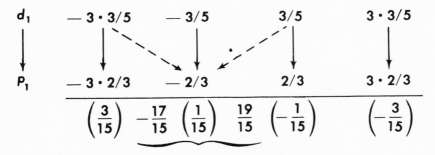

If now, playing with this schema, Landé drew in the diagonal dotted arrows, he would have found, *mirabile dictu,* $-17/15$ $(1/15)$ $19/15$, one of the $\perp - \parallel - \perp$ triplets into which he had originally decomposed the $(p_1 d_1)$ pattern. After that his task would be fairly clear. Each term must be assigned its quota of magnetic energy levels, and then the combinations must be given by a selection rule. In analogy with the normal Zeeman effect, that rule ought to be one restricting the equatorial quantum number, m, to $\Delta m = \pm 1$ or 0, especially as this rule carries with it a theoretical argument for the polarization of the splitting components. To achieve this, while allowing terms to combine symmetrically across the position of the unperturbed line, Landé would be compelled to introduce half integral m's.[150]

150. Consideration of the problem of the dashed (mutually canceling) perpendicular components—schema of 16 Feb., rule 1(a); schema of 23 Feb., rule 3(c)—suggests another intimate connection between Landé's triplet dissections and the precise form for an energy-level construction of the Zeeman patterns whose invention I am reconstructing. Thus if for the sp_2 combination of the doublets one writes:

$$P_2 \qquad [-3/3] \quad -1/3 \quad +1/3 \quad [+3/3]$$
$$\downarrow$$
$$s \qquad\qquad -1 \quad\quad +1$$

then the missing (dashed) perpendicular components are seen to fall, from left and right, at the null point (cf. Fig. 6, appendix, doc. 4). This is likewise the case for the other doublet combination involving equal total angular momenta, $p_1 d_2$, but *only* if one writes $\genfrac{}{}{0pt}{}{d_2}{\downarrow}{p_1}$, not $\genfrac{}{}{0pt}{}{p_1}{\downarrow}{d_2}$, which suggests that this particular ordering of the terms was of some importance in the discovery of the energy level analysis.

On Thursday, 17 March, Landé sent off to Bohr "an essential improvement" of his Zeeman effect rules for doublets:

> In order to construct the splitting pattern out of Sommerfeld's "inner" (better, probably, "total azimuthal") quantum numbers k, and the Runge term denominators r_k, according to the combination principle: There are for k and r_k the $2k$ *energy levels*
>
> $$\pm k/r_k, \quad \pm 3k/r_k, \quad \pm \ldots \pm (2k-1)k/r_k$$
>
> to which one assigns the "equatorial quantum numbers"
>
> $$\pm 1/2, \quad \pm 3/2, \quad \pm \ldots \pm (2k-1)/2 .$$
>
> Then the one and only rule runs thus: If the "equatorial" quantum number changes by ± 1, then $\perp$ components arise, if by 0, then $\|$ components.
> Example $(p_1 d_1)$, $(k/r_k, l/r_l) = (2/3, 3/5)$ thus

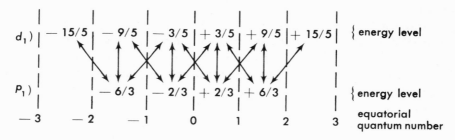

> The diagonal arrows give $\perp$ components, the vertical arrows $\|$ components, e.g., $2/3 - 3/5 = 1/15$, i.e., $\|$ component at $1/15$ the normal Zeeman splitting.[151]

Here then is the solution that Sommerfeld called for, reproducing the known doublet patterns and predicting those as yet unobserved on the basis of quantum numbers, energy levels, and selection rules, and unexpectedly familiar rules at that. The half integral equatorial

---

151. See appendix for text. I have attributed some importance to getting the lower total angular momentum state on the bottom in order that Landé's triplet dissection emerge by subtracting the lower from the upper. Here however Landé appears to contradict me by reading his schema "$2/3 - 3/5 = 1/15$," i.e., reversing the arrow. The explanation may be that he is simply showing Bohr how to read this schema in order to obtain the results as they were expressed in the examples which he had sent Bohr previously.

quantum numbers are put forward very tentatively; in the diagram they are represented as intermediate positions in the sequence of integral quantum numbers. The g factor is not explicitly introduced, and it is not in itself important in the discovery of the schema. Once one has the schema it will intrude itself naturally.

Later the same day Landé wrote a longer letter to Sommerfeld (appendix, doc. 14). Conscious of the different audience, Landé's phrasing is more cautious, less physical. The quotation marks are removed from the "inner" quantum numbers, and the parenthetical physical interpretation, "(better probably 'total azimuthal') quantum numbers," is omitted. Having hidden his interpretation of k as the total angular momentum, he expresses the half integral sequence of angular momentum components even more tentatively as "the ('equatorial' quantum–) numbers." He thus displays a good grasp of the difference between Bohr's approach and Sommerfeld's approach in atomic physics. Landé knows that for the former coherence with a theoretical viewpoint was overriding, while for the latter mere numerical regularity and formal elegance could be quite satisfying and convincing. In this instance he was well aware how little sympathy the *ad hoc* "inner" quantum number would find, as it stood, in Copenhagen, and at the same time he knew that a solicitous concern to extend the range of the inner quantum number, as such, would be met with warm interest in Munich.

At the same time Landé displays a sure grasp of the differences between Bohr's and Sommerfeld's personal styles. Here, in a sense, their roles are reversed, with Bohr insisting always on modest understatement in any suggestion of the ultimate significance of a piece of work, while Sommerfeld was liable to make and accept enthusiastic claims. Thus Landé sends to Bohr merely "an essential improvement" of his previous empirical rules, while he now confidently declares to Sommerfeld that "in the case of the doublets everything is set up in the simplest way for a comprehensible quantum theory of the anomalous Zeeman effects"—perhaps forgetting for a moment that he has kept Sommerfeld in the dark about his physical interpretation.

We might also note that in the letter to Sommerfeld Landé has inverted the diagram, placing the p_1 levels above the d_1. This too could perhaps reflect a lower estimate of Sommerfeld's concern with the physical interpretation of the schema. But as Landé holds to this inverted order in all his subsequent publications, we must simply consider the first form as a clue—soon to be covered over—of Landé's route to this schema.

When writing to Bohr, Landé had to confess that "for the triplets I'm only in a position to set up rules analogous to those I sent you previously." And indeed, superficially, the splitting of the triplet lines presented quite a different case. Because of the curious coincidence that the doublet g factors are numerically equal to $2k/r$, when Landé fastened on this parameter he had hold of an incipient term analysis for the splitting patterns of the doublet lines. But for the triplets k/r could simply not do him any good; and for a time (a very short time) it actually stood in the way of an energy level analysis of the triplet lines. However, in the course of the day he put aside his parameter k/r and obtained empirically the energy splitting of each triplet line. He explained to Sommerfeld that:

> In the case of the triplets one can proceed similarly, but one has to take as equatorial quantum numbers the *whole* numbers $0, \pm 1, \pm 2 \ldots$ (since here $\|$ components appear in the center of the pattern). One does not have a general and explicit rule to put in place of the [doublet] energy levels $(2n - 1) \cdot k/r_k$ given above, but rather each term has its own particular sequence of energy levels which, at present, can only be determined empirically. For example the term d_1 $(k/r_k = 3/3)$ has the (magnetic) energies $0, \pm 4/3, \pm 8/3, \pm 12/3$, with the equatorial quantum numbers $0, \pm 1, \pm 2, \pm 3$. Thus, as its basic energy level d_1 has $4/3$, while d_2 has $7/6$, d_3 has $1/2$; p_1, p_2 and p_3 have $3/2$, s has $2/1$.[152]

We note, parenthetically, that emerging naturally from this schema

---

152. Landé first wrote "statt der Energieniveaus k/r_k," and then inserted the factor $(2n - 1)$, where n is not the azimuthal quantum number, but any positive integer $\leq k$. The p_3 term really has $g = 0/0$, but as the total angular momentum is zero the term does not split at all.

It is interesting that in both these letters of 17 March Landé wrote the Runge denominator of a term with the subscript "k," which strongly suggests that when he wrote Bohr he was already convinced that r was not a function of the azimuthal quantum number alone, i.e., that in general there was not one Runge denominator for a complex term, but that each sublevel (characterized by a value of the inner quantum number or total angular momentum) had its own r. It would seem that *somehow* Landé came to this insight *before* he had acheived an energy level analysis of the Zeeman patterns of the triplet lines, but I am not able to offer a reconstruction of this stage in his thought.

is the "basic energy level" (*Grundenergieniveau*), which functions as a g factor.

Certain, now, that he had the key to the anomalous Zeeman splittings, Landé responds to Sommerfeld's taunt—"Back can predict everything regarding the triplets"—by denying that one can predict anything: "in the case of the triplets the material at hand does not, in my opinion, yet suffice for a general quantitative mastery of the magnetic energy levels of the terms, or for making predictions." Here we can also sense Landé's frustration at being unable to express these splittings by quantum numbers. In both his previous schemes it had been his program to construct the splitting pattern of a line from the quantum numbers and Runge denominators of the terms in question. Thus, as he believed he knew in all cases what these numbers were, there had been no difficulty before in predicting unobserved types. And it certainly must have been annoying to reflect that although he now had the key to the analysis of the triplet patterns, it had been purchased at the cost of that immediate connection with the quantum numbers of the combining states.

It was Back's competition, I have argued, which goaded Landé on to a solution of the problem. Perched now on a theoretical platform well above Back's reach, Landé felt not more, but less, secure. Anxiety is, of course, built into the system, and paradoxically, in such a way that the more unexpected the discovery the more acute the fear of being anticipated. But Landé had better reason; looking backward, he too could now see just how long a detour he had taken, just how short and straight the route could have been.

Continuing his letter to Sommerfeld, Landé describes the "very kind" answer he had gotten from Back, and mentions the "very friendly" letter from Paschen. Therefore, as he assures Sommerfeld, no ill feeing between theory and practice is to be expected, even if he publishes before Back. For, Landé says, "I'm in somewhat of a hurry with this because *Bohr* obviously is thinking about these things; and why should foreign countries forestall us in this." Omitting to tell Sommerfeld that he hoped to keep Copenhagen "fixed" by the same strategy he was using to hold off competition from Munich—namely, full and continuous communication of his results—he plays upon Sommerfeld's expectation that from Bohr would come the new ideas necessary for an explanation of the complex structure and the anomalous Zeeman effect by means of an atomic model.[153] These expectations Landé seeks to convert to apprehensions with the rhetorical question about "das Ausland." The mild mixture of political and scientific nationalism would evoke in almost all German physicists, and certainly in Sommerfeld, concern for the national stake in scientific priorities—a particularly acute concern at a time when the feeling was widespread that "there is one thing which no foreign or domestic enemy has yet taken from us: that is the position which German science occupies in the world."[154] Neither would it in any way serve Landé's present purpose to admit the distinction then commonly made between the northern European scientists who were striving against the exclusion of Germany from international science, and all the rest of the world which had defeated, and was now persecuting, Germany.[155]

Sommerfeld, however, was in the Tyrol on his annual spring vacation ski tour when Landé's letter arrived in Munich on the 18th or 19th,[156] and it came into his hands only when he returned at the end of

153. Cf. note 69. Perhaps we may also see here a suggestion of that particular convention regarding competition which is stated by W. L. Bragg in his "Foreword" to James D. Watson, *The Double Helix*, (New York, 1968), viii: ". . . the story is a poignant example of a dilemma which may confront an investigator. He knows that a colleague has been working for years on a problem and has accumulated a mass of hard-won evidence, which has not yet been published because it is anticipated that success is just around the corner. He has seen this evidence and has good reason to believe that a method of attack which he can envisage, perhaps merely a new point of view, will lead straight to the solution. An offer of collaboration at such a stage might well be regarded as a trespass. Should he go ahead on his own? . . . this difficulty has led to the establishment of a somewhat vague code amongst scientists which recognizes a claim in a line of research staked out by a colleague—up to a certain point. When competition comes from more than one quarter, there is no need to hold back." Thus although neither Landé alone, nor Sommerfeld alone, stood in precisely this relation ("investigator") to Back ("colleague"), together, *qua* German theoretical physics, they did; in any case, it was an acquiescence in this relation which Sommerfeld had been trying to force upon Landé. In reply, then, Landé appealed to the escape clause in the "code" which provided that inasmuch as competition from another quarter (Copenhagen) was imminent, he ought to be released from the obligation to "hold back."

154. Max Planck, addressing the *Preussische Akademie der Wissenschaften*, 14 Nov. 1918. *Sitzungsber.* (1918), p. 993. Also, Hans Hartmann, *Max Planck* (Thun, 1953), p. 47.

155. Brigitte Schröder, *Deutsche Wissenschaft und internationale Zusammenarbeit, 1914–1928* (diss. Geneva, 1966). Paul Forman, "Scientific Internationalism and the Weimar Physicists," *Science, War, and Internationalism, 1900–1939*, ed. Roger Hahn (in press).

156. On the 14th Sommerfeld wrote Einstein: "Jetzt aber gehe ich auf 14 Tage mit Laue, Lenz, Mie, etc. ins Gebirge, um womöglich Ski zu laufen." (Einstein and Sommerfeld, *Briefwechsel, op. cit.* [note 118], p. 80.) Sommerfeld left immediately, for he returned to Munich on the 28th (appendix, doc. 15).

the month. The opening lines of Sommerfeld's reply are best contrasted with his enthusiastic "Bravo" a month earlier in response to the far less significant quasi-empirical rules. "The beginning of your letter is striking proof of my contention that your ideas about the Zeeman effect were, up till now, not ripe for publication."[157] Sommerfeld completely disregards Landé's defenses. "If Paschen and Back write you that they have no objection to your publishing, that's very liberal of them. Formally you are in the right, but from a higher standpoint you may not forestall Back. I know Back would find it very disagreeable, and I will regard it as a direct affront." Sommerfeld discounts Bohr's competition, reminding Landé that Bohr had other things to publish just then, and adding a pointed hint that a real theorist, like Bohr, would not mix in problems of this sort. As for the appeal to German nationalism, Sommerfeld could not, of course, repudiate it, but simply declared, "I don't look upon him [Bohr] as a foreigner."

But there is yet another reason that Sommerfeld is unimpressed by Landé's results, as well as undeterred by his defenses. Knowing the kind of solution he wanted, Sommerfeld had set the problem before one of his students, a brilliant boy just finishing his first semester at the University. And in the course of March 1921 Werner Heisenberg had also come up with an energy level analysis of the anomalous Zeeman patterns: "Your new presentation is identical with that which a student of mine (1st semester) has found, but which has *not* been published."[158]

157. Ca. 31 Mar. (appendix, doc. 15).
158. Heisenberg, interview with SHQP, 1st session, pp. 3–6, 30 Nov. 1962, recollects that Sommerfeld gave him the problem about four weeks after the start of the semester—thus early December—telling him that his own "Allgemeine spektroskopische Gesetze" "was not important at all" and asking him "to find out what the initial and final states were using the selection rules." "So after a very short time, I would say perhaps one or two weeks, I came back to Sommerfeld and I had a complete level scheme." "Then Sommerfeld had received a letter from Landé telling him that he was interested, and in some way he was closer to the experiment, he had worked with Paschen and Back in Tübingen. So I think that I have not published on this and that only Landé has published on it, which was quite all right because he had started earlier." Accepting Heisenberg's dates, we would have to conclude that he had his level scheme by mid-December. And Sommerfeld's draft reply might be read to mean that Heisenberg had indeed found it months, not weeks, before the end of March. Nonetheless, the *sense* of Heisenberg's recollections—not to mention all that we know of Sommerfeld's behavior and reactions in February and early March—is more consonant with the assumption that Heisenberg was given the problem only at the *end* of the first semester, and very likely after Sommerfeld had learned of Landé's schema of 23

7. AFTERMATH, APRIL–SEPTEMBER 1921

Landé pressed on to prepare his work for publication. Now, of course, he must have been willing, and indeed anxious, to forego publication of a summary of his Marburg Lecture. A month earlier Landé felt he had to defy Sommerfeld on this point in order to establish the independence, if not priority, of his own empirical rules. But now, omitting that summary—to be sure, primarily with the intention of covering his tracks—could no longer serve to mollify Sommerfeld. In this situation it was only by *not* waiting for Back's *Annalen* paper that Landé could be sure that no one at Munich would get in ahead; Sommerfeld was obliged to see to that. Moreover, if some "krasser Fuchs" in Sommerfeld's school could solve the problem, then surely there was some substance to the fear that were he to delay he would undoubtedly be scooped. Sommerfeld, through his publication, and thus indirectly through his concern for priority, set in train a social process which he was then simply unable to call back. His inordinate fears for the fragility of the cooperation with Tübingen expressed his own feelings of guilt rather than any realistic analysis of what he could expect from Paschen. Indeed Paschen, who was anxious to get Back's interest and industry into experimental channels again, plainly told Sommerfeld not to overrate Back's contribution, and added: "For this reason I would be sorry if Landé's speculations were not published."[159]

The first paper, or rather the first part of the paper, was received

---

February. Such an interpretation can also be supported on the grounds that Heisenberg shows, in his interviews, a general tendency to recall things as having happened earlier than in fact they did. Despite Heisenberg's recollections, it does not seem possible to say exactly how Sommerfeld put the problem to him, or how Heisenberg cast his solution.

[Upon reading these considerations Professor Landé kindly wrote the author on 27 Nov. 1965 that, "Your conjecture . . . concerning the date of Heisenberg's participation is entirely correct. I positively remember that upon my written objection to Sommerfeld (where is the letter?) that H. found his identical results ('deckt sich') only *after* I had sent to him my very last schema, S. conceded grudgingly that this was indeed so, with the remark 'Why did you send me your results anyway?,' but conceded that the priority was mine (where is this letter?) and added: 'this pupil of mine will find other equally important things,' a very correct prediction."]

159. "Da Landé die Zeeman-Typen tiefer begründet und anders herleitet als Back, so sollte er seine Sachen veröffentlichen. Back ist mehr Experimentator und wird da wichtiges leisten. Seine Componentenregeln sind jedenfalls etwas un-

by the *Zeitschrift für Physik* on the 16th of April.[160] Landé could now afford to be generous toward Back and his index rules, "which make it possible also to predict previously unobserved new types." There is no hint that Landé had attempted anything of that sort. "The present paper is intended now to make the structure of the anomalous Zeeman types comprehensible from the standpoint of the combination principle and the quantum theory." Thus Landé even gives the impression that Back's empirical rules, as well as Back's observations, were the starting point of his own efforts—and, as I have argued, in a peculiar sense that was true.

In this first paper, or first part, the "basic energy level," stripped of its dimensions, made its appearance as the g factor.[161] The problem of expressing the splitting of the triplet terms by means of quantum numbers was only partially solved. Different expressions for the magnetic energy levels were obtained for each of the three members of

---

theoretisch. Landé muss sehr weit sein. Er hat mir Einiges über Neon gesandt, was richtig war.

"Was ich Ihnen früher über die Typen mittheilte, hätten Sie auch wohl mit einiger Kritik aus der damaligen Literatur zusammenfinden können. Back's Verdienst ist, die Ungenauigkeiten und Widersprüche in der Literatur beseitigt zu haben und die dort auch bezweifelte Preston-Regel für die I.N.S. [diffuse series] nachgewiesen zu haben. Popow hat übrigens kräftig mitgeholfen. Wenn Back langsam arbeitet, so liegt das daran, dass das Experimentiren nicht mit Meilenstiefeln geht, besonders, wenn man absolute Sicherheit gewinnen will. Er arbeitet auch jetzt hauptsächlich experimentell, um seine Schlüsse zu erhärten. So wird sein Werk etwas ganz anderes vorstellen, als Landé macht. Aus diesem Grunde würde ich bedauern, wenn Landés Speculationen nicht veröffentlicht würden. In letzter Linie hat ja doch eine wichtige Thatsache sicher begründet mehr Wert als alle Speculationen, die uns wohl zu neuen Auffassungen anregen können, die aber doch bald fadenscheinig werden." Paschen to Sommerfeld, 21 May 1921 (SHQP 33, 1).

160. Landé, "Über den anomalen Zeemaneffekt (Teil I)," *op. cit.* (note 101).

161. It is Professor Landé's impression—letter to the author, 27 Nov. 1965—that he chose the letter "g" as an abbreviation for gyromagnetic. It should be noted, however, that at this time the gyromagnetic ratio was already a well-defined concept, proportional to the reciprocal of the magneto-mechanical factor, g, and represented by the symbol "r" or "ρ." A more likely etymology would point to either one or both of the following sources.

1) The term which Landé used in reporting his discovery to Sommerfeld: "Grundenergieniveau." In *Fortschritte der Quantentheorie* (Dresden, 1922), p. 52, Landé introduces g, bereft of its units, as a "Grundfactor" characteristic of the spectroscopic term.

2) Sommerfeld, in "Allgemeine spektroskopische Gezetze . . . ," *op. cit.* (note 70), p. 253, had written the Runge denominator of a term as $r = g(n)$, g a function to be determined. In "Anomaler Zeemaneffekt und Seriensysteme bei Ne und Hg," *op. cit.* (note 112), which in a number of respects is the publication most revelatory of his earlier efforts, Landé introduces g as "a function of n and k which is to be determined more exactly below," and subsequently writes it $g(n, k)$.

a triplet term; that is, for each of the three possible relations between the values of k and n for the triplets $(k = n, n - 1, n - 2)$. Thus although he lacked a comprehensive formula, Landé must have overcome his doubts about the sufficiency of the material, for he could now predict as yet unobserved combinations. From a theoretical viewpoint this was still not very satisfying. It may well be that a good part of the month which elapsed between the discovery of the quantum-theoretical analysis and the submission of this first paper was spent searching for a comprehensive formula, as well as for a model from which the g factor itself could be derived. In any case, Landé promised that in the second part, which was supposed to follow quickly, he would give an explanation of the physical significance of the g factor and derive the disparate expressions for g from a single unified formula. He was overly optimistic, and soon found himself simply unable to provide a Part II such as he had promised.[162]

Landé's model of coupled angular momenta, although it was critically important in the early stages of his thought about the anomalous Zeeman effect, had slipped into the background before the end of February. Indeed, its initial utility had been in good part due to a "misapplication" of the quantum theory. Correctly interpreted, the only property of the model which could be of any help in the anomalous Zeeman effect was the identification of Sommerfeld's inner quantum number with the total angular momentum. And this was the one point which Landé very tentatively put into print,[163] without, however, explicitly drawing in his picture of coupled valence electrons with separately quantized angular momenta, or suggesting that here lay an explanation of the complex structure of spectral lines. It remained for Heisenberg to invent a particular "misapplication" of mechanics which would make the

162. Landé did submit, after a six-month delay, a paper entitled "Part II" (ZS. f. Phys., 7, 398–405), but it was not the paper which he had envisaged in "Part I." In a real sense "Part II" was "Termstruktur und Zeemaneffekt der Multipletts," ZS. f. Phys., 15 (1923), 189–205, received 5 Mar. 1923; translated as "Term Structure and Zeeman Effect in Multiplets" in W. R. Hindmarsh, Atomic Spectra (Oxford, 1967), pp. 185–205.

163. Landé, "Über den anomalen Zeemaneffekt (Teil I)," op. cit. (note 101), p. 234.

model give doublets and triplets and the anomalous Zeeman effect,[164] even while the one point which Landé ventured to publish continued to be regarded with scepticism.[165] A couple of years later, when the model was generally accepted, the best that Landé could do by way of priority was to cite his ". . . räumliche Orientierung . . ." paper of 1919.

At Tübingen, meanwhile, Paschen and his students were applying Landé's energy level analysis and g factor formulae to a variety of new cases. A particularly interesting one was the forbidden transitions $p_2 d_1 [^2P_{1/2} - {}^2D_{5/2}]$ of the composite doublets, and $p_2 d_1 [^3P_1 - {}^3D_3]$ and $p_3 d_2 [^3P_0 - {}^3D_2]$ of the composite triplets. During the spring of 1921 Paschen had discovered that a moderately strong magnetic field—i.e., strong enough to produce some overlapping of the splitting patterns of the members of the line group—caused these previously unobserved combinations to appear. With some difficulty he managed to involve Back in the measurements, and together they were able to determine the Zeeman patterns of these forbidden lines. "The Zeeman effect," Paschen wrote Sommerfeld on 24 July, "is in all three cases *exactly* that which results from Landé's new rules."[166]

The credit which Landé thereby gained was all the greater in that Sommerfeld—on what we must suppose were personal as well as physical grounds—was unwilling to acknowledge or adopt the key

164. Heisenberg, "Zur Quantentheorie der Linienstruktur und der anomalen Zeemaneffekte," *ZS. f. Phys., 8* (1922), 273–297, received 17 Dec. 1921.

165. Landé discussed the question with Wolfgang Pauli at the Jena *Physikertag* at the end of September. After thinking the question over a good bit, Pauli wrote Landé that the great bulk of the evidence argues for the *old* conception that n, not k, is the total angular momentum quantum number. Pauli to Landé, 5 Oct. 1921 (SHQP 4, 12).

166. "Der Zeeman-Effekt ist in allen 3 Fällen *exact* der, welcher sich aus Landé's neuen Regeln ergiebt sowohl nach Lage wie nach Stärke der Componenten.

"Da ich Landé, der neulich hier war, dieses mitgetheilt habe, er auch die magnetischen Typen hier gesehen hat, möchte ich Ihnen, als dem Vater der inneren Quanten und der Landéschen Regeln dieses ebenfalls mittheilen, bitte Sie aber, jetzt noch nicht darüber zu sprechen, da es erst im Herbst veröffentlicht werden kann (von Back und mir). . . . Es ist also jedenfalls die formale Darstellung von Landé richtig, und man wird seine theoretischen Erläuterungen nur ein wenig besser zu begründen haben, um den Mechanismus der Dublet- resp. Triplet-Bildung zu übersehen." (Paschen to Sommerfeld, 24 July 1921 [SHQP 33, 1].) The delay was due to the commitment of the publication to the issue of the Dutch journal *Physica* commemorating Zeeman's discovery: Paschen and Back, "Liniengruppen magnetisch vervollständigt," *Physica, 1* (31 Oct. 1921), 261–273.

novelties in Landé's energy level analysis. Throughout 1921 Sommerfeld would have none of Landé's half integral equatorial quantum numbers and his identification of the inner quantum number with the total angular momentum. Such proposals, he told Paschen, seemed to him to be "downright silly" [*recht töricht*].[167] More important, he could not accept Landé's triplet *g* factors, for they necessarily involved the abandonment of the idea of a Runge denominator of a complex term, and thus also of Sommerfeld's own splitting rule. Sommerfeld predicted a Runge denominator of 3 for the splitting pattern of the line group resulting from transitions between triplet *d* terms ($r = 3$) and singlet *P* terms ($r = 1$) in the alkaline earths. Landé's rules, on the other hand, predicted 6, the least common multiple of the denominators of the *g* factors of the three *d* terms. Precise measurements were not at hand, but Paschen was fairly certain that Landé was right. Then Paschen's gifted student, Raimond Götze, analyzed the so-called skewsymmetric combination groups in these elements. As Runge denominator of the *dd′* line group he found the 6 which followed from Landé's analysis, not the $3 \times 3 = 9$ demanded by Sommerfeld's splitting rule. Paschen insisted to Sommerfeld in September that "there is no doubt about the correctness of the measurements or the distinctness of the test. *9 is impossible, 6 is positively identified!*" "It is . . . decisive in favor of Landé." "The advance is a great one."[168]

---

167. Draft of Sommerfeld's reply to Paschen's letter of 24 July 1921 (SHQP 33, 1). Sommerfeld avoided both in the third edition of *Atombau* . . . (Braunschweig, 1922), pp. 482–484, preface dated Jan. 1922.

168. "Ich halte überhaupt Alles für richtig, was Landé angegeben hat [concerning *Pd₂*], und zwar auch die Intensitäten. Der Fortschritt ist ein grosser." (Paschen to Sommerfeld, 13 Sept. 1921 [SHQP 33, 1].) "Der Runge-Nenner in der schiefsymmetr. Gruppe, welche eine Combination $d_i d_j$ ist, ist 6 in Uebereinstimmung mit Landé und nicht $3 \times 3 = 9$. Dies hatte ich vergessen, in meinem letzten Briefe zu erwähnen. Es ist aber entscheidend für Landé . . . es ist kein Zweifel an der Richtigkeit der Messungen oder der Deutlichkeit der Versuche. *9 ist unmöglich, 6 ist sicher erkannt!*" (Paschen to Sommerfeld, 15 Sept. 1921 [SHQP 33, 1].) Despite these rebuffs it was still some time before Sommerfeld could bring himself to abandon his Runge denominators. E.g., *Atombau* . . . (3rd ed.; 1922), pp. 480–491, and "Über die Deutung verwickelter Spektren (Mangan, Chrom usw. nach der Methode der inneren Quantenzahlen " *Ann. d. Phys.*, *70* (18 Jan. 1923), 32–62, received 20 Aug. 1922, on p. 39; *Ges. Schr., 3*, 682.

APPENDIX: CORRESPONDENCE OF FEBRUARY AND MARCH 1921
BEARING ON LANDÉ AND THE ANOMALOUS ZEEMAN EFFECT*

	ca.	1 Feb., Landé to Paschen (see nr. 2).
1.		4 Feb., Landé to Bohr.
2.		5 Feb., Paschen to Landé.
3.		14 Feb., Bohr to Landé.
4.		16 Feb., Landé to Bohr.
	ca.	16 Feb., Landé to Sommerfeld (hypothetical).
5.		21 Feb., Landé to Bohr.
6.		[2]3 Feb., Landé to Bohr.
	ca.	23 Feb., Landé to Sommerfeld (see nr. 7, 9, 10, 11).
7.		25 Feb., Sommerfeld to Landé.
	ca.	26 Feb., Sommerfeld to Paschen (see nr. 7, 9, 10, 11).
8.		28 Feb., Kramers to Landé.
	ca.	1 Mar., Back to Sommerfeld (see nr. 9).
		2 Mar., Landé to Paschen (see nr. 10, 11).
9.		3 Mar., Sommerfeld to Landé.
		5 Mar., Landé to Back (see nr. 10, 11).
10.		7 Mar., Back to Landé.
		7 Mar., Back to Sommerfeld (hypothetical).
11.		8 Mar., Paschen to Landé.
12.		8 Mar., Sommerfeld to Born.
	ca.	12 Mar., Born to Landé (hypothetical).
13.		17 Mar., Landé to Bohr.
14.		17 Mar., Landé to Sommerfeld.
	ca.	20 Mar., Sommerfeld to Landé (see nr. 15).
		21 Mar., Landé to Sommerfeld (see nr. 15).
15.	ca.	31 Mar., Sommerfeld to Landé.
		Apr., Landé to Sommerfeld (see note 158).
		Apr., Sommerfeld to Landé (see note 158).

*The numbered items are here printed *in extenso*. The unnumbered items have not been located, and are probably no longer extant.

[1]
A. LANDÉ to N. BOHR   4 February 1921
Handwritten (modern cursive) letter (BSC 4, 3).
Envelope not extant.

4. II. 21.

Sehr verehrter Herr Professor Bohr!

Beiliegend sende ich Ihnen die erste Korrektur meiner Arbeit über
den Diamanten; auf der Seite 8 steht ein Hinweis auf Ihre in Aussicht
gestellte Begründung des Ellipsenzustands.[169] Die Rechnungen, besonders die Störungseinflüsse der Nachbaratome, sind jetzt nicht
mehr so willkürlich, wie ich es damals in Kopenhagen vortrug.[170]
Ausserdem hat sich manches geändert durch genauere Messungen der
Sublimationswärme des Diamanten, und an Stelle der vollkommenen
Einphasigkeit des ganzen Gitters lasse ich jetzt die C-Atome gegen
ihre nächsten Nachbarn 180° Phasendifferenz besitzen. Ich bin sehr
gespannt, ob Sie mit Hilfe der Ellipsenbahnen $n = 1$, $n' = 1$ Fortschritte über Atombau und Spektren gemacht haben, und, vor allem,
ob Resultate zum Verständnis des Zeemanneffekts erreicht sind.
Über die komplizirten Typen des Zeemanneffekts habe ich einige
empirische Regeln gefunden, die über Sommerfelds Zusammenstellung (Ann. d. Phys.) erheblich hinausgehen und Voraussagen für
das Neonspektrum gestatten; aber was diese Regeln bedeuten, ist mir
ganz unfasslich.

Mit den besten Grüssen
Ihr Ihnen sehr ergebener
A. Landé

169. Landé, "Über die Kohäsionskraft in Diamanten," *op. cit.* (note 63), par. 3.
170. On 15 October 1920 Landé addressed the Fysisk Forening on "Würfelatome" (Record book, SHQP 35).

[2]
F. Paschen to A. Landé   5 February 1921
Handwritten (modern cursive) letter (SHQP 4, 11).
Envelope postmarked: Tübingen
                        -6.2.21.8-9V
      addressed: Herrn Professor Dr. A. Landé
                 Frankfurt a.M.
                 Institut f theoret Physik.
                 Robert Mayerstr. 2

                                        Tübingen 5. II. 21
Sehr verehrter Herr Kollege.

Der Zeeman-Typus p_1d_2 der I. N. S. gewöhnlicher Triplets sieht so aus:

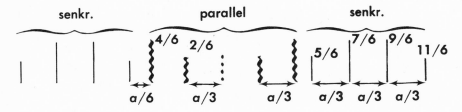

Es ist wohl ausgeschlossen, dass die starke Componente $\perp$ 9/6 bei 13/6 liegt. Der Typ ist so oft und so sicher beobachtet, schon von Runge und mir 1900, dass ein Irrtum kaum möglich sein kann.

Auch die Zeeman-Typen im Neon Spectrum sind nicht zu ändern. Sie wurden neuerdings von T. Takamine und K. Yamada Proc. of the Tokyo Math-Physic. Soc. II Ser. 7 p. 277 1913/14 exact so wiedergefunden. Ein Triplet $3/2\,a$ bei der Schärfe der Neonlinien ist nicht zu verkennen. Obige Arbeit geht nur unwesentlich über Lohmann's[171] hinaus und bringt kaum weitere Gesichtspunkte. Lohmann's Angaben sind fast durchweg bestätigt, und was ich davon verwertet habe,[172] ist nicht zu ändern.

Ich habe gehört, dass Sie auch für das gewöhnliche Heliumspectrum

171. Wilhelm Lohmann, *Beiträge zur Kenntnis des Zeeman-Phaenomens* (Diss. Halle, 1907), 74 pp., as given in the *Jahresverzeichnis der an den dtsch. Hochsch. erschien. Schriften, 22* (1907), 248.

172. Paschen, "Das Spektrum des Neon" and ". . . (Nachtrag)," *op. cit.* (note 58).

Ihr altes Modell verlassen haben und eine Art Würfelsymmetrisches Modell dafür genommen haben.[173] Es würde mich sehr interessiren, ob Ihre Modelle irgend etwas über den Stark-Effect der verschiedenen Heliumserien ergeben resp. schätzen lassen. Soweit ich sehe, würde nach dem alten Modell für alle Serien ungefähr ein gleich starker electrischer Effect herauskommen wenigstens bei gleicher Ordnungs- zahl des subtractiven Terms. Es scheint mir, dass der Stark-Effect wichtige Hinweise für das Modell der verschiedenen Serien giebt, die bisher nicht beachtet sind. Für Helium habe ich die Stark-Effecte qualitativ bis zu hohen Ordnungszahlen und kann sie für die ver- schiedenen Serien vergleichen. Daher würde mich sehr interessiren, was nach Ihrer Theorie ungefähr abzuschätzen ist.

Ich höre, dass Langmuir in America schon lange andere Modelle bearbeitet, als die gewöhnlichen Bohrschen. Näheres allerdings habe ich nicht erfahren.

Mit freudlichem Gruss
Ihr hochachtungsvoll ergebener
F. Paschen.

[3]
N. Bohr to A. Landé   14 February 1921
Typewritten letter (SHQP 4, 1).
Envelope not extant.

UNIVERSITETETS INSTITUT   BLEGDAMSVEG 15,
FOR   KØBENHAVN Ø.
TEORETISK FYSIK   DEN 14. Februar 1921.

Lieber Dr. Landé,

Vielen Dank für Ihren Brief und die Korrektur Ihrer Arbeit über die Kohäsionskraft im Diamanten,[174] die mich natürlich sehr interessiert hat. Zur selben Zeit muss ich Sie um Verzeihung bitten, dass ich nicht früher auf Ihre freundliche Karte[175] geantwortet habe. Auch wir in

173. No application by Landé of the cubic atom model to the helium spectrum is known to me.
174. Doc. 1.
175. Landé to Bohr, 26 Oct. 1920 (BSC 4, 3).

Kopenhagen waren sehr froh über Ihren Besuch und hoffen, dass es gelingen wird Sie zum Herbst wieder hierher zu bringen. Wie Sie wissen, war ich ja selbst besonders in Ihren Arbeiten interessiert, und in der Zwischenzeit habe ich mich viel mit den Problemen beschäftigt, über die wir gesprochen haben. Was die in Ihre Arbeit gemachten Annahmen anbelangt und Ihren freundlichen Hinweis auf einer Publikation von mir, muss ich jedoch gestehen, dass meine Ansichten über diesen Punkt sich in mehrfacher Weise geändert und, wie ich denke, geklärt haben. Ich denke jetzt, dass man nicht allein, wie ich Ihnen damals mitteilte, eine ungezwungene Erklärung für das Auftreten mehrkvantiger Bahnen im normalen Zustand des Atoms erreichen kann, sondern eine Weiterbildung desselben Argumentes hat mich in ungezwungener Weise zu der Annahme einer ganz bestimmten Anordnung dieser Bahnen im Atome geführt. Die Anordnung zeigt wohl eine ausgesprochene räumliche Symmetrie, gehört aber nicht zu der von Ihnen vorgeschlagenen Bewegungstype, sondern möchte am besten dadurch characterisiert werden, dass jedes Elektron im Atome sich im gewissen Masse unabhängig von den anderen bewegt. Ich habe diesen Gesichtspunkt zu einer Theorie des Atombaus entwickelt, wie ich in einem, zur "Nature" eben gesandten Briefe,[176] von dem ich Ihnen eingeschlossen eine Kopie des Manuskriptes schicke, angedeutet habe, und ich beabsichtige die näheren Resultate, die von ziemlich detailliertem Character sind, in einer ausführlichen Publikation baldigst mitzuteilen, die ich Ihnen schicken werde, so bald es fertig ist.[177]

Mit freundlichsten Grüssen von uns allen hier und auch von Professor Franck, der sich hier gelegentlich aufhält um durch seine Ratschläge den Anfang der experimentellen Arbeiten im neuen Institut freundlichst zu unterstützen,

Ihr sehr ergebener,
N. Bohr

---

176. Bohr, "Atomic Structure," *op. cit.* (note 67).

177. Bohr suffered a minor breakdown in the spring of 1921, absented himself from the Solvay Congress in April, and published nothing further until the fall of 1921: "Atomic Structure," *Nature, 108* (13 Oct. 1921), 208–209, dated 16 Sept., and the work cited in note 149. By then Bohr had apparently forgotten this promise to Landé.

[4]
A. Landé to N. Bohr   16 February 1921
Handwritten (modern cursive) letter and enclosures (BSC 4, 3).
Envelope not extant.

Frankfurt 16. II. 21.

Sehr verehrter Herr Professor Bohr!

Bei der Lectüre von Sommerfelds Zusammenstellung „allgemeiner spektroskopischer Gesetze und magnetooptischer Zerlegungssatz" (Ann. d. Phys. 63, 121, 1920) sind mir noch die beiliegenden empirischen Regeln aufgefallen, die wohl durch die Figuren im Vergleich mit Sommerfelds Figuren 4a, 4b und seiner Tabelle 2, 3, 4, 5 verständlich sind.[178] Ich glaube, dass besonders Regel 1, dass sich das komplizierte Zeemannbild in einzelne Gruppen s-p-s auflösen lässt, für eine Theorie des Zeemanneffekts nicht unwichtig ist. Die verschiedenen „inneren" Quantenzahlen eines Terms werden wohl einfach die Gesamtquantenzahlen des Atoms um seine invariable Achse bedeuten bei verschiedener räumlicher Orientierung der Valenzelektronen um das Atominnere, und die Mehrfachheit der Gruppen s-p-s in einem Zeemannbild bedeutet vermutlich mehrere Orientierungen dieser invariablen Atomachse gegen das Magnetfeld an. Ich bin sehr gespannt, wie sich die Entscheidung über das Diamantmodell wenden wird, nachdem die Frage der Bindungsringe röntgenographisch fortwährend schwankt.[179]

Mit den besten Grüssen
Ihr Sie sehr verehrender
A. Landé.

178. See the diagram corresponding to note 92, and Fig. 4.
179. Hans Thirring had made a competing calculation of the lattice energy of diamond, *op. cit.* (note 64), adopting Bohr's original (note 11) hypothesis that homopolar binding arose from the detachment of the valence electrons from both atoms, which electrons then moved in quantized orbits (*Bindungsringe*) between the two nuclei. This mechanism for the cohesive forces in diamond—and, by extrapolation, in other homopolar crystals—had been in disrepute for the past couple of years since Debye and P. Scherrer, "Atombau," *Phys. Zeits.,* 19 (1 Nov. 1918), 474–483, claimed to have proved their nonexistence by X-ray diffraction studies. Recently, however, Dirk Coster, "On the rings of connecting-electrons in Bragg's model of the diamond crystal," *Amsterdam Acad., Proceedings, 22* (1920), 536–541, had criticized Debye and Scherrer's calculations, and had argued that the X-ray diffraction data were better accounted for on the assumption of the presence than of the absence of *Bindungsringe*.

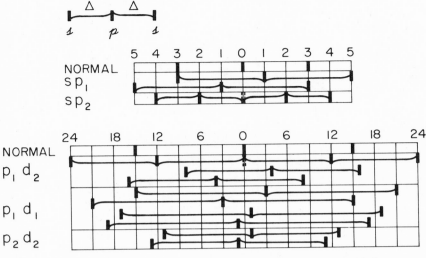

FIGURE 6. Dissection of Zeeman Patterns of Doublet Lines. The splitting pattern of each line is decomposed into a set of Lorentz triplets of equal but anomalous width. *Re* the dashed ("mutually canceling") components see note 150. This and the following figure reproduce in all essentials the diagrams, drawn roughly by hand on a sheet of graph paper, enclosed along with the "empirischen Regeln" in this letter, nr. 4.

1) Ist die Lage aller p-Komponenten und die Lage der äussersten s-Komponente gegeben, so kann man die Lage der übrigen s-Komponenten angeben. Jede p-Komponente ist nämlich rechts und links von zwei s-Komponenten im Abstand $\pm\Delta$ umgeben, wobei Δ der Abstand der äussersten p-Komponente von der äussersten s-Komponente ist.

   a) Scheinbare Ausnahme tritt ein, wenn nach dieser Regel zwei s-Komponenten sich überdecken sollten (vergl. die punktierten s-Komponenten im Zentrum des Zeemann Bildes der (sp_2)- und (p_1d_2)-Dublettlinie). Die beiden s-Komponenten löschen sich dann gegenseitig aus.

   b) Wirkliche Ausnahme: Regel 1 gilt in 13 von 14 Fällen, versagt aber bei der Triplettlinie (p_1d_2), deren Zeemannbild sich nicht in Einzelgruppen s-p-s mit konstantem Δ auflösen lässt.

   c) Dagegen gilt in allen 14 Fällen, dass die Anzahl der s-Komponenten doppelt so gross ist als die Zahl der p-Komponenten, wenn man die nach (a) sich auslöschenden s-Komponenten mitzählt.

243

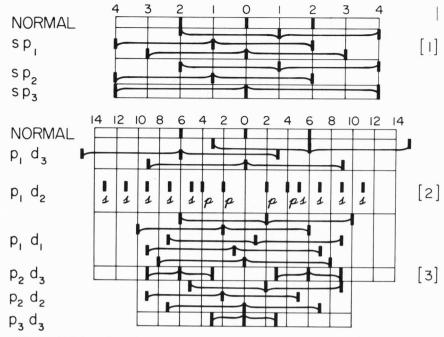

FIGURE 7. Dissection of Zeeman Patterns of Triplet Lines.

1. sp_2 is really (1) 3 4/2 and thus presents the same problem as p_1d_2, namely, the absence of the central | | component, $m = 0 \rightarrow m' = 0$, and the outermost $\perp$ components when $j = j'$.

2. The extra $\perp$ component at $-13/6$ is a residue of earlier attempts to alter the pattern. See document nr. 2.

3. Possible because of a numerical coincidence.

2) Für alle *Triplett*linien gilt die Regel: Die Anzahl der p-Komponenten ist gleich der Summe der „inneren" Quantenzahlen der aufbauenden Terme (doppelt so gross also nach 1c die Zahl der s-Komponenten). Für die Dublettlinien gilt die Regel nur nahezu ($\sim$ statt $=$).

3) Im Zentrum des Zeemannbildes tritt bei den Dubletts eine p-Komponente niemals auf, bei den Tripletts dann und nur dann, wenn die beiden aufbauenden Terme verschiedene „innere" Quantenzahlen haben.

4) Die Grösse Δ erhält man *angenähert* aus den beiden „inneren" Quantenzahlen der aufbauenden Terme, indem man jede von ihnen durch ihren besonderen Runge'schen Nenner dividiert und die beiden Quotienten addiert.

Term-benennung		Innere Qu. zahlen		Summe der inn. Qu. z.		Anzahl der p-Komponenten	Innere Qu. z. durch Runge'schen Nenner dividiert und summiert		$\frac{\Delta}{s}$ $\frac{\Delta}{p}$ $\frac{[180}{s}$	
D u b l e t t s	sp_1	1	2	3	$\sim$	2	$1/1 + 2/3 =$	$5/3$	$\sim$	$4/3$
	sp_2	1	1	2	$\sim$	2	$1/1 + 1/3 =$	$4/3$	$\sim$	$2/3$
	p_1d_2	2	2	4	$\sim$	4	$2/3 + 2/5 = 16/15$		$\sim$	$12/15$
	p_1d_1	2	3	5	$\sim$	4	$2/3 + 3/5 = 19/15$		$\sim$	$18/15$
	p_2d_2	1	2	3	$\sim$	2	$1/3 + 2/5 = 11/15$		$\sim$	$12/15$
T r i p l e t t s	sp_1	1	2	3	$=$	3	$1/1 + 2/2 =$	$6/2$	$\sim$	$3/2$
	sp_2	1	1	2	$=$	2	$1/1 + 1/2 =$	$3/2$	$\sim$	$3/2$
	sp_3	1	0	1	$=$	1	$1/1 + 0/2 =$	$2/2$	$\sim$	$4/2$
	p_1d_3	2	1	3	$=$	3	$2/2 + 1/3 =$	$8/6$	$\sim$	$9/6$
	p_1d_2	2	2	4	$=$	4	$2/2 + 2/3 =$	$10/6$	$\sim$	?
	p_1d_1	2	3	5	$=$	5	$2/2 + 3/3 =$	$12/6$	$\sim$	$8/6$
	p_2d_3	1	1	2	$=$	2	$1/2 + 1/3 =$	$5/6$	$\sim$	$3/6$
	p_2d_2	1	2	3	$=$	3	$1/2 + 2/3 =$	$7/6$	$\sim$	$7/6$
	p_3d_3	0	1	1	$=$	1	$0/2 + 1/3 =$	$2/6$	$\sim$	$3/6$

[5]
A. LANDÉ to N. BOHR   21 February 1921
Handwritten (modern cursive) lettercard (BSC 4, 3).
Lettercard postmarked: Frankfurt (Main)
21.2.21.5-6N
addressed: Herrn Prof. N. Bohr
[following lines excised]

Frankfurt 21. II. 21.

Sehr verehrter Herr Professor Bohr!

Herzlichen Dank für die Übersendung Ihres Manuskripts,[181] das ja für die künftige Atomtheorie von der allergrössten Bedeutung sein wird. Überhaupt scheint es mir, dass bis zum Erscheinen Ihrer ausführlichen Mitteilung es gar keinen Sinn hat, in der Atomtheorie weiter theoretisch zu arbeiten. Deshalb möchte ich Ihnen den Vor-

---

180. Landé's Δ's are equal to the g factors of the higher total angular momentum states. Where the angular momenta of the two terms are equal, Δ is equal to the g factor of the higher azimuthal quantum number term.
181. See document nr. 3.

schlag machen, mir Ihren Brief an die „Nature" einige Zeit zu überlassen, damit ich ihn ins Deutsche übersetze und bald möglichst an die Ztschr. für Physik gelangen lasse, sodass auch die deutschen Physiker möglichst schnell auf Ihre neuen Fortschritte hingewiesen werden. Ich erlaube mir also, Ihr Manuskript bis auf weitere Nachricht noch hier zu behalten, und grüsse Sie herzlich.

Ihr Ihnen sehr ergebener
A. Landé.

Nicht mehr Odenwaldschule,
sondern *Frankfurt am Main*
Institut für theor. Physik, Robert Mayerstr. 2

[6]
A. LANDÉ to N. BOHR [2]3 February 1921
Handwritten (modern cursive) letter and enclosure (BSC 4, 3). Dated
    [13] Feb 21 in *SHQP Report and Inventory*.
Envelope not extant.

Sehr verehrter Herr Professor Bohr!

Als Nachtrag zu den Regeln[182] über Zeemannzerlegung sende ich Ihnen beiliegendes Blatt, auf dem jetzt alles viel einfacher und klarer geworden ist, sodass diese Regeln jetzt das vollständige Bild der *Dublett*linienzerlegungen geben. Die Gesetzmässigkeiten bei den Tripletts scheinen komplizierter zu sein. Übrigens war auf dem Ihnen damals zugeschickten Blatt ein Versehen bei der Triplettlinie (sp_2), deren Bild falsch abgezeichnet war. Mit den besten Grüssen

Ihr sehr ergebener
A. Landé

[2]3. II. 21.[183]

1 *Regel.*   Bei der Zerlegung einer *Dublett*linie kommt eine unverschobene (normale) p-Komponente niemals vor. Also ist die Anzahl der p-Komponenten *grade*.
2. *Regel.* a) Bei der Zerlegung einer Triplettlinie ist die Anzahl der p-Komponenten gleich der Summe der zwei „inneren"

182. See document nr. 4.
183. The first numeral of the date is obliterated.

Quantenzahlen ihrer Terme. Das gilt auch für die Zerlegung einer *Dublett*linie, sofern die Summe der inneren Qu.z. grade ist.

    b) Ist die Summe *ungrade,* so muss wegen Regel 1 eine p-Komponente der *Dublett*linienzerlegung ausfallen.[184]

3. Regel. a) Bei der Zerlegung einer *Dublett*linie gehört zu jeder p-Komponente rechts und links im gleichen Abstand $\pm\Delta$ je eine s-Komponente, und zwar ist Δ für alle Gruppen s-p-s, in welche das Zerlegungsbild *einer* Linie sich auflösen lässt, gleich gross. Dagegen ist Δ bei verschiedenen Linien verschieden.[1]

    b) Diese Regel gilt auch für die Zerlegung einer *Triplett*linie, sofern dieselbe nicht zwei gleichgrosse innere Qu.z. besitzt.

    c) Fallen nach dieser Regel zwei s-Komponenten von rechts und links her aufeinander

$$|s| - |p| - |s|$$
$$|s| - |p| - |s|$$,

so löschen sie sich gegenseitig aus. Zählt man diese sich auslöschenden s-Komponenten mit, so ist ausnahmslos die Zahl der s-Komponenten doppelt so gross wie die Zahl der p-Komponenten.

4. Regel. a) Bei der Zerlegung einer *Dublett*linie folgen die p-Komponenten im gleichen Abstand II aufeinander; jedoch ist II bei verschiedenen Linien verschieden gross.[1]

    b) Wie 3b

5. Regel.   Bei der Zerlegung /innersten \
einer *Dublett*linie (p-Komponente) ... (addiert)[185]
bekommt man

6. Regel.   die Lage der (zugehörigen \
indem man (s-Komponente) . /. (subtrahiert)
ihre inneren Qu.z.
jede durch ihren
Runge'schen Nenner
dividiert und die erhaltenen *Brüche*

---

[1] Beim normalen Zeemanntriplett ist $\Delta = 1$ gesetzt. [Landé's footnote.]

184. See note 115 for justification of this rule, and associated text for this schema in general.

185. Landé here inadvertently interchanges "add" and "subtract."

Durch diese Regeln ist das Zerlegungsbild der Dublettlinien eindeutig durch innere Qu.z. und Runge'sche Nenner bestimmt. Beispiele umseitig. Die Zuordnung der inneren Qu.z. zu den einzelnen Termen ergibt sich aus dem Auswahlprinzip, vergl. *A. Sommerfeld* Ann. d. Phys. **63**, 121, 1920 Fig. 4a und 4b, die empirischen Zerlegungsbilder, vergl. Tabelle 2, 3, 4, 5 l.c.

Beispiele für die Dublettlinienzerlegung.

1) Na-D_2 Linie. (sp_1), innere Qu.z. (1,2), innere Qu.brüche (1/1, 2/3) Zahl der p-Komponenten $= 1 + 2$ (Regel 2a) minus 1 (Regel 2b) $= 2$. Lage der innersten p-Komponente 1/1 $-$ 2/3 $= +1/3$ (Regel 5). Lage der zugehörigen s-Komponente 1/1 $+$ 2/3 $= 5/3$ (Regel 6), also $\Delta = 5/3 - 1/3 = 4/3$. Die zwei p-Komponenten liegen also bei $+1/3$ und $-1/3$, jede ist im Abstand $\Delta = \pm 4/3$ von zwei s-Komponenten umgeben.

2) Na-D_1 Linie. (sp_2) $\qquad\qquad$ (1,1) $\qquad\qquad$ (1/1,1/3)
$1 + 1 = 2$ p-Komponenten
1/1 $-$ 1/3 $= +2/3$ $\quad$ 1/1 $+$ 1/3 $=$ 4/3 $\quad \Delta =$ 4/3 $-$ 2/3 $= 2/3$. Die zwei p-Komponenten liegen also bei $+2/3$ und $-2/3$, jede ist im Abstand $\Delta = \pm 2/3$ von zwei s-Komponenten umgeben. Die beiden im Nullpunkt aufeinanderfallenden s-Komponenten löschen sich aus (Regel 3c).

3) (p_1d_2) (2,2) (2/3,2/5) $2 + 2 = 4$ p-Komponenten 2/3 $-$ 2/5 $= +4/15$ $\quad$ 2/3 $+$ 2/5 $=$ 16/15 $\quad \Delta =$ 16/15 $-$ 4/15 $=$ 12/15. Zwei p-Komponenten liegen bei $+4/15$ und $-4/15$, ihr Abstand II ist 8/15, also liegen die beiden andern p-Komponenten (Regel 4) bei $+12/15$ und $-12/15$. Jede p-Komponente hat zwei s-Komponenten im Abstand $\pm \Delta =$ 12/15. Die beiden s-Komponenten im Nullpunkt löschen sich **aus**.

4) (p_1d_1) (2,3) (2/3,3/5) $2 + 3 = 5$ minus 1 (Regel 2b) $= 4$ p-Komp.
2/3 $-$ 3/5 $=$ 1/15 $\quad$ 2/3 $+$ 3/5 $=$ 19/15 $\quad \Delta =$ 19/15 $-$ 1/15 $=$ 18/15. Eine p-Komponente liegt bei $+$ 1/15, eine bei $-1/15$, II ist also $=$ 2/15 und die andern 2 p-Komponenten liegen bei $+3/15$ und $-3/15$. Jede ist im Abstand $\pm \Delta =$ 18/15 von 2 s-Komponenten umgeben.

5) (p_2d_2) (1,2)   (1/3,2/5) $1 + 2 = 3$ minus $1 = 2$ p-Kompon.
     $1/3 - 2/5 = -1/15$ $1/3 + 2/5 = 11/15$ $\Delta = 11/15 - (-1/15) = 12/15$. Also 2 p-Komponenten bei $+1/15$ und $-1/15$, eingerahmt im Abstand $\pm\Delta = 12/15$ von je 2 s-Komponenten.

[7]
A. SOMMERFELD to A. LANDÉ   25 February 1921
Handwritten (modern cursive) postcard (SHQP 4, 13).
Card postmarked: 25.2.21.11-12N. [place missing]
     addressed: Hn. Dr. A. Landé
               Frankfurt a.M.
               Universität, Robert Mayerstr 2

25. II. 21.

Lieber Landé! Bravo, Sie können hexen! Ihre Construktion der Dublett-Zeeman-Typen ist *sehr* schön. Sie können nun auch den Bergmann-Serien-Typ konstruiren, was für Paschen unmittelbar wichtig wäre.[186] Allerdings können ihn die Spektroskopiker noch nicht beobachten. Aber wenigstens hat Saundars bei Ba Triplett-linien die Dreifachheit des b-Termes (im unzerlegten Zustande) gefunden.[187] Also können Sie auch die Zweifachheit bei Dublett-Linien annehmen. Sie können auch die anomalen Serien (sd) kon-struiren. Für die Triplett-Linien werden Sie das Gesetz auch schon noch herausknobeln. Ihre Diamant-Arbeit habe ich Ewald gegeben. Wir glauben, dass die Thirring'sche überzeugender ist. Dass meine ziemlich auf's Geratewhol geratenen „inneren Quantenzahlen" zu Ehren komen, freut mich.
An Paschen werde ich Ihr Blatt schicken.

Mit besten Grüssen
Ihr ASomerfeld

186. The last clause was an afterthought, squeezed in between the lines.
187. F. A. Saunders, "Revision of the Series in the Spectrum of Barium," *Astrophysical Journal, 51* (Jan. 1920), 23–36.

[8]

H. A. Kramers to A. Landé   28 February 1921

Typewritten letter (original: SHQP 4, 17; carbon copy: BSC 4, 3).
Envelope postmarked: [incomplete and illegible]
    addressed: Herrn Dr. A. Landé.
                 Institut für theoretische Physik
                   der Universität.
                 Robert Mayer Strasse.
                 Frankfurt am Main.
                 Deutschland

UNIVERSITETETS INSTITUT	BLEGDAMSVEJ 15,
FOR	KØBVENHAVN Ø.
TEORETISK FYSIK	DEN 28. Februar 1921.

Sehr verehrter Dr. Landé,

Professor Bohr hat mich gebeten Ihnen einige Zeilen als Antwort auf Ihre letzten zwei Briefe[188] an ihn zugehen zu lassen. Eben in diesen Tagen ist er ausserordentlich beschäftigt mit der Fertigstellung des neuen Instituts, das am 3. März eingeweiht werden soll und hat deshalb keine Gelegenheit selbst seine Korrespondenz zu besorgen.

Was Ihr freundliches Angebot betrifft den Brief an "Nature" auf deutsch zu übersetzen und zum Beispiel in die der Zeitschrift für Physik zu publizieren, glaubt er, dass dieses nicht zweckmässig sein würde. Der Brief darf ja nicht anders aufgefasst werden als ein Einleg in Beitrag zu der Diskussion über Atombau, die eben in dieser Zeit in „Nature" geführt wird, und würde, wenn er für sich allein an anderer Stelle publiziert würde, allzu fordrungsvoll anspruchsvoll erscheinen, zumal weil die Argumente für die in ihm enthaltenen neuen Anschauungen darin so gut wie gar nicht entwickelt werden sind. Ausserdem hofft Professor Bohr innerhalb weniger Wochen eine deutsche ausführliche Artikel über den betreffenden Gegenstand fertiggestellt zu haben, und da wird er Ihnen einen Abdruck des Manuscriptes sofort zugehen lassen.[189]

Was Ihre Bemerkungen über den anomalen Zeemaneffekt betrifft,

188. Actually three letters: documents nrs. 4, 5, and 6.
189. See note 177 to document nr. 3.

bittet er Ihnen zu sagen, dass er natürlich das grösste Interesse an diesen Sachen hat, aber dass er in diesen Tagen, wo er so sehr mit andern Dingen beschäftigt ist, ganz verhindert ist sich darin zu vertiefen und besonders zu prüfen, wie Ihre Resultate sich zu der theoretischen Anschauung verhalten. Auch über diese Fragen wird er aber nach der Fertigstellung des Instituts Ihnen eingehend schreiben.

Mit freundlichen Grüssen von Bohr
und mirselbst
Hochachtungsvoll
H. A. Kramers

[9]
A. SOMMERFELD to A. LANDÉ 3 March 1921
Handwritten (modern cursive) letter (SHQP 4, 13).
Envelope not extant.

München 3. III. 21.

Lieber Landé!

Back schreibt mir, dass Paschen ihm Ihre Dublettregeln nicht gegeben hat, weil er selbst (Back) an denselben Dingen arbeitet, was sehr richtig ist. Ich bitte Sie nun dringend, von jeder Publikation Ihrer Regeln abzustehen, bis Ihnen Back die Correktur seiner in Vorbereitung befindlichen ausführlichen Ann.-Arbeit[190] zuschickt, u. zw. aus folgenden Gründen:
1) Back ist der experimentelle Vater des ganzen Gebietes; er muss die Vorhand haben, die Früchte seines Experimentes selbst zu pflücken. Auch bei meinem Zerlegungssatz habe ich erst seine Einwilligung zur Publikation eingeholt, die er ohne Weiters gab, weil er damals überhaupt keine theoret.[191] Gesichtspunkte hatte; sie sind ihm vielmehr erst durch jenen Satz gekommen.

190. Note 90.
191. Sommerfeld first wrote "keine Q"—perhaps thinking *keine Quantentheorie wusste*—and then recollecting how little quantum theory there actually was in this sort of approach to the problem of the anomalous Zeeman effect, wrote "th" over the "Q."

2) Back arbeitet sehr langsam, ist durch den Krieg körperlich u. seelisch mitgenommen, zumal als Spross einer alten elsässischen, vertriebenen Familie. Er braucht Ruhe zur Arbeit u. fühlt sich durch Ihre Parallelarbeit beunruhigt.

3) Back hat bereits für die Naturwissenschafen eine vorläufige Mitteilung,[192] deren Correktur er mir mitgeschickt hat; er ist also schon[193] auch publicistisch Ihnen voran. Diese Mitteilung beschäftigt sich zwar nur mit der Anzahl der Componenten, lässt aber alles übrige durchblicken.

4) Back scheint auch über die Triplettfragen befriedigendere Regeln zu haben als Sie. Ihre Tripl-Regeln kann ich noch nicht als endgültig ansehen; sie sind mir zu verschieden von Ihren Dubl-Regeln, während doch beide aus der gleichen Wurzel kommen müssen. Back kann auch bei den Tripletts alles vorausprophezeien.[194]

Wenn Sie die Back'sche Ann.-Correktur haben, werden Sie natürlich das, was Ihnen bei Ihrer Darstellung besser scheint, für die Annalen zusammenschreiben (nicht für die Zeitschr. f. Ph., damit Sie auch dann keinen Vorsprung im Erscheinen vor Back gewinnen).[195]

Seien Sie mir nicht böse, dass ich Sie scheinbar in dieser Sache beeinflussen möchte; aber das vertrauensvolle Zusammenarbeiten mit der Praxis, insbesondere derjenigen des Paschen'schen Institutes, darf im Interesse der Theorie nicht gestört werden. Und ich fürchte bei dem sensibeln Charakter von Paschen und der zarten Natur von Back, dass sich ein dauernder Misklang ergeben kann, wenn wir hier nicht mit aller Delikatesse vorgehen.

Ihre Bemerkung, dass Bohr wie eine Bombe eingeschlagen hat, trifft auch für München zu. Ich bekam von Bohr einen Durchschlag seines Nature-Briefes. Wir müssen gründlich umlernen!

Mit den besten Grüssen
Ihr A. Sommerfeld

192. Notes 131 and 133.
193. Struck out.
194. Sentence an afterthought, squeezed in between the lines.
195. The *Annalen der Physik* published papers in three to six months, the *Zeitschrift für Physik* in six to eight weeks. Cf. note 142.

[10]
E. BACK to A. LANDÉ   7 March 1921
Handwritten (gothic cursive) letter (SHQP 4, 15).
Envelope postmarked: Tübingen
                   -8.3.21.3-4N
    addressed: S.H.
                 Herrn Dr. A. Landé
                 Privatdozent.
                 Institut für theoretische Physik
                 Frankfurt a.M.
                 Robert Mayerstr. 2.

                                     Tübingen 7. 3. 21

Sehr geehrter Herr Doktor!

Seien Sie bestens für Ihren freundlichen Brief vom 5.3. und die sehr wohlmeinenden Vorschläge bedankt. Deren Geist und Gesinnung teile ich von ganzem Herzen, die Ausführung ist aber leider nicht mehr ganz in der Weise möglich, wie Sie sie vorschlagen, dazu ist meine Publikation in den Naturwissenschaften schon zu weit gediehen.[196] Das Manuskript wurde in seiner endgültigen Fassung am 10ten Februar von der Redaktion angenommen, abgesandt ist es von hier am 2.II, ein kleiner Schlussnachtrag am 6ten Febr; zu einem Zeitpunkt also, wo ich von Ihren Untersuchungen weder Kenntnis hatte noch haben konnte. Die Korrektur ist längst erledigt und das Heft vermutlich schon beim Buchbinder.

Ich meine, diese Tatsache ist nicht so schlimm, da wir, wie ich glaube, auf ganz verschiedenen Wegen gewandelt sind. Selbstverständlich halte ich es nur für recht und billig, dass Sie ihren Marburger Vortrag drucken lassen, und zwar genau so, wie wenn Sie von meiner Existenz nichts wüssten, denn zum Zeitpunkt Ihres Vortrags wussten Sie ja auch in der Tat nicht das Geringste von meiner Tätigkeit. Ich bin ganz Ihrer Meinung, dass einen etwaigen Prioritätsstreit das physikalische Publikum austragen soll, wenn es Anlass und Bedürfnis dazu hat, ich für meine Person werde ganz gewiss keine Waffen dazu schleifen.

Was Ihren sehr freundlichen und ehrenvollen Vorschlag angeht,

196. Note 131.

eine gemeinsame Veröffentlichung vorzubereiten, so danke ich Ihnen zunächst bestens dafür und bemerke Folgendes: Auf eine selbständige und sehr baldige Publikation in den Annalen kann ich nicht verzichten, weil sie meine Habilitationsschrift sein soll, und ich bei meinen hohen Lebensjahren—dabei aber noch in den ersten Anfängen einer Laufbahn—keine Zeit und keine Frucht zu verschenken habe. Diese Arbeit werde ich Ihnen aber im Manuskript baldigst mitteilen. Ob darüber hinaus noch Stoff zu einer gemeinsamen Publikation verbleiben wird, weiss ich z.Z. noch nicht, wenn es der Fall ist, werde ich sehr gerne Ihrem Vorschlag zur Zusammenarbeit folgen.

Der Stand meiner Arbeit ist dieser: Aus einer „Indexregel" kann ich die Komponentenzahl jeder Kombination und auf Grund einiger weiterer Regeln, die alle in meiner Publik. von 10.2. genau angeführt,[197] wenn auch noch nicht begründet sind, die quantitative Struktur aller Kombinationen aus den 38 Symbolen des Serienschemas angeben. Vor wenigen Tagen, nachdem meine Korrektur längst abgesandt war, machte mir Herr Prof. Paschen Mitteilung von Ihren Briefen. Es ist seltsam und interessant, dass in meinen Regeln die Quantenzahlen keine Rolle spielen (d.h. äusserlich nicht), ich brauche nur die Indizes des Serienschemas, wie sie von Ritz u. Rydberg eingeführt sind und einige Symmetriebetrachtungen. Das zweite Exemplar meiner Korrektur habe ich Herrn Geheimrat Sommerf. geschickt und besitze z.Z. keines mehr, natürlich habe ich nicht nur nichts dagegen, sondern es ist mir geradezu erwünscht, wenn Sie Kenntnis davon nehmen.

Die Aufforderung an Sie, die Veröffentlichung zurückzuhalten, ging ganz spontan von Herrn Geheimrat Sommerfeld aus. Er meinte es damit sehr sehr gut mit mir und ich bin ihm für diese Meinung tief dankbar, aber wie die Dinge liegen, sehe ich keinen Grund, ihnen nach irgend einer Richtung Gewalt anzutun oder Ihnen eine moralische Schweigepflicht zuzumuten.

Mit vorzüglicher Hochachtung und den besten Grüssen—wir kennen uns vom Frankfurter Colloquium 1919—bin ich Ihr sehr ergebener

E. Back.

197. That is not actually the case, and it was even an exaggeration to claim as Sommerfeld did in document nr. 9 that this paper "lässt aber alles übrige durchblicken."

[11]
F. PASCHEN to A. LANDÉ   8 March 1921
Handwritten (modern cursive) letter (SHQP 4, 11).
Envelope postmarked: [missing, excised]
            addressed: Herrn Prof. Dr. A. Landé.
            Frankfurt a.M.
            Institut f. theoret. Physik
            Robert Mayerstr 2.

                                    Tübingen 8. III. 1921

Sehr verehrter Herr Kollege.

Besten Dank für Ihren interessanten Brief vom 2. III, in welchem die vollständige magnetische Aufspaltung eines Dublet-Bergman̄-Gebildes sehr interessant ist. Auch Hr Dr Back hat auf einem, wie es scheint, anderem Wege oder vielmehr nach anderen Regeln säm̄tliche Dublet- and Triplet-Typen construirt, und es ist wichtig, dass die Resultate mit den Ihrigen übereinstim̄en. Nun kennt man erst ein einziges vollständiges Bergman̄-Dublet dieser Art, dessen magnetische Aufspaltung aber nicht genauer zu messen ist, weil es bei zu kurzen Wellenlängen liegt.[198] Immerhin wird es vielleicht gelingen, ein solches Gebilde zu finden und zu analysiren, nachdem man nun einen begründeten Plan seiner Aufspaltung kennt. Es steht experimentell nämlich so, dass die bisher bekannten Bergman̄-Linien ungünstig für die Untersuchung des Zeeman-Effectes sind und meistens im Bergman̄-Term (mb) noch nicht differenziirt auftreten. Aber es könnte sein, dass solche Gebilde besonders in den Spectren höherer Colonnen des period. Systems stark vorhanden sind und nun nach dem Typus erkannt werden können.[199] So ist es ja auch mit den Typen der I Dublet. N.S. gegangen, und so wird es mit den Typen weiterer Gebilde in Zukunft wohl im̄er gehen. Denn man muss sich immer bewusst sein, dass das, was wir über die Gesetze der Spectren kennen, erst ein recht bescheidener Anfang ist.

Bezüglich des gewöhnlichen Helium-Spectrum bin ich sehr ge-

198. Evidently the doublet at 2300Å which S. Popow found in the (spark) spectrum of Ba: "Über eine Gesetzmässigkeit in den Linienspektren," *Ann. d. Phys.*, *45* (1914), 147–175; on p. 172. Cf. Sommerfeld, *Atombau* . . . (3rd ed.; 1922), p. 445.
199. Such fundamental series patterns were indeed found within the year in consequence of the recognition of multiplets in the spectra of these elements. Cf. note 32 and Sommerfeld, "Über die Deutung verwickelter Spektren . . . ," *op. cit.* (note 168).

spannt auf Bohr's Theorie.[200] Ich habe mich weiter damit beschäftigt, und es sind Untersuchungen hier in Gange, welche auf die
Entstehung dieser Spectren Licht werfen können, besonders auch auf
die Entstehung des merkwürdigen Banden-Spectrum des Heliums. Es
scheint, dass J. Fracks Hypothese des metastabilen Zustandes des
einen Theiles des Heliumspectrum mit vielen Thatsachen im Ein-
klang ist.[201] Bezüglich des Stark-Effectes sollte es doch möglich sein,
ganz roh abzuschätzen, ob überhaupt ein solcher Effect in praktisch
realisirbaren Feldern zu Stande kommen kann, und welche Serien
stärkeren Effect zeigen müssten.[202] Schon dadurch würde man sehen
können, ob eine Theorie halbwegs den realen Verhältnissen Rechnung
trägt.

Herr Dr. Back hat mir auch von Ihrem Briefe Mittheilung gemacht. Soweit ich es beurtheilen kann, besonders nach Ihren Hrn
Somerfeld mitgetheilten Regeln, besteht eine bedeutende Verschiedenheit in den zur Construction herangezogenen Elementen. Danach
dürfte es sich empfehlen, wenn jeder seine Sache für sich weiter führt.
Das Gemeinsame könnte nachher wohl herausgeschält werden. Jedenfalls ist Back schon längere Zeit ziemlich fertig. Er hat auch die
Bergman-Typen der Triplets und behauptet jeden Combinations-
Typus construiren zu können. Da er aber Gewicht auf völlige Abge-
schlossenheit der Sache legt, bevor er publiciren will, so will er doch
auch das experimentell jetzt Mögliche noch erledigen, um seine
theoretischen Regeln endgiltig durch Thatsachen beweisen zu können.
Er war schon 1913 und 1914 mit diesen Studien beschäftigt und ist
durch den Krieg an der Ausführung verhindert. Die Sicherstellung
der Typen der I. N. S. der Dublets und Triplets war schon damals
hauptsächlich durch seine experimentellen Arbeiten erfolgt. Diese
Typen wurden bekannt, indem ich sie Hrn Voigt und Sommerfeld
mittheilte. Denn es schien mir nützlich, sie bekannt zu geben, weil es
damals unsicher war, ob Hr Back jemals wieder daran arbeiten
würde. Ausserdem hätte Jeder sich die Typen auch aus der Literatur

200. The results of Bohr and Kramers' helium calculations, although adverted
to by Bohr in general terms on several occasions after he learned of Landé's
work in the spring of 1919, were detailed only in 1923: H. A. Kramers, "Über das
Modell des Heliumatoms," *Zeits. f. Phys., 13* (Jan. 1923), 312–341.

201. I.e., metastable states (see note 46) justify hypothesizing He_2 molecules,
which are necessary to explain the band spectrum of helium.

202. Cf. Paschen's previous queries, document nr. 2.

zusam̅enstellen können, allerdings mit einigen Unsicherheiten, die eben erst Back endgültig behoben hat.

Mit freundlichen Grüssen
Ihr hochachtungsvoll ergebener
F. Paschen

[12]

A. SOMMERFELD to M. BORN   8 March 1921

Carbon copy of typewritten letter without complimentary closing or signature (SHQP 29, 11). Date given erroneously as 3 Mar 21 by SHQP.

München, den 8. März 1921.

Lieber· Born!

Ich bin ernstlich böse auf Lande. Er ist wieder dabei etwas gänzlich Unreifes zu publizieren. Aber schlimmer noch, er ist dabei, etwas zu tun, was sich nicht schickt. Es schickt sich nicht, dem Experimentator die Schlussfollgerungen aus seinen Versuchen vorweg zu publizieren. Zumal bei Herrn Back der seelisch so vom Krieg mitgenommen ist. Alle unsere Weisheit (meine und Lande's) gründet sich auf Back's unpublizierte Messungen. Als ich ihn seinerzeit darum anging war er einverstanden, dass ich sie benutzte, jetzt aber ist er es gar nicht. Jetzt braucht er Ruhe und Lande ist durch seinen unruhigen Ehrgeiz dabei, ihn in seinen Kreisen zu stöhren.

Sie müssen durchaus verhindern, dass Lande in den Verhandlungen von seinem Marburger Vortrag berichtet. Desgleichen, dass er Back mit dem Vorschlag einer gemeinsamen Publikation beunruhigt. Das Publizieren von Halbfertigfabrikaten, die in der nächsten Publikation abgeändert werden, ist wirklich unerträglich. Ausserdem ist die natürliche Folge, dass Paschen uns aus seinem Institut überhaupt nichts mehr wissen lässt, wenn wir seine oder Back's Liberalität missbrauchen.

Ich hätte wirklich nicht gedacht, dass Lande auch in diesem Falle schon wieder beim Publizieren ist und nochweniger, dass er aus meinen ernstlichen Vorstellungen heraus nicht das Unschickliche gegenüber Back eingesehen hat. Er ahnt offenbar gar nicht, wie sehr er sich selbst damit schadet.

Ich habe nichts dagegen, wenn Sie diesen Brief Lande zeigen; ich habe ihm ganz dasselbe schon direkt geschrieben, wenn auch zunächst in höflicherer Form.

257

[13]

A. Landé to N. Bohr   17 March 1921
Handwritten (modern cursive) postcard (BSC 4, 3).
Card postmarked: Elberfeld
                 17.3.21.3-4N.
    addressed: Herrn Professor N. Bohr
               Dänemark Kopenhagen
               Blegdamswej 37

Sehr verehrter Herr Professor Bohr! Verzeihen Sie, dass ich Ihnen schon wieder schreibe, ohne Ihre Antwort auf meine vorigen Briefe abzuwarten. Aber ich möchte Ihnen doch eine wesentliche Verbesserung meiner Zeemannregeln mitteilen, um aus Sommerfelds „inneren" (besser wohl „gesamten azimutalen") Quantenzahlen k und den Runge'schen Termnennern r_k die Zerlegungsbilder nach dem Kombinationsprinzip zu konstruiren: Es gibt zu k und r_k die $2k$

$$\textit{Energieniveaus} \quad \pm \frac{k}{r_k}, \quad \pm 3\frac{k}{r_k}, \quad \pm \ldots \pm (2k-1)\frac{k}{r_k}$$

denen [203] man zuordne die

$$\textit{äquatorialen Quantenzahlen} \quad \pm\tfrac{1}{2}, \quad \pm\tfrac{3}{2}, \quad \pm \ldots \pm \frac{2k-1}{2}$$

Dann heisst die einzige Regel so: Ändert sich die „äquatoriale" Qu.z um ± 1, so entstehen s-Komponenten, um $0 \ldots$ p-Komponenten.—

Beispiel: $(p_1 d_1)$, $\left(\dfrac{k}{r_k}, \dfrac{l}{r_l}\right) = (\tfrac{2}{3}, \tfrac{3}{5})$ also

Die schrägen Pfeile geben s-Komponenten die graden Pfeile p-Komponenten, z. B. $2/3 - 3/5 = 1/15$ d.h. p-Komponente bei $1/15$

203. The "d" is obliterated.

der normalen Zeemannaufspaltung. Für die Triplettzerlegungen bin ich erst auf dem Stand, Regeln nach Analogie der Ihnen früher gesandten aufzustellen. Mit den besten Grüssen

Ihr Sie sehr verehrender

Elberfeld   17. III. 21.   A. Landé

[14]
A. Landé to A. Sommerfeld   17 March 1921
Handwritten (gothic cursive) letter (SHQP 32,1).
Envelope not extant.

Elberfeld   17. III. 21
Luisenstr. 15 [?]

Sehr verehrter Herr Geheimrat!

Die Zeemannzerlegungsregeln, die ich Ihnen neulich schrieb, lassen sich noch sehr vereinfachen, auf *Term*energien und eine Auswahlregel reduzieren, zunächst bei den Dubletts: Ein Zustand mit der inneren Quantenzahl k und dem Termnenner r_k soll im Magnetfeld die $2k$ (magnetischen Zusatz-)

—Energieniveaus

$$\pm 1 \cdot \frac{k}{r_k}, \quad \pm 3 \cdot \frac{k}{r_k}, \quad \ldots \pm (2k-1)\frac{k}{r_k}$$

haben, denen man die („äquatorialen" Quanten-) zahlen

$$\pm \tfrac{1}{2} \qquad \pm \tfrac{3}{2} \qquad \pm \frac{2k-1}{2} \quad \text{zuordne.} \qquad \text{Dann}$$

bekommt man alle Dublettzerlegungen richtig heraus mit Hilfe der

einen *Regel:* $\begin{cases} \text{p-} \\ \text{s-} \end{cases}$ Komponenten entstehen bei Änderung der äquatori-

alen Quantenzahl um $\begin{matrix} \cdot 0 \\ \pm 1 \end{matrix}\Big\}$. Beispiel: $(p_1 d_1)$, $\left(\dfrac{k}{r_k}, \dfrac{l}{r_l}\right) = \left(\tfrac{2}{3}, \tfrac{3}{5}\right)$

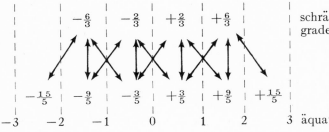

schräge Pfeile: s-Komp.
grade   ″   : p- ″

259

Dadurch kommt erstens die Regel: Anzahl der p-Komponenten $=$ $k + l$ mit der Ausnahme $k + l - 1$, zweitens: Das Verschwinden der s-Komponenten im 0-Punkt selbsttätig heraus u.s.w.[204] (auch die Intensitäten stimmen einigermassen mit den Schätzungen.)[205] Bei den Tripletts kann man ähnlich vorgehen, muss aber als äquat. Qu.z die *ganzen* Zahlen 0, ± 1, ± 2, .. nehmen (da hier p-Komponenten im 0-Punkt auftreten), und statt der obigen Energieniveaus $(2n - 1) \cdot k/r_k$[206] hat man hier kein so allgemein und einfach angebbares Gesetz, sondern jeder Term hat seine besondere, vorläufig nur empirisch zu bestimmende Energieniveaureihe, z.B. der Term d_1 ($k/r_k = 3/3$) hat die $2k + 1$ (magnetischen Zusatz-) energien

0, $\pm 4/3$, $\pm 8/3$, $\pm 12/3$ mit den äqu. Qu.z 0, ± 1, ± 2, ± 3 also d_1 hat 4/3, ferner d_2 hat 7/6, d_3 hat 1/2; p_1, p_2 und p_3 haben 3/2, s hat 2/1 als Grundenergieniveau.[207] Während also bei den Dubletts alles in einfachster Weise auf eine vernünftige Quantentheorie der anomalen Zeemanneffekte vorbereitet ist, genügt meiner Meinung nach bei den Tripletts das vorhandene Material noch nicht, um allgemein die magnetischen Termenergieniveaus zahlenmässig zu beherrschen oder Voraussagen zu machen.

Nun zu der Angelegenheit von Dr. Back. Ich schrieb an ihn einen ausführlichen Brief, auf den ich eine sehr liebenswürdige Antwort bekam,[208] in der stand, dass er Ihnen für Ihr Interesse ganz ausserordentlich dankbar sei, aber keinen Grund sehe, warum ich mit der Publikation meiner Ergebnisse auf ihn Rücksicht nehmen sollte, zumal sein Weg (eine „Indexregel") offenbar ganz anders sei als der meinige. Auch Paschen schrieb mir einen sehr freundlichen Brief,[209] sodass also eine Verstimmung zwischen Theorie und Praxis nicht zu erwarten ist, auch wenn ich meine Arbeit, wie ich an Back schrieb, früher als seine publiziere. Ich habe es damit etwas eilig, weil *Bohr*[210] offenbar über diese Dinge nachdenkt; und warum

204. Notes 114 and 150.
205. The sentence here enclosed in parentheses was an afterthought inserted in the letter as a footnote: ("u.s.w.¹").
206. See note 152.
207. See commentary in text, section 6.
208. Document nr. 10.
209. Document nr. 11.
210. Modern cursive in MS.

soll das Ausland uns darin zuvorkommen. Born habe ich seit seiner Berliner Reise, auf der Sie ihn vielleicht gesprochen haben, noch nicht gesehen. Mit den besten Grüssen

Ihr Sie sehr verehrender
A. Landé.

[15]
A. SOMMERFELD to A. LANDÉ   ca. 31 March 1921
Stenographic draft on reverse of document nr. 14, as transcribed in 1963 by Miss S. Hellmann, Niels Bohr Institutet, Copenhagen (SHQP 32, 1).

L.L.

Der Anfang Ihres Briefes ist ein schlagender Beweis für meine Behauptung, dass Ihre Z.E. Ueberlegungen bisher nicht reif zur Publikation waren. Ihre neue Darstellung deckt sich mit dem, was ein Schüler[211] von mir (1.Sem.) gefunden hat, was aber *nicht* veröffentlicht worden ist. Die Platten [Typen?] lassen sich analysieren, wenn auch weniger schön darstellen. Dass Sie B. [Bohr] zuvorkommen wollen, ist kein Grund für Ihre Eile. B. hat jetzt andere Dinge zu publizieren und ist ja Theoretiker. Ich sehe ihn nicht als Ausländer an.

Wenn B. [Back] und P. [Paschen] Ihnen schreiben, dass sie gegen Ihre Publikation nichts einwenden, so ist das sehr nobel von ihnen; Sie stellen sich auf den formellen Standpunkt, von dem aus Sie recht haben. Aber von einem höheren Standpunkt aus, dürfen Sie B. [Back] nicht zuvorkommen. B. wäre es sehr unangenehm, wie ich weiss, und ich werde es Ihnen direkt übelnehmen.

Meine Karte aus Ehrw. war vor Ihrem Brief vom 21. geschrieben, Letzteren habe ich hier erst am 28. vorgefunden.[212]

211. Heisenberg; see note 158.
212. Neither the "card from Ehrwald," which Sommerfeld evidently sent while on ski tour (cf. note 156), nor Landé's letter of the 21st (The arabic numeral is unmistakable in the stenographic draft), is known. *Perhaps* Sommerfeld meant to say that although his card from Ehrwald was written (on the 21st?) after Landé's letter (of the 17th), he received that letter only upon returning to Munich (on the 28th).

Helmholtz' "Kraft": An Illustration of Concepts in Flux

BY YEHUDA ELKANA*

"In the world of human thought generally and in physical science particularly, the most fruitful concepts are those to which it is impossible to attach a well-defined meaning."[1]

1. INTRODUCTION

The history of the principle of conservation of energy has been told many times. The most fruitful years in the development of the principle, 1820–1850, have been examined from various points of view; but whatever the point of view, there is unanimous agreement that the first mathematical formulation of the principle in all its generality was given by Hermann von Helmholtz in 1847.[2] What Helmholtz actually did was to formulate clearly the principle of the conservation of mechanical energy and then show that all the various

* Department of History and Philosophy of Science, The Hebrew University, Jerusalem, Israel.

1. H. A. Kramers, as quoted by L. Tisza, *Generalized Thermodynamics* (Cambridge, Mass., 1966), p. 1.
2. "Ueber die Erhaltung der Kraft," read before the Physical Society of Berlin on 23 July 1847. As I shall be quoting from this work extensively, a few remarks about it will be in order. It was published privately that same year by G. Reimer, Berlin. The essay is fifty-three pages long and is subdivided in the following manner:

    Introduction
  I. The Principle of Conservation of *Vis Viva*
  II. The Principle of Conservation of Force
  III. The Application of the Principle to Mechanical Theorems
  IV. The Force-Equivalent of Heat
  V. The Force-Equivalent of Electric Processes
  VI. The Force-Equivalent of Magnetism and Electromagnetism

"forces of nature" can be subsumed under this principle. He thus created the general concept of energy as the one entity that is conserved under all circumstances in a fundamentally mechanical world. The generality of his treatment is underlined by the fact that, for the first time, the principle is regarded not only as a law of nature covering all phenomena, but, to use an expression of Maxwell, as a "science-producing doctrine": "The object of the present memoir is . . . to draw further conclusions regarding laws which are as yet but imperfectly known, and thus to indicate the course which the experimenter must pursue."[3] It is argued here that Helmholtz' paper "On the Conservation of Force" serves as an illustration of a general feature of the way in which scientific concepts develop: they are in a state of flux while the scientist is struggling to clarify his thoughts; that is, while the discovery is being made.[4] I suggest that Helmholtz believed in a vaguely formulated conservation law, and that only after he had arrived at his proof of the conservation of "Kraft" did he distinguish between his concepts of energy and force in our sense of the terms. (Later energy became for him more than just the effect of force—it gained primary status.) It is widely accepted that scientific theories and concepts are first vague and later specific. What I emphasize here is that the theory and concept crystallize[5] simultaneously out of the vague early formulations. The official view of the scientist is

3. This passage appears at the beginning of Section I. I am using here, and in all other quotations (except where otherwise indicated), the translation of Helmholtz' essay by John Tyndall in Taylor's *Scientific Memoirs, 1*, Pt. II (London, 1854), 114–162.

4. Here several reservations have to be made. These statements are not to be understood on a logical level; i.e., they do not relate to recent controversies on the logical status of concepts vs. theories. I am dealing with that stage of the creation of a new theory where logical analysis is not yet known to apply. My thesis is that concepts originate in metaphysical principles in the form of vague ideas. These are not the concepts that we analyze after a theory has been formulated, but rather the vague conceptual tools which the scientist uses while struggling to formulate a theory. What the logical relation is between the two kinds of concepts I do not claim to know. What I am doing with the physical concepts of "force" and "energy" has been admirably done for mathematical concepts by I. Lakatos in his "Proofs and Refutations," Pts. I–IV, *Br. J. Phil. Sci., 14* (1963–1964), 1–117. Lakatos distinguishes between the original vague stage of concepts and their clarified version in the context of a theory. He calls the transition "concept-stretching," and he emphasizes that concepts "grow" while theories are refuted and replaced.

5. I do not mean to imply that concepts are ever "fixed." I am only momentarily interrupting the historical development to get a clearer view. I am grateful to my friend J. Ravetz, who pointed out the necessity of this clarification.

opposed to this position, and it was even more strongly so in the nineteenth century.

To put the problem in historical context it is useful to see what Helmholtz' contemporaries thought of his "Kraft." Maxwell, in an article in *Nature,* says: "Helmholtz' essay 'Ueber die Erhaltung der Kraft' . . . we must now (and correctly, as a matter of science) translate as Conservation of Energy, though in the translation which appeared in Taylor's scientific memoirs, the word 'Kraft' was translated as 'Force,' in accordance with the literary usage of that time."[6] In Maxwell's time the English scientific journals still used the terms "force" and "energy" interchangeably. Even as late as 1895 the substitution approach was widely accepted. (By this I mean that it was thought that one could simply substitute "energy" or "force" for "Kraft.") Thus, in that year, T. C. Mendenhall wrote in an essay on Helmholtz' conservation paper that "its excellence is shown by the fact that if rewritten today it would be changed only a little in its nomenclature."[7] Such examples could be quoted by the hundreds. One of the few historians of science to comment on this is J. Agassi in his *Towards a Historiography of Science.*[8] Agassi mentions that Helmholtz follows Faraday in advocating a "conservation of force," but that later Helmholtz "unscrupulously changes his tune and also relabels his 'On the Conservation of Force' as 'On the Conservation of Energy.' " I do not agree with his interpretation of Helmholtz' volte-face in terminology as "unscrupulous." The volte-face was unconscious. It is exactly the change from "Erhaltung der Kraft" to "Constanz der Energie" which has to be investigated. I hope to show that this change was natural and inevitable.

Let me make it very clear that I am not engaged in any sort of historical debunking of the value of Helmholtz' contributions to science. I consider this case as an illustration of that general aspect of

6. J. C. Maxwell, "Scientific Worthies: Hermann von Helmholtz," *Nature, 16* (1877), 389.

It seems to have been a characteristic of Maxwell to attribute to other great physicists only opinions which seemed to him the correct ones. I shall give below an even more convincing example of this attitude: Maxwell assumes that Faraday knows exactly the difference between "force" and "energy" and only uses a confused terminology.

7. T. C. Mendenhall, "Helmholtz," *Ann. Rep. Smiths. Inst., 1895,* p. 787.

8. J. Agassi, *Towards a Historiography of Science,* in Beiheft 2, *History and Theory* (The Hague, 1963), p. 91.

scientific change which I shall call "concepts in flux." By this I mean that at the time that Helmholtz derived the general law of conservation of energy, the two concepts—or, rather, what are for us the two concepts—of force and energy were not at all clearly defined, separate entities, but rather different guises of a vaguely defined "Kraft" that Helmholtz thought of as being conserved. The concept of energy as it is fixed in our minds now (the "now" covers classical, pre-relativity physics) implies the conservation of energy principle as formulated by Helmholtz and others; but it is not the case that these discoverers had a clear idea of energy for which they tried to derive a conservation law. To re-create the conceptual framework of physicists before the new concept emerged requires more than historical precision; it requires an effort in sympathy. As a result of the rapid development of energetics[9] in the late nineteenth century, the energy principle penetrated philosophy and psychology on all levels. We think of kinetic energy, potential energy, and heat as instances of one entity—energy. For the laymen and even for the physicists of the eighteenth and nineteenth centuries, this was not the case. Only long after Lagrange and his "final" formulation of analytical mechanics did the mechanical philosophers realize that the *vis viva* of a moving particle, the potential function of Green, and various other concepts were the same. It is not accidental that what certain Germans called "Leistung," the English and certain other Germans called "effect of force," Smeaton called "mechanical power," and Lazare Carnot called "moment d'activité" or (for different cases) "force vive virtuelle."[10] All this simply means that for different problems—collisions of elastic bodies, collisions of hard bodies, vibrations of springs, or motions along inclined planes—the conserved quantity was thought to be different. There were philosophers

9. The science of energetics incorporates the first law of thermodynamics, but rejects the directional character of the second law. In other words, it does not admit any fundamental difference between heat and other forms of energy, and thus raises the concept of energy to the same mystical level as *Naturphilosophie*'s "Kraft." Energetics is the link between late nineteenth-century *Naturphilosophie* and its heirs, psychophysics and psychoanalysis.

10. In my book in preparation, "The Emergence of the Energy Concept," I show that each of the mechanical philosophers who developed the mechanical energy conservation principle was concerned with one typical case, and that each invented a name for the conserved quantity which he was considering, not realizing that the others were treating the same quantity. It is certainly true that they knew of each other's work and that there was no necessity for such diversity in terminology.

of nature who talked about some vaguely conceived "forces" that were conserved in nature; but it took Helmholtz, with his thorough knowledge of physics and mathematics, to realize what mathematical form this conserved entity must take. The substantiation of this thesis necessitates various kinds of indirect evidence. The only document by Helmholtz which could serve as direct evidence is lost; I am referring to the "philosophical introduction" which he attached to his essay and which, on du Bois-Reymond's advice, he threw into the wastepaper basket.

I shall endeavor to show for the case of Helmholtz that the concept in flux is compatible with an exact mathematical formulation, that Helmholtz' background and the various intellectual influences on him support this approach, and that the internal evidence in the text of his paper shows that the very ambiguity in the word "Kraft" is indispensable for the comprehension of his achievement. By intellectual influences, I mean the double tradition in mechanics (the Newtonian-vectorial and the Lagrangian-analytical), the cross-fertilization between physics and physiology which, in his case, was very strong, and the philosophy of Kant.[11] The philosopher-scientists of the eighteenth and nineteenth centuries used "force" intuitively. The English expressions "force of a muscle," "force of a machine," "force of gravitation," "electric forces," "magnetic forces," "galvanic forces," "mesmeric forces," "vital forces," and "forces of nature" (not to mention all those expressions which had a theological overtone) have their parallels in all the other European languages. All natural philosophers who followed the Cartesian tradition of seeking general principles in nature felt strongly that something in nature had to be conserved. The principle of conservation of force is that vaguely formulated principle to which Helmholtz was committed and which, together with the other influences, led him to a generalization of the law of conservation of *vis viva* and to the creation of the concept of energy. From the 1880's Helmholtz himself talked of "Constanz der Energie."

I hope to show, on the basis of my analysis of Helmholtz' use of words, that the conservation principle must have been primarily an *a priori* one and that there is ample evidence in Helmholtz' works

11. The Kantian influence on Helmholtz' metaphysics will be treated in a subsequent paper.

to justify disregarding his proclaimed extreme inductivism; I intend to work out in a subsequent article the implications for the conservation principle of the admitted influence on Helmholtz of Kant's mechanistic, causal theory.

2. MECHANICS

In his early youth Helmholtz, according to his own evidence, read the works of Newton, Euler, d'Alembert, and, naturally, Lagrange. Before his professional training in medicine, he read mainly in physics, philosophy, and mathematics. The nineteenth century inherited two basically different traditions in mechanics: Newtonian-vectorial mechanics, with its emphasis on forces, and the Leibniz-Euler-Lagrange formulation of analytical mechanics, with its emphasis on the scalar quantities of *vis viva* and the potential function. The major concern of vectorial mechanics, as formulated by the Newtonians, was to measure the action of a force by its momentum. This approach originated with Descartes; Descartes' momentum, however, is undoubtedly a scalar quantity and, as such, serves as a foundation for both traditions. The basic concepts in Newtonian mechanics were space, time, mass, and force. A drawback of the Newtonian formulation was that for cases where constraints occurred, the treatment became rather tedious. Besides, the action-reaction law does not embrace all cases; it proves to be sufficient only for the dynamics of rigid bodies. The great advantage of the Newtonian approach was in treating forces which are not derivable from a work-function, i.e., forces which are not conservative but of frictional origin and which therefore cannot be dealt with by the mechanical energy conservation principle. The basic concepts in the Euler-Lagrange formulation were space, time, mass, and energy. This formulation is applicable only to Lagrangian forces that are conservative (it excludes, for example, frictional forces); here the conservation of mechanical energy holds. The great generality of the scalar treatment of mechanics becomes evident only if Hamilton's principle is introduced; this principle covers all cases, even where the work function is a function of time. If action is defined as the integral of the difference between the kinetic energy and the work function, the principle says: the actual motion realized in nature is that for which the action assumes

268

the smallest value. In all probability Helmholtz did not know of Hamilton's work at this stage. But the reasons for not having taken this into account are deeper. The science of energy had a conceptually different structure from that of classical mechanics. The former was founded on *two* basic principles: the principle of conservation of energy and the principle of least action. The two principles taken together teach us how to select from the multitude of conceptually possible motions the actual one taking place in nature. But here, as usual, the logical development of a science is not parallel to its historical development. To realize the importance of excluding processes which do not take place, it was absolutely necessary to have understood the two fundamental principles of thermodynamics (insofar as Carnot understood them). The principles are the conservation of energy and the entropy-conserving character of complete, reversible cycles.[12]

In 1847 Helmholtz, then twenty-six years old, was still struggling with the Lagrangian formulation of mechanics and trying to reconcile it with the vectorial force-treatment. Although he was a Newtonian "mechanical philosopher," he must have realized and appreciated the methodological advantages of the analytical approach. In the latter, constraints are dealt with in an elegant and easy manner and there is no need for hypotheses concerning forces of constraint. Mathematically, however, the two formulations of mechanics are equivalent, and Helmholtz and his contemporaries must have asked the fundamental question which Cornelius Lanczos raised and answered in his *The Variational Principle of Mechanics:*

> Since motion by its very nature is a *directed* phenomenon, it seems puzzling that two scalar quantities should be sufficient to determine motion. The energy theorem which states that the sum of the kinetic and potential energies remains unchanged during the motion, yields only one equation while the motion of a single particle in space requires three equations; in the case of mechanical systems composed of two or more particles the discrepancies become even greater. And yet it is a fact that these two fundamental scalars contain the complete dynamics of even the most complicated material systems, provided they are used as the basis of a principle rather than of an equation.[13]

12. This topic is discussed at length in my forthcoming book.
13. Cornelius Lanczos, *The Variational Principle of Mechanics,* 2nd ed. (Toronto, 1962), p. xviii.

269

The principle referred to is that of least action—a unifying principle which implicitly includes all the equations of motion of a system. This must have raised the question whether the two scalars and the mechanical energy equation in which they occur cannot be generalized beyond mechanics.

What Helmholtz actually did was the only thing he could have done, logically speaking, in view of his premises. He was looking for a unification of the two traditions in mechanics, knowing their individual advantages and insufficiencies. He was very much committed, *a priori,* to two fundamental beliefs: all phenomena in physics are reducible to mechanical processes, and there is a basic entity in nature that is conserved. And in his physiological studies he was a thorough reductionist: all organic processes, he believed, are reducible to physics. If Helmholtz' premises are collected systematically, his reconstructed argument will appear as follows:

i) Newtonian "force" is a fundamental concept in mechanics.
ii) Physics is reducible to mechanics.
iii) The fundamental concept in physiology is "force of life"; physiology is reducible to physics, i.e., to mechanics.
iv) There is a basic entity in nature that is conserved.

Conclusion: the basic conserved entity must be "force" or "Kraft." For Helmholtz' final formulation of the conservation law another premise must be added:

v) The Lagrangian formulation of mechanics is equivalent to the Newtonian formulation mathematically. The fundamental entity of the Lagrangian formulation is mechanical energy, and this entity is conserved.

Conclusion: the basic entity "Kraft" that is conserved in nature must be equivalent in dimension and mathematical form to mechanical energy. This is the generalized conservation of energy principle. After the appearance of Helmholtz' paper, the German "Kraft" came to mean simply energy (in the conservative context) and later gave place slowly to the expression "Energie." The Newtonian "Kraft," with its dimensions of mass times acceleration, became simply "force."

This systemization requires a few qualifying statements:

a) Such an analysis can be done, naturally, only *ex post facto*. The separation of the two conclusions presupposes an awareness of the difference between the conserved "Kraft" and the later "energy"; this again presupposes that the concept has passed the stage of being in flux; i.e., it presupposes the final conclusion.

b) I did not refer to the impossibility of perpetual motion as one of Helmholtz' premises because I do not think it filled that role in his thought-world. Naturally, he knew that a perpetuum mobile was impossible; i.e., it had been known for almost a hundred years without its having resulted in a clear formulation of the principle of conservation of energy in all its generality. The meaning of this impossibility for Helmholtz was that any conservation law must imply the impossibility of perpetual motion.

c) If one still feels inclined to ask why Helmholtz did not say plainly after 1847 that now one had to distinguish clearly between the concepts of force and energy, the answer is simply that he was not aware of the fact that there had ever been any confusion. In making a new discovery or creating a new concept, one has to think in, or work with, a set of older concepts. The logical connection between the new and the old concepts cannot be formulated in the same conceptual framework; one has to be outside of it. This is one sense in which one can understand the thesis that a "logic of discovery" is not possible.[14] The set of concepts before Helmholtz' discovery included "vis viva," "Spannkraft," "living force," etc.; these concepts were not regarded as instances of one conserved entity.

To see how important Helmholtz' training as a mathematical physicist was, it is helpful to compare his case to that of Faraday. Faraday was also committed to the great principle of conservation, but he did not have the guideline afforded by Lagrangian mechanics.

---

14. Anyone who uses the expression "logic of scientific discovery" in such a context is rightly understood to mean that he accepts Popper's unequivocal "no" to the question: "Is there a logic of scientific discovery?" The sense in which I use it here, though naturally Popperian, is best expressed by Emile Meyerson in his *Identity and Reality* (New York, 1962): our reason is competent to scrutinize everything except itself. When one reasons one is really powerless to observe the action of one's reason. This makes a logic of discovery impossible. Not all introspection is denied, but one has to be out of one's own train of thought in order to realize what has been happening. Some individuals never reach this point, while for others it is only a question of time.

271

He taught himself the principles of Newtonian mechanics, but, for lack of mathematical background among other reasons, he was not well acquainted with the Euler-Lagrange formulation of analytical mechanics. For this reason his quest for a great unifying principle of nature centered around the concept of force. In view of the investigations of L. Pearce Williams, I do not have to convince anyone how deeply Faraday was committed to a metaphysical belief in the conservation of force. There is no question whatsoever that when Faraday used "force," he did not mean "energy";[15] in his famous essay "On the Conservation of Force," published in the *Philosophical Magazine* in 1857 (ten years after Helmholtz' paper was published and several years after it had become available in English in John Tyndall's translation), Faraday wrote:

> The case of a force simply removed or suspended, without a transferred exertion in some other direction, appears to me absolutely impossible. . . . The principle of conservation of force would lead us to assume that when A and B attract each other less, because of increasing distance, then some other exertion of power, either with or without them, is proportionately growing. . . . When the equivalents of various forms of forces, as far as they are known, are considered, their differences appear very great; thus a grain of water is known to have electric relations equivalent to a very powerful flash of lightning. It may therefore be supposed that a very large apparent amount of force causing the phenomena of gravitation, may be the equivalent of a very small change in some unknown condition of the bodies whose attraction is varying by change of distance.[16]

These passages illustrate beyond doubt that the entity dealt with is not energy. Faraday's "force" is directional; it is equivalent to some "extension of power." Faraday, who knew of Helmholtz' paper, must have been impressed much more by the verbal part of it than by the mathematical. If one compares his use of the word "force" with Helmholtz' use of "Kraft," the similarity is immediate. There is a letter from Maxwell to Faraday, which Pearce Williams published for the first time, that sheds light on the whole problem. Fara-

---

15. See L. Pearce Williams, *Michael Faraday* (New York, 1965), p. 457.
16. Michael Faraday, "On the Conservation of Force," *Proc. Roy. Soc., 2* (1857), 352–365, republished in the *Phil. Mag., 13* (1857), 225–239.
Faraday's theory of the conservation and correlation of forces underlies his charming Royal Institution lectures for "a juvenile audience," *On the Various Forces of Nature* (New York, 1960).

day had asked Maxwell what he thought of his essay "On the Conservation of Force." Maxwell replied: "Now first I am sorry that we do not keep our words for distinct things more distinct and speak of the 'conservation of Energy' as applied to relations between the amounts of 'vis viva' and of 'tensions' in the world."[17]

Maxwell assumes (according to his usual habit) that Faraday has distinct concepts, and that Faraday is just not careful enough with his choice of words. In the same letter, he defines force and energy very clearly: "Energy is the power a thing has of doing work arising either from its own motion or from the tension subsisting between it and other things. Force is the tendency of a body to pass from one place to another and depends upon the amount of change of 'tension' which that passage should produce."[18] Faraday disagreed basically with this definition: "I perceive that I do not use the word 'force' as you define it, 'the tendency of a body to pass from one place to another.' When I mean by the word is that source or sources of all possible actions of the particles or materials of the universe."[19] If Faraday had had Helmholtz' mathematical training, he either would have realized that "force" is not conserved or he would have discovered the principle of the conservation of energy. But this is idle speculation. As it was, his ideas were in a state of flux; his conservation principle was vague and metaphysical. To the end of his days, his ideas were similar to those with which Helmholtz began, and from which Helmholtz evolved his ideas on conservation of energy.

Many years after his discovery of the fully general principle of conservation of energy, Helmholtz returned to the fundamentals of mechanics and to the role of mechanics in physics. Never abandoning his fundamentally mechanistic philosophy, one of his last original researches was his paper on the principle of least action, published in Crelle's journal in 1887. At this stage, he was as deeply influenced by the generality and beauty of Hamilton's principle, as were de Broglie, Schroedinger, and Feynman in our century. I intend to discuss at length the importance and place of the least-action principle in Helmholtz' philosophy of physics in another place.

17. L. P. Williams, *op. cit.* (note 15), p. 511.
18. *Ibid.*
19. *Ibid.*, p. 514.

3. PHYSIOLOGY

Helmholtz was a physician by training, and, beginning in 1839, he spent several years in Johannes Mueller's laboratory, where the most brilliant thinkers in physiology and medicine were working;[20] these thinkers included du Bois-Reymond, Bruecke, Virchow, and the somewhat older Ludwig. Most of them were considered to be thorough reductionists. At the same time, while under Mueller's guidance, they tried to solve the problem of the sources of animal heat, a problem formulated in Liebig's "vitalistic" language. What I wish to emphasize is that they tried to reduce to mechanics a problem which belonged to biology and which was formulated with the help of concepts like "vital force" or "forces of life." When assessing the influence of his background on Helmholtz' "On the Conservation of Force," we have to bear in mind that this was his first paper in physics. His previous papers were all connected with the question of animal heat, and they all show a preoccupation with the balance of "forces" in the basic life processes. One of these was a review article, "Bericht ueber die Theorie der physiologischen Waermeerscheinungen fuer 1845,"[21] published in the *Fortschritte der Physik im Jahre 1845*. It is notable that Helmholtz, when editing his collected scientific papers in 1881, grouped this among the papers headed "Zur Lehre von der Energie."

Especially revealing is the article "Waerme, physiologisch" that Helmholtz wrote in 1845 for the *Encyklopaedisches Hand-Woerterbuch der medicinischen Wissenschaften*. It reads like an expanded version of all the above premises; only the final conclusion is not drawn. Helmholtz explains at length both the caloric and the mechanical theories of heat. He considers the motion theory as the correct one, but emphasizes its incomplete state. He states explicitly his reductionist views—both of physics to mechanics, and of life processes to physics. He states clearly that there must be a unique connection between the mechanical conservation law and the "life

20. See Everett Mendelsohn, "The Biological Sciences in the Nineteenth Century: Some Problems and Sources," *Hist. of Sci., 3* (1964), 39–59. See also Mendelsohn, "Physical Models and Physiological Concepts: Explanation in Nineteenth-Century Biology," *Boston Studies in the Phil. Sci., 2* (1965), 127–150.
21. All three are in Helmholtz' *Wissenschaftliche Abhandlungen*, 3 vols. (Leipzig, 1882–1895). For further treatment of these papers see my forthcoming book.

forces." His belief in a basic conservation law in nature is not stated in so many words, but it can be read between the lines.

In later years, Helmholtz repeatedly said that his "On the Conservation of Force" was written primarily for physiologists. Though he had read in his early youth the physical writings of Newton, Euler, d'Alembert, and Lagrange, he was active in the 1840's in physiological research, and he formulated his research problems in biological language. It is natural for Rueckert to have said of Helmholtz in 1894:

> The study of medicine led him to the problem of the nature of "vital forces." He convinced himself that if—as Stahl had suggested—an animal had the power now of restraining and now of liberating the activity of mechanical forces, it would be endowed with the power of perpetual motion. This led to the question whether perpetual motion was consistent with what was known of natural agencies. The Essay on the Conservation of Force was, according to von Helmholtz himself, intended to be a critical investigation and arrangement of facts which bear on this point for the benefit of the physiologists.[22]

The controversy in which Helmholtz took such a prominent part was over Liebig's "vitalistic" theory. The question was whether the forces responsible for the production of physiological heat were "vital forces" or well-known physico-chemical processes. (The discussion among historians of physiology as to what extent Liebig can be labeled a "vitalist" is of no concern here.) The ambiguity of the terms "vital forces" and "Kraefte" in physiology is well illustrated in June Goodfield's *The Growth of Scientific Physiology*. Goodfield quotes the following passage from Liebig: "In the animal ovum as well as in the seed of a plant, we recognize a certain remarkable force, a source of growth . . . this force is called the vital force, vis vitae or vitality."[23] She continues:

> No explanation is given here of force: the paragraph as it stands is the description of a potentiality. But if we read further, we find that he does mean more; he believes that there is a unique agency operating in living material, different in nature and manifestation from any other. . . . His vital force, like Bichat's vital properties, has no connec-

22. A. W. Rueckert, "Helmholtz," *Ann. Rep. Smiths. Inst., 1894*, p. 709.
23. J. Goodfield, *The Growth of Scientific Physiology* (London, 1960), pp. 135–145.

tion with the soul or mind, and is directly open to experimental study. . . . Nor is this vital force *immaterial*. . . . Liebig introduces the word "force" into his physiological framework . . . treats it as a kind of central force, capable of giving rise either to resistance or to motion: "The chemical force, which kept the elements together acted as a resistance, which was overcome by the active vital force." Several things must be noted about this passage; first of all two ambiguities. To begin with the word "resistance" appears sometimes to mean actual physical "pushes and pulls" and sometimes the capacity to slow down or prevent the occurrence of "physical or chemical processes." Of course, for anyone thinking in terms of central forces, the "invincible resistance" of a compound "to the action of the decomposing agent" would seem explicable only by supposing the decomposing force to be weaker than the binding force. . . . [24]

Full justice is done here to the difficulty, and the conceptual muddle is clearly explained. However—and this is the reason I have been quoting at such length—I cannot accept the treatment of the problem merely as a verbal one.

Discussing Liebig, Goodfield says: "Suppose for that word 'force' with its precise twentieth century associations, we here substitute the word 'energy.' The passage at once becomes clear and acceptable. Liebig's word 'Kraft' in fact carried the meaning both of 'force' and of our 'energy.' In 1840, the distinction between the two had not yet been made clear, and the word 'Energie' did not appear in scientific papers until some twenty years later."[25] It is precisely this sort of substitution which cannot be carried out consistently. Not only was the distinction not made between force and energy, but many sentences have meaning only if "Kraft" is left intact; similarly, Faraday's "force" must be understood in the framework of his own definition, otherwise his essay "On the Conservation of Force" becomes meaningless. It was not the unavailability of the term "energy" that prevented its being used. Thomas Young introduced it with a clearly defined meaning, denoting "vis viva" as "energy" in his 1807 *Lectures on Natural Philosophy,* and Thomson and Tait used it in articles and in their textbook. Helmholtz used it sometimes in the same vague sense as he did "Kraft": "weil wir die Proteinverbindungen ueberall als Traeger der hoechsten Lebens-

24. *Ibid.*
25. *Ibid.*

276

energien finden..." ("because we find the protein combinations every-where to be the carriers of the highest energies of life . . . ").[26] Here, even more than in discussions of physics, the ambiguity of the word "Kraft" was the very condition of intelligibility. The physiological problem of the sources of animal heat could have been resolved only after the physical law of the conservation of energy had been formulated. The moment that the physical balance of energies had been measured for living bodies, a thorough vitalist had to shift the idea of a special vital something from the question of physiological heat to a new and as yet unsolved question. In the same physiological essay, Helmholtz writes:

Eine der hoechsten, das Wesen der Lebenskraft selbst unmittelbar betreffenden Fragen der Physiologie, naemlich die, ob das Leben der organischen Koerper die Wirkung sei einer eigenen, sich stets aus sich selbst erzeugenden, zweckmaessig wirkenden Kraft, oder das Resultat der auch in der leblosen Natur thaetigen Kraefte, nur eigenthuemlich modificirt durch die Art ihres Zusammenwirkens, hat in neuerer Zeit, besonders klar in Liebig's Versuch, die physiologischen Thatsachen aus den bekannten chemischen und physikalischen Gesetzen herzuleiten, eine viel concretere Form angenommen, naemlich die, ob die mechanische Kraft und die in den Organismen erzeugte Waerme aus dem Stoffwechsel vollstaendig herzuleiten seien, oder nicht.[27]

One of the most important questions of physiology, one that touches on the very essence of "Lebenskraft," is the question whether the life of organic bodies is the action of a purposefully acting "Kraft," which has the power to create itself, or the effect of "Kraefte," which are active also in lifeless nature, and are specially modified through their interactions; this question has recently achieved a much more concrete form in Liebig's attempt to derive the physiological facts from known chemical and physical laws. This new form of the question is: can the mechanical "Kraft" and heat created in the organism be obtained in the metabolic process?

It is not really necessary to show again, word by word, that the above passage is untranslatable into modern terminology. There is, however, another point in this passage worth noting. It seems as if Helmholtz denies that Liebig is a vitalist; he clarifies his position

26. Helmholtz, "Ueber den Stoffverbrauch in der Muskelaction," *Mueller's Archiv* (1845), pp. 72–83, or in *Wiss. Abh.*, 2, 735–744.
27. Helmholtz, *loc. cit.*

in a review article discussing Liebig's work: "Liebig, in his Animal Chemistry . . . lays down the theoretical demand, that the origin of heat, as a principle that is to correspond to a 'Kraftaequivalent,' can be derived from other 'Kraefte' and not out of nothing."[28] Liebig's "vital forces" were neither immaterial nor metaphysical in the sense of having something in common with soul or mind; this seemed to Helmholtz to be in line with his reductionist demands. I would rather overemphasize this point than allow it to pass unnoticed; in 1845 Helmholtz' concepts were still in a state of flux to the extent that he could accommodate in his conceptual framework "vital forces" while believing that they could be reduced to physico-chemical terms.

It cannot have been an accident that Julius Robert Mayer, who came to realize the energy conservation principle as generally as did Helmholtz, arrived at his result through physiology. Mayer was no mere "hunch" philosopher, though it is difficult to find the logical path from the change in the color of venous blood in the tropics to a full annunciation of the principle of conservation of energy. His work gives us a real flavor of mid-nineteenth-century biology, when concepts were indeed in a state of flux. I will quote a passage from Mayer's first paper where he formulates his theory of the mechanical equivalent of heat. In all fairness to him I will say in advance that, though his first version was full of mistakes, his next was not, having enlisted in the meantime the help of his physicist-friends. The longer, corrected version appeared in 1845, and it is on this that his fame justly rests. However, for our purpose, the first paper is much more useful, as it reveals the muddled terms from which the principle of energy conservation emerged:

Der Zweck folgender Zeilen ist, die Beantwortung der Frage zu versuchen, was wir unter 'Kraeften' zu verstehen haben und wie sich solche untereinander verhalten. Waehrend mit der Benennung 'Materie' einem Objekte sehr bestimmte Eigenschaften, als die der Schwere, der Raumerfuellung zugeteilt werden, knuepft sich an die Benennung Kraft vorzugsweise der Begriff des Unbekannten, Unerforschlichen, Hypothetischen. Ein Versuch, den Begriff von Kraft ebenso praecis als

28. Helmholtz, "Bericht ueber die Theorie der physiologischen Waermeerscheinungen fuer 1845," *Fortschritte der Physik fuer 1845* (1847), p. 347, or in *Wiss. Abh., 1,* 1.

den von Materie aufzufassen, und damit nur Objekte wirklicher Forschung zu bezeichnen, duerfte, mit den daraus fliessenden Konsequenzen, Freunden klarer, hypothesenfreier Naturanschauung nicht unwillkommen sein.

Kraefte sind Ursachen; mithin findet auf dieselben volle Anwendung der Grundsatz: *causa aequat effectum*. Hat die Ursache c die Wirkung e, so ist c = e; ist e wieder die Ursache einer andern Wirkung f, so ist e = f, usf. c = e = f = ... = c. In einer Kette von Ursachen und Wirkungen kann, wie aus der Natur einer Gleichung erhellt, nie ein Glied oder ein Teil eines Gliedes zu Null werden. Diese erste Eigenschaft aller Ursachen nennen wir ihre Unzerstoerlichkeit.[29]

I shall use the English translation from the collection *Correlation and Conservation of Forces* (Faraday's influence is evident in the title), edited by Youmans in 1865:

The following pages are designed as an attempt to answer the questions, what are we to understand by "Forces"? and how are different forces related to each other? Whereas the term *matter* implies the possession, by the object to which it is applied, of very definite properties, such as weight and extension, the term force *conveys* for the most part the idea of something unknown, unsearchable, and hypothetical. An attempt to render the notion of force equally exact with that of matter, and so to denote by it only objects of actual investigation, is one which, with the consequences that flow from it, ought not to be unwelcome to those who desire that their views of nature may be clear and unencumbered by hypotheses. . . . Forces are causes, accordingly, we may in relation to them make full application of the principle—*causa aequat effectum*. If the cause c has effect e, then c = e; if, in its turn, e is the cause of a second effect f, we have e = f, and so on: c = e = f ... = c. In a chain of causes and effects, a term or part of a term, can never, as plainly appears from the nature of the equation, become equal to nothing. This first property of all causes we call their *indestructibility*.

Mayer sees the term "force" to have been hopelessly muddled and unscientific. By announcing his conservation principle, by defining forces as causes, by taking recourse to the old scholastic argument *causa aequat effectum*, Mayer hopes to have made the concept of

29. J. R. Mayer, "Bemerkungen ueber die Kraefte der unbelebten Natur," *Ann. der Chem. Pharm.*, *42* (1842), 233–241. Reprinted in *Raum und Kraft*, ed. E. Wildhagen (Berlin, n.d.), pp. 357–363. The essay is translated under the title "The Forces of Inorganic Nature," *The Correlation and Conservation of Forces*, ed. E. L. Youmans (New York, 1865), pp. 251–258.

"force" well-defined and less hypothetical. What Mayer could not have done was to give the principle the mathematical form that defines our concept of energy. It becomes clear that for the period when the concepts were in a state of flux, we cannot follow logically what happened between the stage when those vaguely formulated principles were conceived and the point when their final, general form was realized. Even when Mayer's conservation principle served him as a "science producing" tool for further research, it was not yet in its final form; the concept of energy had yet to be defined.

4. THE LANGUAGE OF THE "ERHALTUNG DER KRAFT"

In Helmholtz' "Ueber die Erhaltung der Kraft" we find the Kantian matter-force dualism, a vague conception of what a "Kraft" is, and a deep conviction of its indestructibility.[30] It is imperative to show now, in a word-by-word textual criticism, that the substitution theory, which did not work in the case of Liebig and Faraday, will not apply here either; again the ambiguity of the term "Kraft" is essential for understanding the text. But before testing whether one can or cannot merely replace Helmholtz' terms—"Arbeitskraft," "Bewegungskraft," ' 'bewegende Kraft," "lebendige Kraft," "wirkende Kraft"—sometimes by "force" and sometimes by "energy," a procedure sanctioned by such an authority as Maxwell, there is a prior question to be raised: why should Helmholtz, one of the most didactic writers, hesitate to create a new word for a clearly defined concept if he is aware of any ambiguity? He coined the phrase "Spannkraft" for the clearly defined mechanical entity that we call "potential energy"; he created the concept of "circulation" in his hydrodynamical vortex theory; and he invented the expression "cyclid variables" in his mechanical writings. Moreover, the word "Energie" was known to and used by him; besides, he had read Young's *Lectures on Natural Philosophy*, where the word had been introduced, denoting the concept that we call "kinetic energy."

And then, from the late 1870's, Helmholtz began to use the expression "Constanz der Energie," no longer "Erhaltung der Kraft." Especially noteworthy is that in his 1881 notes (prepared for the edi-

30. Helmholtz' philosophical development and his metaphysics will be treated briefly in section 6; an extended treatment will be given in a forthcoming article.

tion of his collected scientific papers[31] and appended to the original 1847 paper) he used the new expression. In his 1887 article, "On the Principle of Least Action," which contains a short review of the energy principle, the old expression never occurs. Helmholtz was famous for his scrupulous honesty in acknowledging errors and priorities; e.g., after some authors had tried to show that he had ignored Mayer's contributions, he did his utmost to repair the damage and to show Mayer's true priority. Had he thought that his 1847 paper had been erroneous or confusing in any respect, he would have certainly remarked on it in his 1881 notes. The only explanation I can see for his not having done so is that he honestly considered the change in terminology from "Erhaltung der Kraft" to "Constanz der Energie" to be merely a verbal change. He was not aware of the conceptual changes accompanying the emergence of the concept of energy. He certainly did not think that anything had been wrong with his concept of "Kraft" prior to his own proof of its conservation. In other words, he was not aware that his concepts had been in a state of flux, and that the emergence of the concept of energy in his own mind and in the minds of his contemporaries was due to his proof of the conservation of energy in 1847.

In the following, I shall give several passages from Helmholtz' 1847 paper, numbering them for easy identification.

1) . . . das alle Wirkungen in der Natur zurueckzufuehren seien auf anziehende und abstossende Kraefte, deren Intensitaet nur von der Entfernung der auf einander wirkenden Punkte abhaengt.

. . . that all actions in nature can be ultimately referred to attractive or repulsive "Kraefte," the intensity of which depends solely upon the distances between the points by which the "Kraefte" are exerted.

2) . . . wenn wir von verschiedenartigen Materien sprechen, so setzen wir ihre Verschiedenheit immer nur in die Verschiedenheit ihrer Wirkungen, d.h. in ihre Kraefte.

. . . for when we speak of different kinds of matter we refer to difference of action, that is, to differences of the "forces of matter."

31. Helmholtz, *op. cit.* (note 43).

3) Es ist einleuchtend, dass die Begriffe von Materie und Kraft in der Anwendung auf die Natur nie getrennt werden duerfen.

It is evident that in the application of the ideas of matter and "Kraft" to nature, the two former should never be separated.

4) Eine reine Kraft waere etwas, was dasein sollte und doch wieder nicht dasein, weil wir das Daseiende Materie nennen. Ebenso fehlerhaft ist es, die Materie fuer etwas Wirkliches, die Kraft fuer einen blossen Begriff erklaeren zu wollen, dem nichts Wirkliches entspraeche . . . wir koennen ja die Materie eben nur durch ihre Kraefte, nie an sich selbst, wahrnehmen.

A pure "Kraft" would be something which must have a basis, and yet which has no basis, for the basis we name matter. It would be just as erroneous to define matter as something which has an actual existence, while force is a mere idea which has no corresponding reality. . . . Matter is only discernible by its "Kraefte," and not by itself.

5) . . . die aeusseren Verhaeltnisse, durch welche die Wirkung der Kraefte modificirt wird, koennen nur noch raeumliche sein, also die Kraefte nur Bewegungskraefte.

. . . the only alteration possible to such a system is an alteration of position, that is, of motion; the "Kraefte" can only be "Bewegungskraefte."

So far, everything seems extremely simple. "Forces" could probably be substituted for "Kraefte." Yet, even here, the resulting expressions like "forces of matter" and "forces of motion" would be deceptive; although they sound perfect in English, they do not have any clearly defined meaning. The influence of the nineteenth century is still strong enough on our language so that these expressions sound extremely clear. But if we listen carefully, we will soon realize that they convey a meaning only if "force" means "a source of all possible actions of the particles or materials of the universe," which is Faraday's definition. Let us continue:

6) Bewegungskraft . . . ist also zu definiren als das Bestreben zweier Massen, ihre gegenseitige Lage zu wechseln. Die Kraft aber, welche zwei ganze Massen gegen einander ausueben, muss aufgeloest werden in die Kraefte aller ihrer Theile gegen einander.

"Bewegungskraft" . . . is therefore to be defined as the endeavor of two masses to alter their relative position. But the "Kraft" which two masses exert upon each other must be resolved into those exerted by all their particles on each other.

7) Eine Bewegungskraft, welche sie gegen einander ausueben. . . .

A "Bewegungskraft" therefore, exerted by each upon the other. . . .

8) Die Kraefte, welche zwei Massen auf einander ausueben, muessen nothwendig ihrer Groesse und Richtung nach bestimmt sein.

But the "Kraefte" which two masses exert upon each other must be given by their intensity and their direction.

The translations are John Tyndall's (with the exception of passage [8]), from Taylor's *Scientific Memoirs* (1854). This is a noteworthy fact because Tyndall held very similar ideas to those of Faraday on the nature of force and felt a strong intellectual affinity between Faraday's and Helmholtz' work. (In view of this intellectual relationship it is not surprising that Helmholtz wrote the introduction to the German edition of Tyndall's *Faraday as a Discoverer*.) Concerning the translation, let me note only a few features: passage (8) is my translation, not having been translated in full by Tyndall; the full translation makes it explicit that the "Kraefte" have both intensity and direction. Another feature is that the concept "Bewegungskraft" is once translated as "moving force," whatever this may mean, and at other places as "force, which originates motion," which sounds more like an explanation than a translation. Could it be possible that the translator in 1854, in the years when the use of the word "energy" in mechanics was gaining ground through the publications of Thomson and Tait, felt somewhat uneasy? "Bewegungskraft" for Helmholtz is a "tendency" or, as Tyndall puts it, an "endeavor"; it is a tendency exerted by two masses having both intensity and direction! But this is not all:

> 9) Es bestimmt sich also endlich die Aufgabe der physikalischen Naturwissenschaften dahin, die Naturerscheinungen zurueckzufuehren auf unveraenderliche, anziehende und abstossende Kraefte. . . . Die Loesbarkeit dieser Aufgabe ist zugleich die Bedingung der vollstaendigen Begreiflichkeit der Natur. Die rechnende Mechanik hat bis jetzt diese Beschraenkung fuer den Begriff der Bewegungskraft nicht angenommen, einmal weil sie sich ueber den Ursprung ihrer Grundsaetze nicht klar war. . . .

> Finally, therefore, we discover the problem of physical natural science to be, to refer natural phenomena back to unchangeable attractive and repulsive forces. . . . The solvability of this problem is the condition of the complete comprehensibility of Nature. In mechanical calculations this limitation of the idea of "Bewegungskraft" has not yet been assumed. . . .

The last clause of the quoted German passage is missing in the translation. It should read: "On the one hand, because it was not yet certain about the origin of its fundamental theorems. . . ." But this is the crucial point. What was the origin of the fundamental theorems that had not been clearly established in mechanics at this stage? Helmholtz answers this question at the beginning of the next section of his essay:

> 10) Wir gehen aus von der Annahme, dass es unmoeglich sei durch irgend eine Combination von Naturkoerpern bewegende Kraft fortdauernd aus nichts zu erschaffen.

> We will set out with the assumption that it is impossible by any combination whatever of natural bodies, to produce "bewegende Kraft" [here Tyndall uses "force"] continually from nothing.

For Helmholtz, the principle of the impossibility of a perpetual-motion machine is the origin of one of the fundamental theorems, namely, the basic conservation law.[32] Such language could sound offensive enough to the ears of the English inductivists in 1854 so that the last phrase of quotation (9) could have been left out on purpose. In the last passage quoted, the translator has also given up the battle with the "bewegende Kraft"; he simply translates the expression as "force." In passage (9) we have seen that "Bewegungskraft" must be closer to our "force" than to our "energy," having both intensity and direction. Could we reasonably say that the same word in passage (10) should be translated as "energy" to sound more correct in modern terms? The very formulation of the principle of the impossibility of a perpetuum mobile uses the vague term "bewegende Kraft" and is thus difficult to translate exactly; but in the German original it all sounds very smooth. And even if we did translate "Kraft" as "Energie," how could we possibly talk about "bewegende Energie"? This, even more than "Bewegungskraft," sounds like an active entity; it should mean something active and having direction, and it is not exactly the same as "Kraft der Bewegung" ("bewegende Energie" could possibly be translated as "energy of motion").

The difficulties of passage (10) are compounded in the sentence immediately following it:

11) Aus diesem Satze haben schon Carnot und Clapeyron eine Reihe theils bekannter, theils noch nicht experimentell nachgewiesener Gesetze . . . theoretisch hergeleitet.

By this proposition [impossibility of a perpetuum mobile] Carnot and Clapeyron have deduced theoretically a series of laws, part of which are proved by experiment, and part not yet submitted to this test.

If the concept of energy had had a fixed meaning for Helmholtz he would have read it into Carnot. But Carnot was talking about "force

32. Although in his 1847 paper Helmholtz posits the impossibility of perpetual motion as a fundamental assumption for the derivation of the principle of conservation, in his 1863 popular lecture on "The Conservation of Force" he says: "The possibility of a perpetual motion was first finally negatived by the law of the conservation of force . . ." (*Popular Scientific Lectures* [New York, 1962], p. 222).

vive" and "force vive virtuelle," both clearly defined terms applying solely to conservative systems; Clapeyron and Carnot both used the expression "puissance motrice." Whatever difficulties Clapeyron or Carnot may have had (whether heat or entropy is what they took to be conserved),[33] they certainly did not have the "force" versus "energy" difficulty. Helmholtz should have read them as if they were writing about energy, had this concept already had a fixed meaning for him. Clearly it had not.

I shall end this textual criticism with two more examples:

12) Denken wir uns ein System von Naturkoerpern, welche in gewissen raeumlichen Verhaeltnissen zu einander stehen, und unter dem Einflusse ihrer gegenseitigen Kraefte in Bewegung gerathen, bis sie in bestimmte andere Lagen gekommen sind: so koennen wir ihre gewonnenen Geschwindigkeiten als eine gewisse mechanische Arbeit betrachten und in solche verwandeln. Wollen wir nun dieselben Kraefte zum zweiten Male wirksam werden lassen, um dieselbe Arbeit noch einmal zu gewinnen, so muessen wir die Koerper. . . .

Let us imagine a system of natural bodies occupying certain relative positions towards each other, operated upon by "Kraefte" mutually exerted among themselves, and caused to move until another position is attained; we can regard the velocities thus acquired as a certain mechanical work and translate them into such. If we now wish the same "Kraefte" to act a second time, so as to produce the same quantity of work, we must in some way, by means of other "Kraefte" placed at our disposal, bring the bodies. . . .

Should we now say that *velocities are* mechanical work? And finally:

13) Nennen wir nun die Kraefte, welche den Punkt m zu bewegen streben, so lange sie eben noch nicht Bewegung bewirkt haben, Spannkraefte, im Gegensatz zu dem, was die Mechanik lebendige Kraft nennt, so wuerden wir die Groesse $\int_r^R \phi dr$ als die *Summe der Spannkraefte* zwischen den Entfernungen R und r bezeichnen koennen.

33. Much has been written concerning this fascinating topic, but it is still not fully realized that the caloric theory was a far more sophisticated and theoretically satisfactory structure than the mechanical theory in its initial stages. Moreover, the mathematical treatment of Lavoisier and Laplace, of Fourier, and in great part of Carnot and Clapeyron, does not really presuppose the mechanism or the nature of heat. The connection between the two theories of heat and the development of the concept of energy will be treated at length in my book.

Calling the "Kraefte" which tend to move the point m before the motion has actually taken place, "Spannkraefte," in opposition to that which is meant in mechanics, "vis viva," then the quantity $\int_r^R \phi dr$ would be the *sum of the "Spannkraefte"* between the distances R and r.

All this sounds very clear in German. The English translation with "forces" for "Kraefte" and "tensions" or "potential energy" for "Spannkraefte" is also without difficulty, at first glance. But in passages (6) and (8) we found that "Kraft" in this context, where it constitutes a "tendency," should be rendered as "force" (in our sense); the "Spannkraefte" in passage (13), however, as is clearly seen from the mathematical expression, must be understood as our energy or, rather, our potential energy.

Planck, who also read Helmholtz "correctly" in attributing the use of "Kraft" instead of "energy" to the different terminology of an earlier period, had the following to say:

As long as there was no clear notion connected with the word "Kraft" any dispute on the quantity of this "Kraft" was without a proper theme. Yet it must be admitted that this dispute had a much deeper content at its foundation; for, the parties to the dispute were to some extent united, even if they did not express this very clearly and often, as to what they wanted to understand under the word "Kraft." Descartes, as well as Leibniz, had certainly some, even if not very precise, notion about a principle, which expresses the unchangeability and indestructibility of that of which all motion and action in the world consists.[34]

I read this statement as another formulation of my claim that new ideas and discoveries can and often do grow out of a pool of vague concepts and that this vagueness is essential for the formulation of the problem.

To sum up the argument: these passages, taken altogether, are intelligible only if we leave "Kraft" unchanged. The Helmholtzian "Kraft" is a vague entity—something which must be conserved in nature. "Kraft" is a concept in flux that emerged as our "energy" only after Helmholtz had completed his mathematical reduction of all "Kraefte" to mechanical energy.

34. Max Planck, *Das Prinzip der Erhaltung der Energie* (Leipzig, 1913), p. 9.

5.   HELMHOLTZ AND HIS ENGLISH CONTEMPORARIES

Helmholtz' 1847 essay was technical; he addressed it expressly to physicists. Physicists either detected immediately the *a priori* flavor of the work, and in view of their strong inductivist philosophy discarded it completely, or they accepted it immediately, because of its mathematical argument. To this second group belonged the mathematicians Jacobi and du Bois-Reymond. Du Bois-Reymond had worked with Helmholtz in Johannes Mueller's laboratory on the problem of physiological heat, and he understood immediately what Helmholtz' work meant for that problem. It is clear from du Bois-Reymond's scattered remarks that he realized the nonempirical basis of Helmholtz' work, but it did not deter him from examining its implications. Jacobi would have been drawn to Helmholtz' work through a common concern with analytical mechanics. The majority of physicists, however, ignored it. Poggendorff, whose *Annalen* had rejected Mayer's work a few years previously, also rejected "Ueber die Erhaltung der Kraft," and Helmholtz had to publish it privately. Considering that it was regarded as a major task of science to eliminate all nonempirical elements, it is small wonder that Mayer's and Helmholtz' works seemed too speculative.

The confusion between the concepts of force and energy was stronger in England than in Germany. One reason was that Faraday's "On the Conservation of Force"—a nontechnical, popular lecture read by scientist and layman alike—was so influential. Especially revealing of its influence is a debate, conducted in the pages of *Nature,* between John Moore and the well-known physicist Charles Brooke. Moore had written an article, "The Heresies of Science," which Brooke answered with a long article "Conservation of Energy a Fact —Not a Heresy of Science." Brooke wrote that "there is probably no term employed in physics that has been more misapplied, and in its misuse has led to greater confusion of ideas than 'force'."[35] He implicitly accepted Faraday's definition of force, characterizing it as the "source or sources of all possible actions on the particles or materials of the universe." He considered his definition an amplification or explanation of Faraday's: "Force is a mutual action between the atoms or molecules of matter, by which they are either attracted

35. Charles Brooke, "Force and Energy," *Nature, 6* (1872), 122.

towards or repelled from, each other; and by this action energy is imparted to the matter put in motion. It may be further remarked that force is essentially either an attraction or a repulsion." That for Brooke there were no vague concepts involved and that he knew exactly what he was talking about may be seen from the following passage: "If the above definition of force and energy be accepted, it is obvious that the term 'force,' as used by Grove, Tyndall and many others, means sometimes force and sometimes energy."

Brooke, like Maxwell, read Faraday "correctly." But Moore had read his Faraday too, and in his first paper he wrote that "a given motion viewed as a cause is force, while the very same motion thought of as an effect is energy."[36] In his reply to Brooke's sharp criticism of his first paper, Moore wrote:

> Mr. Brooke says that "energy is the power of doing work." He does not tell us what he means by work. If he means motion in any of its modes, then he confounds what he holds to be distinct realities viz. Force and Energy. The "Theory of Conservation of Energy" as now maintained by physicists is opposed in several respects to the doctrine of the "Conservation of Force" as held by Faraday. Stewart Brooke and others teach most explicitly that energy is not only constantly changing its form, but always shifting about from one position of matter to another. If I mistake not, Faraday asserts the very opposite respecting force. He seems to teach that each material particle, into whatever combinations it may enter, retains all its original forces.

Stewart Brooke, the well-known author of a physics textbook, and "the others" are all those clear-thinking physicists who, in the tradition of Thomson and Tait's textbook, had very clear ideas about force and energy in mechanics and had no great concern with the unifying principles of nature. Moore quoted Faraday's famous "Conservation and Correlation of Forces": "Holding as I do, that forces are both conserved and correlated, I feel no difficulty whatever in accepting the facts established by Dr. Joule. . . . But the molecular actions or motions are the effects of force but not the force itself. In no instance whatever can force be resolved into molecular action." The last sentence again is a reflection of Faraday's definition of force. In short, those physicists who were dealing with specific problems

36. John Moore, "The Conservation of Energy not a Fact, but a Heresy of Science," *Nature, 6* (1872), 180.

had very clear ideas about energy and force, and they either ignored Faraday's work as speculative or simply read him "correctly." Some others, whose questions were of the same kind and breadth as Faraday's, understood him on a vague, metaphysical level, and utter confusion was the result.

Holding with Meyerson[37] that the depth of understanding of new ideas (by the society at large of a period) is best judged by seeing what the nonprofessionals have to say, I will quote from J. Nicolson's paper on the conservation of force, which appeared in *Nature* in 1871. Claiming to have spent much thought on Faraday's "On the Conservation of Force," he asked:

> What then, does the "conservation of forces" doctrine amount to in plain English? It amounts to the simple admission that the tendency to move is a property of matter inseparable from it and coexistent with it, and it is this tendency to move which is the cause of all the changes which we observe around us. There is however nothing new under the sun, for the old doctrine of Argan in Le Malade Imaginaire is revived again; when Argan answers his examiner for a license to practise in medicine, he says:
>
> > "Mihi a docto Doctore
> > Domandatur causum et rationem quare
> > Opium facit dormire
> > Quiae est in eo
> > Virtus dormitiva
> > Cujus est natura
> > Sensus assoupire."
>
> Many a clever student has laughed at this answer who little thought that research and experience would confirm it so strongly as they do now.[38]

I think that this passage speaks for itself; Faraday is considered to have revised the old scholastic occult qualities.

I have chosen to treat the reaction to Faraday's paper rather than to Helmholtz' for two reasons. First, Helmholtz was very reluctant to participate in any scientific debate if he thought it fruitless. In his preface to the German edition of Thomson and Tait's *Natural Phi-*

37. Meyerson, *op. cit.* (note 14), p. 196.
38. J. Nicolson, "The Conservation of Force," *Nature, 4* (1871), 47–48.

losophy, Helmholtz wrote: "I have as a rule considered it necessary to reply to criticisms of scientific propositions and principles only when new facts were to be brought forward or misunderstandings to be cleared up, in the expectation that, when all data have been given, those familiar with the science will ultimately see how to form judgement even without the discursive pleadings and sophistical arts of the contrary parts."[39] Moreover, he did not think that there was anything to be cleared up; whenever anyone exhibited confusion, Helmholtz classified him as one who was not familiar with the science.

My other reason for dealing with Faraday is that Helmholtz was preoccupied with Faraday's achievements. In the preface to Tyndall's *Faraday as a Discoverer,* Helmholtz wrote:

> It is moreover by no means for the philosopher only that such an insight possesses interest. His interest certainly is the most immediate, for it has hardly been the lot of any single man to make a series of discoveries so great and so pregnant with the weightiest of consequences as those of Faraday. Most of them burst upon the world as surprises, the products apparently of an inconceivable instinct; and Faraday himself, even subsequently was hardly able to describe in clear terms the intellectual combinations which led to them. These discoveries, moreover, were all of a kind calculated to influence in the profoundest manner our notions of the nature of Force, and in the presence of Faraday's electromagnetic and diamagnetic discoveries more particularly, it was impossible for the old notions of force acting at a distance to maintain themselves without submitting to essential expansions and alterations.[40]

This is a very revealing passage indeed. Helmholtz acknowledges fully the vague, loosely formulated terms with which Faraday's mind worked. It may be that he thought this to have been a peculiarity of Faraday's, but he certainly admits how very much those hypothetical, unproved ideas resulted in a change in our conception of force. This is an indirect admission by Helmholtz that while discoveries are being made the concepts are vague. However, a few passages later, probably not having realized the implications of his praise, he goes on to reprimand Faraday for his misapprehension of the law of con-

39. Translated in *Nature, 11* (1874), 149.
40. Published in *Nature, 3* (1870), 51–52. See also Helmholtz' "Faraday Lecture," delivered to the Chemical Society at the Royal Institution in 1881 and published in *Nature, 23* (1881), 535–540.

servation: "More especially, he [Faraday] opposed the action of forces at a distance, the assumption of two electric fluids and of two magnetic fluids, and in like manner, all hypotheses which contradicted the law of conservation of force, of which he had an early presage, though he singularly misapprehended its mathematical form." I find this change of tune most natural, in view of the interpretation suggested here. Helmholtz' concepts having crystallized, he did not realize that his own conception of the nature of "Kraft" had ever been in a state of change. He realized that Faraday had contributed important discoveries involving his peculiar concept of force. To suppose that Faraday did not really understand the nature of force would have been absurd to Helmholtz; thus, the only explanation he could find was Faraday's mathematical misapprehensions.

6. HELMHOLTZ' PHILOSOPHY OF SCIENCE

Helmholtz was a typical nineteenth-century scientist in that he tried to eradicate the last remnants of *Naturphilosophie,* which, in his opinion and in the opinion of his contemporaries, did enormous harm to the development of German science. Along with this went a loud empiricism, a repetitious denunciation of all hypotheses and conjectures.[41] Helmholtz said that his discovery of the principle of conservation was rooted in the inductive generalization that it was impossible to construct a perpetuum mobile: "In this way I came to the question: 'What must be the relations between the various forces of nature if no perpetuum mobile should be possible?' "[42] And in his 1881 notes to his 1847 paper, he added: "Besides, this law, like all knowledge about the proceedings of the real world, has been found inductively; that no perpetuum mobile can be built, i.e., no motive power can be gained endlessly without corresponding expenditure, had become, through many unsuccessful attempts to build one, a slowly gained induction."[43]

Naturally Helmholtz was convinced that no perpetual-motion

41. I deal here only with the question as to whether or not Helmholtz was an inductivist as he claimed to be. Elsewhere I shall discuss Helmholtz' epistemology and metaphysics, and show the influence of Kant and *Naturphilosophie* on his whole system of thought.
42. Helmholtz, "Ueber die Erhaltung der Kraft," *Wiss. Abh., 1,* 12.
43. Helmholtz' notes added to the original 1847 paper, "Ueber die Erhaltung der Kraft," in the 1881 edition of his *Wiss. Abh., 1,* 68. The translation is mine.

machine could be built. So was the French Académie des Sciences as early as 1775, when it decided not to consider any further attempts to construct one. Helmholtz did not actively seek a law that would in some sense be a generalization of this impossibility. He only knew that whatever law he found must not contradict this impossibility. But in this there was nothing new. Every well-educated physicist in the last two centuries avoided any theory that resulted in a perpetuum mobile. The difficulty, naturally, was that until the connection between the impossibility of a perpetuum mobile and the general principle of the conservation of energy became clear, there was no way to foresee whether or not a perpetuum mobile was implied by a new theory. Rumford, for example, certainly knew that it was useless to try to construct a perpetuum mobile, and yet he did not use that principle as a crucial argument against the caloric theory. Rather, he argued that he had exhibited an inexhaustible source of heat, and that heat could not then be matter because matter obeyed an all-embracing conservation law. The implication of his argument is that if heat is not matter it cannot be conserved. Although Rumford advocated a mechanical theory of heat, he had no idea at all of the conservation of energy.[44]

Besides, the impossibility of gaining work from nothing, even if it is no less general than the principle of conservation of energy, is only half the story, as Planck pointed out. The conviction of this impossibility worked psychologically only in one direction. Scientists never realized that one could not *lose* work indefinitely, and they even thought that Rumford's famous cannon experiment taught them the contrary. However, for a clear understanding of the conservation of energy principle, one has to see that the impossibility works in both directions. It was certainly clear to Helmholtz that whatever form the law that connects the forces of nature acquires, it must satisfy this self-evident and long-familiar condition.

Moreover, it was not the case that Helmholtz inferred correctly that energy had to be conserved, while others inferred wrongly that

44. There is no direct dependence of the conservation of energy principle on the mechanical theory of heat. The historical sequence seems not to be from the mechanical-heat theory to the conservation of energy principle to the second law of thermodynamics. Rather, the sequence began with the caloric theory, which led to Carnot's principle; then, after the independent discovery of the principle of conservation of energy, the foundation of thermodynamics was separated into two fundamental laws. For more on that topic see my forthcoming paper, "Energy Conservation: A Case of Simultaneous Discovery?"

it is "force" (in our sense of the word) that is conserved. There were not two clearly formulated, mathematically expressible conservation laws: a law of conservation of force, and a law of conservation of energy. In other words, it was not the case that due to the work of Helmholtz and others, the proof of the second law was found to be correct while the proof of the first was not satisfactory. In fact, no proof of a conservation of force, as we understand force, can even be formulated. One could talk of the "conservation of force" in English as long as the definition of "force" was on the level of imprecision of Faraday's. In German one could speak of the "Erhaltung der Kraft," before the clear statement of the conservation law rendered the concepts precise; afterwards "Kraft" became simply "force," and the conserved quantity was called "Energie." Helmholtz, a mathematician of the first rank, aimed at a mathematical formulation of the conservation principle; the mathematical proof and the clarification of concepts emerged together. It is sufficiently clear that a conservation law formulated with vague concepts of force and lacking mathematical clarity could not have been drawn as an inductive inference. Certainly, one could arrive inductively at the conclusion that a perpetuum mobile is impossible, or that the sum of the "tension" and "life force" is conserved, but one could not arrive inductively at the science-producing principle that "something in nature must be observed." The principle of the conservation of force was not disproved; it was labeled a "different verbal formulation" of the true principle, and then the whole thing faded away, as generally happens with superseded scientific theories. Recently more attention has been given to the stage in which scientific concepts are only vaguely formulated. Laszlo Tisza has propounded a theory that bears some resemblance to what I have been trying to prove: concepts undergo differentiation and integration when great scientific discoveries occur. The implication of his theory for the case treated here is that through the establishment of the principle of conservation of energy the concept "Kraft" differentiated into "force" and "energy," while the concepts of "puissance motrice," "Spannkraft," "Arbeitskraft," "vis viva," "vis mortua," and many others were integrated into the concept of "energy."[45]

That the concept of energy could have been clearly defined only

45. Laszlo Tisza, "The Conceptual Structure of Physics," *Rev. Mod. Phys., 35* (1963), 343.

after the law of its conservation had been established was clearly seen by Planck: "The concept of energy gains its meaning for physics only by the principle [of its conservation]."[46]

Helmholtz tried to convince himself and others of his thorough empiricism, which, for him, meant an antimetaphysical and antihypothetical approach. For instance, he had this to say about Mayer's discovery of the mechanical equivalent of heat in his 1881 notes on his essay on the conservation law:

> Lately, the followers of metaphysical speculations have tried to stamp the law of the Conservation of Energy as an *a priori* valid one, and they are celebrating J. R. Mayer as a hero in the field of pure thought. What these people consider as the zenith of Mayer's contribution, namely, the metaphysically formulated would-be proofs for the *a priori* necessity of this last, will be considered the weakest point of his arguments, in the eyes of anyone who is used to the severe scientific method of the natural scientist; this must have been, no doubt, the grounds for his work having remained unknown in the circles of natural scientists.

Helmholtz was carried away by his battle against the metaphysical speculators. He had known the value of Mayer's work for three decades when writing these lines. Why, then, had he not commended Mayer's originality before? The answer does not lie in petty priority resentments, but rather in Helmholtz' lip-service commitment to expurgate metaphysics from science. Helmholtz never attacked Joule or any of the other discoverers of the conservation law. His only severe critcisms are against Mayer and Faraday, both of whom were outstanding scientists who did not conform to Helmholtz' avowed empiricist principles.

It is instructive to compare what Helmholtz has to say about the status of the law of conservation with what others say. The *a priori* nature of the principle was seen even by some none-too-eminent experimental physicists like Charles Brooke. He wrote in 1872 that "one proposition only shall be alluded to as having been by some writers rather overstrained, viz., that the amount of energy in the world is unchangeable, the sum of the actual of kinetic and potential energies being a constant quality. This may be taken as a postulate,

---

46. M. Planck, *op. cit.* (note 34), p. v.

and is probably true but it is a proposition that is equally incapable of proof or disproof."[47] Meyerson in his *Identity and Reality* dealt profoundly with the methodological status of the law of conservation of energy and tried to understand what Helmholtz could have meant by an experimental proof of it. He wrote:

> What really would be a valid experimental demonstration of the conservation of energy? We should need a considerable series of experiments showing that through all kinds of change, under the most varied conditions, different forms of energy transform themselves into one another according to equivalents remaining constant within the limits of error of measuring instruments. It seems to be a demonstration of this kind that Helmholtz was thinking of in 1847 when after having furnished the double deduction . . . ended by "the complete confirmation of the law must be considered one of the principal tasks which physics will have to accomplish in the years to come."[48]

That Helmholtz did not practice what he preached can be shown by positive evidence. When he did not speak as a methodologist of empirical science, he expressed himself quite differently. In the original 1847 essay, he presented the following program: "The final aim of theoretical natural science is therefore to discover the ultimate and unchangeable causes of natural phenomena."[49] Or again, a few paragraphs later, he wrote that "we have seen above that the problem before us is to refer back the phenomena of nature to unchangeable final causes. This requirement may now be expressed by saying that for final causes unchangeable forces must be found."[50] It is clear from here that he does not take the empirical character of the conservation law very seriously. Having thus formulated the task of physical science, Helmholtz went on to investigate how general he could render the old law of conservation of *vis viva*. He applied his generalized law to all other branches of physics and to the science of living organisms. Both the formulation of the program and the generalization of the conservation law are typical of a very general *a priori* principle. Even René Dugas, who otherwise represents the con-

47. Charles Brooke, *op. cit.* (note 35), p. 122.
48. Meyerson, *op. cit.* (note 14), p. 196.
49. Helmholtz, "Ueber die Erhaltung der Kraft," *Wiss. Abh., 1,* 14.
50. *Ibid.*

servative positivist attitude, admits the nonempirical nature of Helmholtz' argument:

> In 1847 Helmholtz published at Berlin a paper on the Conservation of Force. From the philosophical, not to say the metaphysical point of view, Helmholtz assigned to the theoretical sciences the task of inquiring into the "constant causes" of phenomena. . . . In some things Helmholtz appears a Cartesian . . . , it is well to remember his intention "to reduce all natural phenomena to invariable forces. . . ." There is some illusion in the pursuit of such an ideal, and this, to Helmholtz himself, had, at least in part, the character of a wish.[51]

What Dugas calls Cartesian was, in Helmholtz' philosophy, Kantian in origin. I shall deal with this topic in detail in another place where I discuss Helmholtz' philosophy. The only conclusion I wish to draw here is that his implicit claim to be the "self-elected high-priest" of inductivism[52] should be treated with proper skepticism.[53]

7.   AFTERMATH

By the 1880's the principle of the conservation of energy had become one of the most fundamental principles in all of natural science. In the physical literature and the semipopular press in England, the concept of "Conservation of Force" had faded out of memory. Maxwell summed up the situation very clearly in his preface to *Matter and Motion* in 1877:

51. René Dugas, *A History of Mechanics,* trans. J. R. Maddox (Neuchâtel, 1955), p. 435.

52. The reference is to R. Graves' description of Wordsworth as a "self-elected high priest of Nature."

53. Not only did Helmholtz formulate the law of conservation of energy, and not only did he in the last phase of scientific activity return to basic mechanical principles by expounding the principle of least action, but he also indulged in brilliant speculative deductions; some of these were proved to be wrong and were forgotten as, for example, his electromagnetic theories, and some proved to be astonishingly correct. In the same Faraday lecture he wrote: "now the most startling result, perhaps, of Faraday's law is this: If we accept the hypothesis that the elementary substances are composed of atoms we cannot avoid concluding that electricity also, positive as well as negative, is divided into definite elementary portions, which behave like atoms of electricity. As long as it moves about in the electrolytic liquid each atom remains united with its electric equivalent or equivalents. At the surface of the electrodes decomposition can take place if there is sufficient electromotive power, and then the atoms give off their electric charges and become electrically neutral" (*Nature, 23* [1881], 538).

Physical science, which up to the end of the eighteenth century had been fully occupied in forming a conception of natural phenomena as the result of forces acting between one body and another, has now fairly entered on the next stage of progress—that, in which the energy of a material system is conceived as determined by the configuration and motion of that system, and in which the ideas of configuration, motion and force are generalized to the utmost extent warranted by their physical definitions.[54]

This summary might have been somewhat premature, but it would certainly apply fifteen years later. In Germany, also, the term "Energie" was accepted, and no one used the term "Kraft" any more except for "force." In 1887 the philosophical faculty of the University of Goettingen announced a prize essay, the purpose of which was to clarify whether or not the well-known principle of conservation of energy was identical with Helmholtz' "Erhaltung der Kraft." As they put it:

> Ever since Thomas Young (*Lectures on Natural Philosophy*, London, 1807, Lecture VIII) the physicists ascribe *energy* to bodies, and ever since William Thomson (*Philosophical Magazine* and *Journal of Science* IV Series, London, 1855) it is claimed that the Principle of the Conservation of Energy is valid for all bodies. It seems that what is meant here is the same principle which was announced earlier by Helmholtz as the Principle of the Conservation of Force.[55]

The second prize was awarded to Max Planck, who wrote an exhaustive historical and philosophical survey on the principle, attributing to Helmholtz the clear concept of "energy" before its real emergence through the proof of its conservation. It is ironic that Planck was the one whose contributions set in motion a new philosophy of nature with a new concept of energy.

54. J. C. Maxwell, *Matter and Motion* (New York, n.d.). Preface.
55. M. Planck, *op. cit.* (note 34), p. ii. The original is as follows: "Seit Thomas Young (*Lectures on Natural Philosophy*, London, 1807, Lecture VIII) wird den Koerpern von vielen Physikern *Energie* zugeschrieben, und seit William Thomson (*Philosophical Magazine* and *Journal of Science*, IV Series, London, 1855) wird haeufig das *Prinzip der Erhaltung der Energie* als ein fuer Koerper gueltiges ausgesprochen worunter das selbe Prinzip verstanden zu werden scheint, was schon frueher von Helmholtz unter dem Namen der Erhaltung der Kraft ausgesprochen war."

Clausius and Maxwell's Kinetic Theory of Gases

BY ELIZABETH WOLFE GARBER*

In May 1859 James Clerk Maxwell wrote at length to George Stokes,[1] describing a theory of gases he had developed as a result of reading a paper of Rudolf Clausius on the subject.[2] In four months Maxwell had produced a theory which not only encompassed all previously described gas properties, but also included transport coefficients and a powerful statistical method for describing the state of a gas. Clausius provided him with critically important ideas for both of these developments.

Clausius severely criticized the parts of Maxwell's 1860 publication[3] relating to transport properties, and he published a counter-

* Physics Department, State University of New York, Stony Brook, New York, 11790.

1. James Clerk Maxwell to George Gabriel Stokes, *Memoir and Scientific Correspondence of the Late Sir George Gabriel Stokes,* ed. Joseph Larmor (Cambridge, 1909), *2,* 8–11. See also Lewis Campbell and William Garnett, *The Life of James Clerk Maxwell* (London, 1882), 328, 332, and 335–336. [*Life.*]

2. Rudolf Julius Emmanuel Clausius, "On the Mean Lengths of Paths described by separate Molecules of gaseous Bodies," *Annalen der Physik, 105* (1858), 239–258. [*Ann. Phys.*] Maxwell saw it in translation in the February issue of the *Philosophical Magazine, 18* (1859), 81–91. [*Phil. Mag.*]

3. Maxwell, "Illustrations of the Dynamical Theory of Gases. Part I. On the Motions and Collisions of Perfectly Elastic Spheres," *Phil. Mag., 19* (1860), 19–32; "Part II. On the Process of the Diffusion of two or more Kinds of moving Particles among one another," *Phil. Mag., 20* (1860), 21–37; also in *Scientific Papers of James Clerk Maxwell,* ed. W. D. Niven (New York, 1960), *1,* 377–409. [*Scientific Papers.*] The paper had been read before the August 1859 meeting of the British Association for the Advancement of Science and was abstracted in *British Association for the Advancement of Science, Report, 1859,* Transactions of Section A, p. 9, indicating that Maxwell had derived the distribution function and the simple gas laws. The following year he concentrated upon transport phenomena and the equipartition theorem. (*B.A.A.S. Report, 1860,* Transactions of Section A, p. 15.)

theory in 1862.[4] Maxwell responded to this in 1867,[5] fully acknowledging the acuteness of Clausius' remarks and incorporating Clausius' method. In spite of this promising start, the productive relationship between Maxwell and Clausius was not sustained. After 1867 Maxwell turned, for several years, to other problems, while Clausius continued to pursue his own ideas on gases and thermodynamics.

By 1857, when he first began to publish papers on gases, Clausius had already established a reputation as a leading theoretical physicist with his clear, incisive papers on thermodynamics.[6] These latter papers reveal his ability to separate the essential physical principles from the special matter-theory assumptions that he had used in his development of the subject[7] (all of his thermodynamics papers published after 1860 were based upon his concept of the structure of matter[8]). In 1857 Clausius was still cautious about revealing these assumptions; he only published the ones needed to construct a correct theory of gases after seeing similar ideas expressed, incorrectly he felt, by August Krönig.[9]

4. Clausius, "On the Conduction of Heat by Gases," *Phil. Mag., 23* (1862), 417–435 and 512–534.

5. Maxwell, "On the Dynamical Theory of Gases," *Philosophical Transactions of the Royal Society of London, 157* (1867), 49–88. [*Phil. Trans.*] Also in *Scientific Papers, 2,* 26–78.

6. Details of Clausius' life are hard to find, as no full-length biography exists. But there are several obituaries: George F. Fitzgerald, "Rudolf Julius Emmanuel Clausius," *Proceedings of the Royal Society, 48* (1890), i–vii; Josiah Willard Gibbs, "Rudolf Clausius," *Proceedings of the American Academy of Arts and Sciences, 14* (1890), 438–465; F. Folie, "Rudolf Clausius, sa vie, ses travaux et leur portée métaphysique," *Revue des questions scientifiques, 27* (Brussels, 1890), 419–487; Eduard Riecke, *Rudolf Clausius* (Göttingen, 1888).

7. Clausius, "On the Moving Force of Heat and the Laws which can be deduced therefrom," *Ann. Phys., 89* (1850), 368–397 and 500–524; trans. in *Phil. Mag., 2* (1851), 1–21 and 102–119; "On a Modified form of the Second Law of Thermodynamics," *Ann. Phys., 92* (1854), 481–506; trans. in *Phil. Mag., 12* (1856), 81–98. These papers do not discuss his ideas on matter, which were first made public in "On the Application of the Laws of the Equivalence of Transformations to Inner Work," *Ann. Phys., 114* (1862), 73–112; trans. in *Phil. Mag., 24* (1862), 81–97 and 201–213. William Thomson accepted Clausius' theory of heat, as he presented it in 1850, because it was free of assumptions about matter; see William Thomson, "On the Dynamical Theory of Heat," *Phil. Mag., 4* (1852), 8–21, 105–117, 168–176, and 424–434.

8. For the molecular theory underlying Clausius' thermodynamics, see Edward E. Daub, "Atomism and Thermodynamics," *Isis, 58* (1967), 293–303, and Martin J. Klein, "Gibbs on Clausius," *Historical Studies in the Physical Sciences, 1* (1969), 127–149.

9. Karl August Krönig, "Grundzüge einer Theorie der Gase," *Ann. Phys., 99* (1856), 315–322.

From his elastic-sphere model of the molecule, Krönig was able to deduce the simple gas laws. He mentioned the applicability of probability arguments to gas theory, but he did nothing to develop them himself: "the wall facing the gas atoms is regarded as very rough, the path of each molecule must be irregular and its calculation impossible. Yet according to the laws of the theory of probability one is able to assume, in spite of this chaos, complete uniformity."[10] As Stephen Brush has remarked, Krönig was a chemist rather than a physicist, and the loose arguments that he gave in his derivation of the gas laws probably satisfied him.[11] Clausius disliked Krönig's oversimplified arguments, and his criticisms are well founded.[12]

From his criticism of Krönig's work, Clausius constructed a more rigorous theory of his own, which included all of Krönig's results. In addition Clausius developed a criterion for distinguishing atoms from molecules,[13] and he demonstrated that gas molecules were complex bodies with moving constituent parts.[14] Two statements that

10. *Ibid.*, 316.

11. Stephen G. Brush, "The Development of the Kinetic Theory of Gases. III Clausius," *Annals of Science, 14* (1958), 185–196.

12. Clausius, "On the Kind of Motion which we call Heat," *Ann. Phys., 100* (1857), 353–380; trans. in *Phil. Mag., 14* (1857), 108–127. Clausius claimed in this paper that he had developed his ideas before reading Krönig's paper. For details of Clausius' paper and for his contributions to the development of gas theory, see Brush, *op. cit.* (note 11), and Brush, *Selected Readings in Physics: Kinetic Theory* (New York, 1965), *1*, 23–25.

13. To establish this distinction, Clausius used a result that Krönig had derived: equal volumes of gases under the same conditions of temperature and pressure contain the same number of molecules. He addressed himself to the problem of the structure of the molecules of oxygen and nitrogen. Taking equal volumes of oxygen and nitrogen at the same temperature and pressure and allowing them to combine, Clausius concluded that if the volume of the combined gases remained the same as the volume of each gas when separated, the compound contained the same number of molecules as either of the gases. The compound, therefore, contained one atom of oxygen and one of nitrogen. If, however, one volume of nitrogen mixed with two of oxygen, and if the reaction was accompanied by a change in volume in the ratio of 3:2, then the molecules of the constituent gases contain two atoms each. This diminution of volume was not predicted by current chemical theory, for chemists did not adopt Avogadro's hypothesis until after 1860.

14. Clausius had already shown in his first thermodynamics paper of 1850 that, for an ideal gas, the specific heats at constant pressure and volume, C_p and C_v, respectively, were constants. He now used this result to determine the ratio K/H of the translational (K) to the total (H) kinetic energy of a gas. For an ideal gas, $K/H = (3/2) [(C_p/C_v) - 1]$, which, if the gas molecules were simple material points, should be 1. But for air, C_p/C_v was already known, and $K/H = 0.63$; i.e., some of the kinetic energy of a gas molecule was taken up in internal motions. $K/H < 1$ indicated that a gas molecule was a complex body whose constituent parts were in motion with respect to its center of mass.

301

Clausius made in this paper were important for the later development of the theory. The first was his assumption that while molecular collisions with the container were not necessarily elastic, they could be treated as such:

> Although it is not actually necessary that a molecule should obey the ordinary laws of elasticity with respect to elastic spheres and a perfectly plane side, in other words, that when striking the side, the angle and velocity of incidence should equal those of reflection, yet according to the laws of probability, we may assume that there are as many molecules whose angle of reflection falls within a certain interval, e.g., between 60° and 61° as there are molecules whose angles of incidence have the same limits, and that on the whole, the velocities of the molecules are not changed by the side. No difference will be produced in the final result, therefore, if we assume that for each molecule the angle and velocity of reflection are equal to those of incidence.[15]

The other important statement was his explicit recognition that the velocity he used in his expressions for pressure, etc., was an average velocity "which for the totality of the molecules gives the same *vis viva* as would their actual velocities. At the same time it is possible that the actual velocities of the several molecules differ materially from their mean value."[16] However, in this paper he did not examine the divergence from the mean value.

Clausius now included concrete results in his theory: he estimated the mean speeds for the molecules of some common gases. These speeds were immediately challenged by Christoph Buys-Bullot,[17] who pointed out that the observed rates of diffusion for gases were far too low to be compatible with Clausius' mean-speed calculations.

In his published answer to this criticism, Clausius introduced the idea of the mean free path of a molecule.[18] The path traveled by the molecule between collisions was very small in absolute measure, although it was large compared with the size of the molecule. The

15. Clausius, "On the Kind of Motion which we call Heat," *Phil. Mag.*, *14* (1857), 120.
16. *Ibid.*, 124.
17. Christoph H. D. Buys-Bullot, "Über die Art der Bewegung, welche wir Wärme und Electrizität nennen," *Ann. Phys.*, *103* (1858), 240.
18. Clausius, "The Mean Path of Molecules," *Phil. Mag.*, *18* (1859), 81–91. For details of his answer, see Brush, "Development of Kinetic Theory. III Clausius," *Annals of Science, 14* (1958), 194–195.

path-length was an inaccessible quantity; one could only ask: "how far, on the average, can the molecule move, before its centre of gravity comes into the sphere of action of another molecule?"[19] Yet, despite all of his arguments for the reasonableness of his mean-free-path idea, Clausius had no way of actually answering Buys-Bullot's objections; for he could measure neither the sizes of the molecules nor their number.[20] Although he had introduced the mean free path into the theory, Clausius was thus unable to take advantage of it to defend his ideas on gases. When Maxwell began his study of gases in 1859, he found a theory based upon a sophisticated molecular model from which the known gas laws could be derived. Yet the theory was unsatisfactory because no new phenomena could be predicted with it and because it seemed to lead to a result contrary to known observations. The one new quantity introduced into the theory, the mean free path, could not be estimated. Maxwell was the first to extend gas theory beyond the already empirically derived gas laws; in so doing, he introduced as many difficulties as he solved.

Clausius' motive for studying the properties of gases was to explore the nature of heat. Maxwell's was quite different, as he used kinetic theory to study the structure of matter. The primary purpose of his first paper (1860) was to find a numerical value for the mean free path of a molecule. He says this, albeit obliquely, in the first paragraph of his paper:

It is not necessary to suppose each particle to travel to any great distance in the same straight line; for the effect in producing pressure will be the same if the particles strike against each other; so that the straight line described may be very short. M. Clausius has determined the mean length of path in terms of the average distance of the particles, and the distance between the centres of two particles when collision takes place. We have at present no means of ascertaining either of these distances; but certain phenomena, such as the internal friction of gases, the conduction of heat through a gas, and the diffusion of one gas through another, seem to indicate the possibility of determining accu-

19. Clausius, "The Mean Path of Molecules," 84.
20. Clausius writes the mean free path l in terms of other molecular quantities in the form $l/\sigma = (3/4)\lambda^3/\pi\rho^3$, where σ is the radius of the molecule, λ is the average intermolecular distance, and ρ is the density of the gas. He considers one molecule as moving through a cubical mesh of stationary molecules. None of the quantities σ, λ, or l is known.

rately the mean length of path which a particle describes between two successive collisions.[21]

Maxwell used Graham's experimental data on the diffusion of gases with his own expression for the diffusion coefficient to obtain a numerical value for the mean free path.[22] Similarly from Stokes's data on the amplitude decay of pendulum swings, he calculated the coefficient of viscosity and the mean free path again. He found the two values in "reasonable agreement."

Because of his interest in the transport properties of gases, Maxwell had to consider intermolecular collisions. To avoid the impossible task of describing the positions and velocities of every molecule in the gas, Maxwell introduced statistical methods following Clausius' suggestion. To obtain the expressions for the pressure, and for the rate of transport of molecules, momentum, and kinetic energy through a gas, Maxwell used the mean free path.[23]

Maxwell fully understood his debt to Clausius, and he spoke of it in a letter to William Thomson in 1870. Among his other contributions, Clausius had introduced "The relation between their [the molecules'] diameter, the number in a given space and the mean length of path. . . . Mathematical methods . . . for dealing *statistically* with immense numbers of molecules by arranging them in groups according to their directions, velocities etc."[24] Maxwell also pointed

21. Maxwell, "Illustrations of the Dynamical Theory of Gases," *Scientific Papers, 1,* 377.

22. *Ibid.,* 403. For details of Maxwell's derivation of the diffusion and viscosity coefficients in this paper, see Ian B. Hopley, *Clerk Maxwell's Contribution to Physics,* 2 vols. (unpub. Ph.D. diss., University of London, 1956), and Brush, "The Development of the Kinetic Theory. VI Viscosity," *American Journal of Physics, 30* (1962), 269–281. Maxwell repeated this calculation of the mean free path in his review of Loschmidt's work on gases. He then ventured "on more hazardous ground" and used his value to estimate the sizes of the molecules of different gases. Maxwell, "On Loschmidt's Experiments on Diffusion in Relation to the Kinetic Theory of Gases," *Nature, 8* (1873), 298–300; also in *Scientific Papers, 2,* 343–350.

23. For a detailed discussion of Maxwell's derivation of the distribution function and his introduction of intermolecular collisions, see Brush, "Kinetic Theory. IV Maxwell," *Annals of Science, 14* (1958), 243–254, and Hopley, Ph.D. diss., *1,* Part 2, Sect. 1. Hopley includes a detailed discussion of how Maxwell derived the expressions for pressure, etc., using the mean free path, and he fills in many analytical steps which Maxwell omitted in his published account.

24. Henry T. Bernstein, "J. Clerk Maxwell on the History of the Kinetic Theory of Gases," *Isis, 54* (1963), 206–216. The italics are Maxwell's. The occasion for the letter was Thomson's request for a brief history of kinetic theory for his address to the British Association in 1870.

out his own additions; he had shown that the "velocities of the molecules range through all values, being distributed according to the same law which prevails in the distribution of errors of observation and in general in all cases in which general uniformity exists in the mass amid apparent irregularity in individual cases."[25]

Maxwell's was the first fluid transport theory where the assumption of a molecular structure was a necessary supposition. Using molecular properties, he constructed transport equations equivalent to the macroscopic hydrodynamic equations containing transport coefficients. Comparing the two forms of the equations, Maxwell expressed the transport coefficients in terms of molecular parameters. To obtain the viscosity coefficient μ, he compared the macroscopic equation for the net momentum passing in the x-direction through unit area in unit time, $F = \mu \, du/dx$, to the same quantity derived from microscopic considerations, $(1/3) \, l\sigma\bar{c} \, du/dx$, from which, $\mu = (1/3) \, l\sigma\bar{c}$.[26]

The controversial part of Maxwell's original transport theory was his derivation of the conductivity coefficient. Maxwell sketched, in less than two pages, a derivation of an expression for the energy crossing unit area of a plane in the gas perpendicular to the temperature gradient. Writing the kinetic energy of translation of a particle as $(1/2)m\beta c^2$,[27] Maxwell obtained an energy-transport-formula,

$$J \cdot H = -\tfrac{1}{3}\frac{d}{dx}\left(\tfrac{1}{2}\beta m\overline{c^2}N\bar{c}l\right),$$

where J is the mechanical equivalent of heat, N the number of molecules in unit volume, and H the heat transferred across the plane. He deduced the mean-square and mean velocities from his distribution function. For the conductivity coefficient κ he found[28]

$$\kappa = \tfrac{3}{4}\beta Pl\bar{c}\frac{1}{T},$$

where P is the pressure and T the temperature of the gas.

Maxwell's gas theory went far beyond Clausius' in method, con-

25. *Ibid.*
26. l is the mean free path, σ the diameter of the molecule, and $\bar{c}$ its mean speed.
27. β is the ratio of the total to the translational kinetic energy of the molecule. Maxwell followed Clausius, letting experiment dictate the value of β. He used Clausius' ratio $1/\beta \equiv K/H = (3/2)\,(C_p/C_v - 1)$.
28. The correct form for the thermal conductivity coefficient is $\kappa = (1/6)N\bar{c}kl$, where k is Boltzmann's constant.

cept, and scope, but Clausius' prior work had been essential. Clausius had given Maxwell two crucial ideas: the mean free path and the suggestion of a statistical method. The distribution function, which was the most important innovation for the ultimate statement of the theory, appeared fully developed, though ill defined, in Maxwell's first paper;[29] but we should perhaps discount the form in which it was first introduced because it was of secondary interest to Maxwell in 1860. He was casual about it in his letter to Stokes and was more excited about the behavior of the viscosity coefficient. He became interested in the function itself only after Boltzmann's extensive analysis of it.

Maxwell's first formulation of kinetic theory has been criticized[30] both with respect to his derivation of the distribution function and his treatment of the transport coefficients. However, the main thrust of the commentary came several years later, and, apart from Clausius', immediate reactions were few.[31] Until Maxwell's own experiments on viscosity, the discussion was whether or not a kinetic theory could describe the behavior of a gas. Afterwards the discussion shifted to the relative merits of the contending kinetic theories.[32] The initial lack

29. Maxwell earlier confronted the problem of a large number of particles in motion and colliding with one another in his analysis of the stability of Saturn's rings. He analyzed the problem without considering interparticle collisions, because "when we come to deal with collisions among bodies of unknown number, size and shape, we can no longer trace the mathematical laws of their motion with any distinctions. All we can now do is collect the results of our investigations and to make the best use we can of them in forming an opinion as to the constitution of the actual rings of Saturn." ("On the Stability of the Motion of Saturn's Rings," *Monthly Notices of the Astronomical Society of London, 19* [1859], 297–304; also in *Scientific Papers, 1,* 288–378, esp. 336.) The question of the origin of Maxwell's ideas on probability has come under scrutiny lately; it is still an open question. See Charles C. Gillispie, "Intellectual Factors in the Background of the Analysis by Probabilities," *Scientific Change,* ed. A. C. Crombie (New York, 1963), 431–453; Brush, "Foundations of Statistical Mechanics, 1845–1915," *Archive for the History of Exact Sciences, 4* (1967), 145–183, esp. 152; and Elizabeth W. Garber (work in progress).

30. See Brush, "Development of Kinetic Theory. IV Maxwell," *Annals of Science, 14* (1958), 243. In addition there are algebraic and numerical errors, some of which were noted by Clausius and, later, by Boltzmann and Burbury. See C. W. F. Everitt, "Maxwell's Scientific Papers," *Applied Optics, 6* (1967), 639–649, and his forthcoming biography of Maxwell.

31. For contemporary reactions to the kinetic theory, see Brush, *ibid.,* 249–255. There were some physicists who accepted the theory immediately, e.g., Thomas Graham and J. Stefan.

32. After the appearance of Maxwell's second gas theory (1867), several others advanced theories that included transport properties. See O. E. Meyer, *Kinetic Theory of Gases* (Breslau, 1877).

of interest is not surprising when one considers the uncertainty, which Maxwell emphasized in this first paper, of the existing data on transport properties. The only completed experiments on gases were Graham's on diffusion, performed almost forty years previously, and Stokes's on pendulums, which were done to examine the decay of amplitudes and not to explore gases. It was not until 1860 that Magnus showed that gases possessed a thermal conductivity coefficient.[33] Maxwell's theoretical discussion followed hard on the heels of this work and provided the first theoretical framework against which experimentally determined transport coefficients could be interpreted. This aspect of his theory was welcomed, and by 1873, when he returned to gas theory, Maxwell was able to draw upon an increasing body of literature on transport coefficients.[34]

The other major innovation of Maxwell's gas theory was not so easily assimilated. His statistical description was ignored or dismissed as incorrect. Of the people who were uncomfortable with this aspect of his kinetic theory, Clausius was the most important and the first to respond.

While Clausius regarded Maxwell's theory as a serious attempt to solve the problems of gaseous behavior, he criticized it on basic physical grounds. He considered Maxwell's use of a spherically symmetric distribution function in analyzing transport properties as incorrect. He developed his own theory, taking into account the additional *vis viva* associated with motion in the direction of the gradient within the gas. The factor, ignored by Maxwell, was the cause of the phenomenon, and, in the case of thermal conduction, the gas molecules possessed additional momentum in the direction of the temperature gradient; therefore, the distribution of molecular motions could not be spherically symmetric. Even if this added momentum was the same for all molecules, "with reference to the distribution of molecules among the various directions of motion, it is easy to see that if the original system of motion were such that an equal

33. For a discussion of the history of thermal conductivity, see Alexander Burr, "Notes on the History of the Thermal Conductivity of Gases. Part 2," *Isis, 21* (1934), 169–186; for later developments in transport theories, see the historical sketch in Chapman and Cowling, *The Mathematical Theory of Non-Uniform Gases* (Cambridge, 1952).

34. By 1873 Loschmidt had published the results of his experiments on diffusion. O. E. Meyer had independently confirmed Maxwell's results on viscosity (1861), and Stefan had performed his experiments on conductivity (1872).

number of atoms moved in each direction this could no longer be the case in the modified system of motion."[35] A weighting factor favored motion in the direction of the temperature gradient; all directions of motion were not equally probable, and Maxwell's distribution function did not apply.

As the molecules moved irregularly, a normal spread in velocities was produced by intermolecular collisions. Superimposed on this "accidental" spread was the weighted motion in the direction of the thermal gradient. Clausius used only this weighted motion to find the mean speed and its direction due to the thermal gradient in the gas. To obtain an absolute value for the mean velocity of the molecules, Clausius had to consider not only this mean speed but also that due to the "accidental" variation. But Clausius gave no details for calculating the latter, nor would he use the value given by Maxwell's distribution function; yet he clearly recognized the statistical nature of the problem: the effects of the "accidental" variations cannot be determined by following the motions of the individual molecules, "but the rules of probabilities enable us to establish certain general principles for a *large number* of molecules. Maxwell has thus deduced a formula purporting to represent the manner in which the various existing velocities are distributed among the molecules. It is not, however, necessary for our present purpose to enter upon this."[36] Maxwell's derivation of the distribution law depended on the mean free path, l. Clausius commented:

> In order to be able rightly to calculate these mean values, we must know the law which regulates the various velocities which occur. As I have already stated above, such a law was established by Maxwell, and it might perhaps be employed for calculating mean values. I prefer, however, not to discuss this subject here, as a few remarks concerning this law would be required which would lead us too far at present; and I feel more justified in leaving this point, since the numerical value of l is so imperfectly known that the accurate numerical calculation of a formula in which it occurs is not possible. I will therefore content myself, in the calculation of the conduction of heat, with employing the above formula, which is deduced without the consideration of the accidental variation, a mean value for the velocity which though not

35. Clausius, "On the Conduction of Heat by Gases," *Phil. Mag., 23* (1862), 417–435 and 512–534, esp. 427.
36. *Ibid.,* 428.

strictly accurate, may be regarded as sufficiently so, considering the uncertainty which still prevails in regard to the value of l.[37]

The statistical method might be correct in principle, but Maxwell's particular method was inaccurate.[38] Clausius' acceptance of a statistical method was only tentative and he finally rejected it altogether.[39] His hesitancy in using the distribution function reflects the difference between his and Maxwell's concept of matter and gases.

From his preliminary criticism of Maxwell's use of the distribution function, Clausius developed his transport theory. He rejected Maxwell's method of deriving the transport coefficients, which depended on the mean free path, and introduced a new method that Maxwell later adopted. Clausius considered the alteration, due to collision, of an unspecified property of a gas molecule; the property was defined individually for each transport coefficient. He wrote the transport equations in the form

$$E = \tfrac{1}{2}mN\int_{-1}^{1} I \bar{c} \mu d\mu \, ,$$

$$F = \tfrac{1}{2}mN\int_{-1}^{1} I(\overline{c^2}) \mu d\mu \, ,$$

$$G = \tfrac{1}{2}mN\int_{-1}^{1} I(\overline{c^3}) \mu d\mu \, ,$$

where E is the mass, F the momentum, and G the *vis viva* (heat) carried across a unit area of a plane perpendicular to the x-axis in unit time; μ is the cosine of the angle that the velocity of the molecule makes with the x-axis; c is the velocity of the molecule; I, which is not spherically symmetric, is the proportion of molecules whose velocities lie at angles whose cosines are between μ and $\mu + d\mu$. The

37. *Ibid.*, 512.
38. Although Clausius was obviously uncomfortable with Maxwell's use of the mean free path, he did not attack Maxwell's method on physical grounds; he only remarked on his misgivings about the use of statistics.
39. Clausius did not use the statistical distribution even after Maxwell derived it independently of the mean free path. He only used the distribution function once, to find the mechanical expression for the heat added to a gas, in "On the Theorem of the Mean Ergal and its Application to the Molecular Motions of Gases," *Sitzungsberichte der Niederrheinischen Gesellschaft für Natur- und Heilkunde zu Bonn* (1874), 183–231; trans. in *Phil. Mag., 50* (1875), 26–46, 107–117, and 191–200.

velocities $\bar{c}$, $\overline{(c^2)}$, and $\overline{(c^3)}$ are averages that Clausius did not calculate absolutely; he considered only the effects of the thermal gradient. The averaging process takes the nonuniform distribution of mole-cules and momentum into account. In the case of thermal conduc-tivity, Clausius assumed that the additional momentum in the x-direction was directly proportional to the equilibrium mean free path, l. He thus could not escape a correlation between the average velocity and the mean free path. The actual path also depends on l and on the angle at which the molecule is projected with respect to the x-axis; all averaging processes are finally reduced to averaging over μ or l. The numerical factors in Clausius' final equations were different from Maxwell's, not only because he used the nonspherical weighting factor I, but also because he insisted that $l = (4/3)\pi\sigma^2 N$ for the equilibrium mean free path.[40]

Clausius' method of averaging over I was an involved chase through functions of l, μ, and c, all written in the form $\alpha\xi + \beta\xi^2 + \gamma\xi^3 \dots$ (ξ substituting for l, μ, or c). To avoid unwieldy results, he ne-glected terms in ξ of second and higher orders, in which case it is not clear how different I was from a spherical distribution. This is not, however, the important point here. What is important is that Clau-sius saw the physical problems involved in calculating the transport coefficients and said so clearly. He said it clearly enough that Max-well, despite his difficulty in understanding Clausius' thought,[41] could identify the problems and improve his understanding of them.

Maxwell's first attempts to answer Clausius' criticisms remained

40. Clausius obtained a 4/3 rather than a $\sqrt{2}$ factor in his expression for the mean free path because he did not use the distribution function. He had pre-viously published a short justification of his 4/3 factor in "On the Dynamical The-ory of Gases," *Phil. Mag.*, *19* (1860), 434–436. Maxwell wrote to Tait rejecting Clausius' expression for the mean free path:

> As for Clausius he pointed out *gross* mistakes in M[axwell] I have no doubt he has some of his own but I have not had the patience to find them out, except that he stuck to uniform velocity in the molecules though I proved it impossible and pointed out the only . . . possible distribution of velocity. Clausius' uniform velocity leads (by sound mathematics) to an expression for the mean relative velocity which is unsym-metrical with respect to the components so that you need to know which is the greater of the two velocities and put it in the right place in the formula (Maxwell to Tait, March 1868, Maxwell-Tait Correspondence, Cambridge University).

41. Maxwell acknowledged the difficulty that he had in understanding Clausius. For further details, see Elizabeth W. Garber, "Maxwell and Thermodynamics," *American Journal of Physics, 37* (1969), 146–155.

unpublished.[42] In the introductory remarks of his manuscript, Maxwell acknowledged the force of Clausius' criticism:

> M. Clausius has recently published an investigation of the particular case of the conduction of heat through a gas which was very imperfectly treated by me in the paper referred to ["Illustrations of the Dynamical Theory of Gases," *Phil. Mag., 19* (1860), 19–32; *20* (1860), 21–37]. I have reexamined it and found some others the influence of which extends to other parts of my investigation. I shall therefore state here so much of my former results as will make the requisite corrections intelligible and I shall retain the methods used in my former paper except when obliged to compare them with those of M. Clausius.[43]

Maxwell was here only concerned with transport phenomena. He therefore gave only brief sketches of derivations for the distribution function and the simple gas laws, obtaining his previous results. He also included a discussion of the specific heat ratio for a gas of billiard-ball molecules. He then turned "to the question which I neglected to consider in my former paper. When the density and the temperature of a gas, or the composition of a system of mixed gases vary from one place to another, what is the proportion of particles, which, starting from one given place arrive at another given place without collision."[44]

The corrections Maxwell made in his initial transport theory were to rederive two propositions of his 1860 paper. The first was the expression for the quantity of matter transferred across a plane, and the second was the net effect of all collisions that could occur in a layer of gas. Using the modified derivations, he eliminated the nonconservation of mass, which was a consequence of his expression for the thermal conductivity coefficient, thereby correcting a major physical error that Clausius had pointed out in his first paper.

42. Maxwell, "On the Conduction of Heat in Gases," Maxwell Manuscripts, Scientific Papers–6, Cambridge University. For a transcript of this paper, see Elizabeth W. Garber, *Maxwell, Clausius and Gibbs: Aspects of the Development of Kinetic Theory and Thermodynamics* (Ph.D. diss., Case-Western Reserve University, 1967), Appendix A, 279–297. This paper is to be published in the collected papers of Maxwell on kinetic theory, thermodynamics, and statistical mechanics, ed. C. W. F. Everitt, Stephen G. Brush, and Elizabeth W. Garber.
43. Maxwell Manuscripts, Scientific Papers–6, Cambridge University; see Garber, *ibid.,* 280.
44. Maxwell Manuscripts, *ibid.,* and Garber, *ibid.,* 288.

To take the density and temperature gradients within the gas into account, Maxwell used a path length that depended on the end points of the actual path of the molecule. Apart from this correction, his derivations followed his former ones closely. The coefficients of diffusion and viscosity were of the same form as before, but the method for deriving them "was now rendered strict." The only coefficient which was actually changed was that of thermal conductivity.[45]

Using the device of a mixture of two kinds of particles of densities ρ_1 and ρ_2, Maxwell examined the mean free path λ_1 of the first kind of particles. The velocity of these particles was c', which was slightly different from c_1, the velocity they had if ρ_1, ρ_2, c_1, and c_2 were all constant.[46] He found:

$$\frac{1}{\lambda_1} = s_1^2 N_1 \left(1 + \frac{c_1^2}{(c')^2}\right)^{1/2} \pi + \pi s'^2 N_2 \left(1 + \frac{c_2^2}{(c')^2}\right)^{1/2} ,$$

where N_1 and N_2 are the numbers per unit volume of the two kinds of particles. As c' and c_1 are only slightly different, Maxwell wrote c_1 as $c_1 = c' + dc$, thereby obtaining

$$\frac{1}{\lambda_1} = \pi s_1^2 N_1 \left(1 + \tfrac{1}{2} \frac{dc}{c_1}\right) \sqrt{2} + \pi s'^2 N_2 \left(1 + \frac{c_1^2}{(c')^2}\right)^{1/2} \left[1 + \frac{c_2^2}{c_1^2 + c_2^2} \cdot \frac{dc}{c_1}\right] .$$

He used this expression to rederive the mass crossing a unit area in unit time. To take the varying mean path length into account, he actually used an average $\Lambda = 1/2(\lambda + l_1)$, where l_1 is the path at the beginning of the excursion and λ that at the end, assuming that the molecules had been projected with their actual velocity c_1. The number of molecules that are projected from a layer and pass through a plane perpendicular to the x-direction, dropping the device of two kinds of particles, now becomes

$$\frac{Nc(x \pm n\Lambda)}{2nl\Lambda} e^{-n} dx dn ,$$

45. His final expression for the thermal conductivity coefficient was $\kappa = 5/3$ $(\gamma - 1)$ $(P/\sigma T)$ μ/ρ, where γ is the ratio of the specific heats, T the temperature, and ρ the density of the gas.

46. This is a modified form of Eq. (12) in Maxwell, "Illustrations of Gas Theory," *Scientific Papers, 1,* 388. The notation in both versions is the same, s_1 being the distance of separation of the centers of parties of the first kind and s' being the same distance in a collision of unlike particles.

where x lies between $\pm n\Lambda$.[47] For the net flow of mass, Maxwell now obtains

$$q = -\tfrac{1}{3}\frac{d}{dx}\left[\rho c \frac{\Lambda^2}{l}\right],$$

which can be approximated by $q = -(1/3)d/dx(\rho c\lambda)$.[48] His next step was to examine the rate of change of momentum in the layer, using his new average value Λ for the mean free path. This expression was compatible with the ordinary laws of hydrodynamics, and as before he obtained $(1/3)(d/dx)(\rho c^2) = dP/dx$.

The total quantity of mass transferred across the boundary must be zero, i.e., $q + PV = 0$, V being the translational velocity of the gas as a whole; in addition, the condition $dP/dx = 0$ can be applied to the situation. The energy transferred across the boundary due to the translational motion of the gas is[49]

$$E_{\text{trans}} = \tfrac{1}{2}\beta\rho c^2 V \,.$$

The energy transferred by the agitational motion of the molecules is

$$E_{\text{ag}} = -\tfrac{1}{3}\frac{d}{dx}\left[\tfrac{1}{2}\beta\rho c^2 \cdot cl\right].$$

The total energy transferred across the plane is

$$E = -\tfrac{1}{12}\beta\rho\frac{l}{c}\frac{dc}{dx}(c^2 \cdot c + 3c^3) \,.$$

The velocities in the final expression are the mean, mean-square, and mean-cube velocities. To go further and to use this expression to find the mean free path, Maxwell needed a value for these velocities. In this manuscript he could only use the values given by his homogeneous distribution function.[50] He later tried at least twice to investi-

47. This is a correction to Proposition xiv of the 1860 paper. (*Ibid.*, 393.) Maxwell is here clearly only counting particles, not masses.

48. This expression has the same form as his previous $q = -(1/3)(d/dx)(\rho cl)$, but λ differs from l.

49. Maxwell used β in the same way that he used it in his 1860 paper.

50. To find an expression for energy in terms of temperature, pressure, and density, Maxwell assumed that c, (c^2), (c^3) were given by the normal distribution function. He obtained

$$E = -\tfrac{5}{4}\left(\frac{2}{\pi}\right)^{1/2}\beta l_0 P_0^{3/2}T_0^{-3/2}\rho^{-1/2}T^{1/2}\frac{dT}{dx},$$

where P is the pressure and T the temperature.

313

gate a distribution function of a more general form than that of his first paper, but he was unsuccessful in carrying the problem of the transport properties any further at this time.[51] This is perhaps why the paper was never published.[52]

An equally important objection to this whole theory from Maxwell's point of view was its use of the elastic-sphere model of the molecule. Even in his first paper he had treated this model only as an hypothesis: "If the properties of such a system of bodies [hard, elastic spheres] are found to correspond to those of gases, an important physical analogy will be established, which may lead to a more accurate knowledge of the properties of matter."[53] He was more specific in his unpublished paper where he demonstrated that the elastic-sphere model could not account for the known specific heat of steam.

The equipartition theorem indicated that the energy per unit volume of a gas of elastic spheres should be ρc^2, rather than $1/2 \, \rho c^2$. For a medium consisting of a collection of perfect spheres and other bodies, the total kinetic energy can be written as $1/2\beta\rho c^2$, where β is different from 1 and can be written as $\beta = 1 + q$, where q is the ratio of the mass of the nonspherical particles to the total mass. The specific heat ratio can then be written

$$\gamma = 1 + \tfrac{2}{3}\beta .$$

In the case of steam, $\gamma = 1.304$, and therefore β is 2.19. Then q is 1.19 and $1 - q = -0.19$; "that is, we must suppose a negative quantity

51. In one unpublished attempt, Maxwell wrote the rate of change of some property Q, whose velocity distribution was not symmetric, as

$$(h_1 h_2 h_3/\pi^{3/2}) \exp \left[- (h_1^2 u^2 + h_2^2 v^2 + h_3 w^2 + 2g_1 vw + 2g_2 wu + 2g_3 uv)\right],$$

without defining the g's or h's. To use this function, he had to eliminate the g's. In his second published paper, he modified the distribution function for the case of a gas that is compressed or expanded:

$$f(u, v, w) = (N/\alpha\beta\gamma\pi^{3/2}) \exp \left[-(u^2/\alpha^2 + v^2/\beta^2 + w^2/\gamma^2)\right],$$

where α, β, and γ are all different. However, he could not use it in any physical problem and could not justify its form. There are other fragments of notes on the problem in Maxwell Manuscripts, Scientific Papers–6, Cambridge University.
52. The unpublished answer to Clausius' remarks is written in the form of a paper with title, footnotes, etc., as opposed to a series of notes for private digestion. In the text, Maxwell outlines the problem before launching into his solution and methods. The whole manuscript reads as a rough draft of a paper intended for publication.
53. Maxwell, "Dynamical Theory of Gases," *Scientific Papers, 1,* 378.

of spherical particles exist, or in other words our theory fails to explain how the value of γ can be so low as 1.304."[54]

Maxwell therefore had very good reasons to drop this whole approach to the theory and initiate another. A more immediate avenue lay open to test the elastic-molecule hypothesis: the experimental determination of the coefficient of viscosity. He had already found that the viscosity for a gas of elastic molecules would vary as the square root of the temperature and would be independent of pressure. The pressure variation tested the kinetic theory as a whole, while the temperature dependence followed from the molecular model assumed. For a gas of center-of-force molecules, the viscosity changed directly with temperature. Maxwell's experiments confirmed his earlier prediction that viscosity was independent of pressure[55] and that it varied directly with temperature. He could drop the elastic-sphere model completely and turn to Clausius' suggestion of a center-of-force molecule.

Maxwell's account of a new kinetic theory based upon Clausius' concept of the molecule contained his reply to Clausius' criticism of his first paper.[56] He dropped the mean free path as the method for deriving his equations for pressure. Instead he presented a theory based upon Clausius' method of considering some property Q that could change during a collision. The rate of change of Q is

$$\frac{\delta Q}{\delta t}\, dN_1 = \int (Q' - Q)Vb\, db d\phi dN_1 dN_2 \, .$$

where Q' is the value of Q after the collision, and V is the velocity of the colliding particles with respect to one another; b is the distance of closest approach of the two interacting molecules, and ϕ is the angle between the directions with which they approach each other; dN_1 and dN_2 are the numbers of particles with velocities between

54. Garber, *op. cit.* (note 42), 285; *Maxwell Manuscripts, Scientific Papers–6,* Cambridge University.

55. Maxwell's experiment was to measure the viscous drag on a physical pendulum at varying pressures and temperatures. The pendulum consisted of three thick metal disks clamped to the same vertical wire. Although the moment of inertia of the system remained constant, the surface area of the pendulum could be varied. Maxwell read his results to the Royal Society in his Bakerian Lecture, 1865. ("On the Viscosity of Air and Other Gases," *Phil. Trans., 156* (1866), 249–268; also in *Scientific Papers, 2,* 1–25.)

56. Maxwell, "On the Dynamical Theory of Gases," *Phil. Trans., 157* (1867), 49–88; also in *Scientific Papers, 2,* 26–78.

c_1 and $c_1 + dc_1$ and c_2 and $c_2 + dc_2$ respectively. To complete the integration and find the rate of change of Q for the whole gas, Maxwell had to find specific functions for Q and for the dependence of dN on V and b. In the final form of the integral, V entered as $V^{(n-5)/(n-1)}$, where n is an integer deriving from Maxwell's assumption that the force law was k/r^n.

Maxwell now brought the results of his viscosity experiments to bear on the determination of n. With $n = 5$, the viscosity coefficient varied directly with the temperature. Maxwell said that "we have reason from experiments on the viscosity of gases to believe that $n = 5$. In this case V will disappear from the expressions . . . and they will be capable of immediate integration."[57] To derive expressions for the different transport coefficients, Maxwell used different functions for Q. In the case of thermal conduction,

$$Q = M(u + \xi)[u^2 + v^2 + w^2 + 2u\xi + 2v\eta + 2w\zeta + \beta(\xi^2 + \eta^2 + \zeta^2)],$$

where u, v, w are the components of the mass motion of the gas and ξ, η, ζ are those of the agitational motion of the molecules. Q also had to obey the normal hydrodynamical equations of continuity and equilibrium, so that

$$N \frac{\partial \overline{Q}}{\partial t} + \frac{d}{dx} (\overline{\xi Q N}) + \frac{d}{dy} (\overline{\eta Q N}) + \frac{d}{dz} (\overline{\zeta Q N}) = N \frac{\delta Q}{\delta t},$$

from which Maxwell could obtain the equation of conduction and the coefficient of conductivity.

In this paper Maxwell met Clausius' criticism of the mean-free-path method, and for this particular model the form of the distribution function need not be specified.[58] Therefore he met Clausius' criticism of his use of the distribution function for finding the thermal conductivity. However, Maxwell did not see his k/r^5 model as meeting this criticism, and he argued that "when one gas is diffusing through another, or when heat is being conducted through a gas, the distribution of velocities will be different in the positive and negative directions, instead of being symmetrical, as in the case we had con-

57. For the details of Maxwell's method, see Brush, "Statistical Mechanics," *Archive for the History of Exact Sciences, 4* (1967), 154–155. See also Maxwell, *ibid.*, p. 46.

58. Sir James Jeans, *The Dynamical Theory of Gases*, 4th ed. (New York, 1954), Chapter 8.

sidered. The want of symmetry however may be treated as very small in most actual cases."[59]

After these first years of fruitful work on gas theory, Maxwell returned to the study of electrodynamics. He was among the group appointed to measure the absolute value of the ohm. Following that assignment he retired to complete his treatise on electricity and magnetism.[60] Clausius continued to work steadily on gas theory in an effort to express the second law of thermodynamics in mechanical terms. In 1862 he had revealed the molecular ideas that had guided his thermodynamics, and his later papers were all directed to the problem of expressing disgregation and entropy in mechanical form. Unlike Boltzmann, Clausius never fully appreciated the power of the statistical method, and he continued to work within the traditional, yet sophisticated, field of analytical mechanics.

When Maxwell returned to gas theory in 1873, Clausius' methods had hardened; they stood in sharp contrast to even the first papers of Maxwell. Both were now approaching gas theory from entirely different points of view. Maxwell's reaction to Clausius' efforts to express thermodynamic quantities in purely mechanical terms was amusement mixed with a little smugness:

> Have you been introduced to Virial and Ergal? Ergal is an old friend $=$ Potential of a system . . . or $\Sigma m_1 m_2 \phi_{12}$. Virial is $m_1 m_2 \phi_{12}/dr_{12}$. He appeared in dp/dt on Reciprocal frames etc. p. 13 near the bottom.[61] Clausius is now working along with these eminent artistes at the second law of $\theta \Delta \eta$cs [thermodynamics]; but as far as I can see they have not yet furnished him with the dynamical condition of the equilibrium of temperature.[62]

To Clausius' efforts to relate the second law to Hamilton's principle of least action, Maxwell had the same reaction—Clausius "not being

59. Maxwell, "Dynamical Theory of Gases," *Scientific Papers, 2*, 46.
60. Maxwell, *A Treatise on Electricity and Magnetism*, 3rd ed. (New York, 1954); the first edition was published in 1873.
61. Maxwell is referring to his paper "On Reciprocal Figures, Frames, and Diagrams of Forces," *Transactions of the Royal Society of Edinburgh, 26* (1870). However, the expression is for the statical case, and Maxwell did not recognize the physical significance of the virial theorem until he read Van der Waals's thesis. The notation dp/dt refers to a thermodynamic equation, $J. C. M. = dp/dt$ with which Maxwell usually signed his letters.
62. Maxwell to Tait, February 1871, Maxwell-Tait Correspondence, Cambridge University.

a constant reader of the R.I.A. [Royal Irish Academy] transactions and knowing nothing of H[amilton] except (lately) his Princip[le of least action] which he and others try to degrade into the 2nd law of $\theta\Delta\eta$cs, as if any pure dynamical statement would submit to such an indignity."[63] Maxwell regarded the second law as statistical even before the appearance of Boltzmann's H-theorem in 1872. Again, writing to Tait, Maxwell used his demon to illustrate the statistical nature of the second law.[64]

After 1873 Maxwell's work was not directed to proving the statistical nature of gases, but to defending gas theory where it was most vulnerable. He was concerned with the implications of the equipartition law and the ergodic hypothesis. From his first kinetic-theory paper Maxwell was aware of the problem the equipartition law raised for the kinetic theory, and much space in his later papers is taken up with ways of avoiding it. These consisted mainly of constructing molecular models (center-of-force, elastic-sphere, or vortex-ring complexes) that would bypass the law, none of which was entirely satisfactory. Neither of the above problems concerned Clausius; his theory did not contain statistical analyses, and he allowed experiment to dictate the ratio of the specific heats. Both men were by now working on entirely different theories, and, therefore, Clausius had little to give Maxwell and Maxwell little to offer in return.

Another reason why Maxwell could not understand Clausius, especially when the latter was speaking of subjects in which he had little interest and less sympathy, was one of style.[65] Maxwell's style of thought was geometrical, and he frequently used geometrical images in his papers.[66] Not that he was not a master mathematician when he chose to be, as anyone reading his gas-theory papers will realize, but he preferred to think in structural rather than algebraic terms.[67]

63. Maxwell to Tait, 1 December 1873, Maxwell-Tait Correspondence, Cambridge University.

64. Maxwell to Tait, 11 December 1867, quoted in Cargill G. Knott, *Life and Scientific Work of Peter Guthrie Tait* (Cambridge, 1911).

65. Maxwell obviously continued to read Clausius' papers and to use those which contained material he found relevant to his own work. He used Clausius' virial theorem to analyze, unsuccessfully, Andrews' experiments on carbon dioxide; see Maxwell, "Van der Waals on the Continuity of the Gaseous and Liquid States," *Scientific Papers, 2,* 407–415.

66. The differences in style between Maxwell and Clausius are considered in more detail in Garber, "Maxwell and Thermodynamics," *op. cit.* (note 41).

67. *Ibid.*

In contrast, Clausius thought in algebraic terms and seemed to delight in the complexity of analysis. His thermodynamics was an algebraic subject, not the geometrical one that Gibbs's was. Maxwell always found Clausius difficult to understand and evaluate. Speaking about the early editions of his *Theory of Heat*, Maxwell remarked to Tait: "Observe how my invincible ignorance of certain modes of thought has caused Clausius to disagree with me (in the digestive sense) so that I failed in my attempts to boil him down and he does not occupy the place in my book on heat to which his other virtues entitle him. . . ."[68]

Although Clausius had been critically important to Maxwell and had supplied him with two very fundamental ideas for his initial kinetic theory, when Maxwell returned to gas theory it was Boltzmann rather than Clausius who was closest to him and who served as his main stimulus. Clausius had been, indeed, as Maxwell said, a "prime source" for his kinetic theory, but only for a short time, and Clausius could hardly have approved of the final form that his seminal ideas took.

68. Maxwell to Tait, undated postcard (but after 1871, the date of the first edition of *Theory of Heat*), Maxwell-Tait Correspondence, Cambridge University.

Entropy and Dissipation

BY EDWARD E. DAUB*

Entropy and the dissipation of energy are as inseparable as Siamese twins in the thought of every student of thermodynamics, and the mere mention of Clausius' famous phrase, "Die Energie der Welt ist constant; die Entropie der Welt strebt einem Maximum zu," inevitably brings to mind those two seemingly disparate images of energy behavior: conservation in quantity and increasing unavailability with time. The modern thermodynamic mind would initially suffer cultural shock, therefore, if whisked back a century to watch Peter Guthrie Tait penning his *Sketch of Thermodynamics* and writing in italics, ". . . the *Entropy of the Universe tends to zero.*"[1]

Tait was not trying to stand the thermodynamic world on its head, for the word entropy and the idea of dissipation were not yet intimately bound in Britain. Tait, a faithful follower of William Thomson and his dynamical theory of heat, provided the first systematic presentation of Thomson's thermodynamic thought in his book. Prior to this, one had to go to scattered numbers of the *Philosophical Magazine* of the early 1850's to read Thomson's own papers. In keeping with Thomson's approach, Tait wished to deal with the availability of heat rather than with its unavailability and chose, rather presumptuously, to adopt what he called Clausius' "excellent term Entropy" to designate that quantity. Tait was simply saying that the available energy of the universe tends to zero.

Actually, his choice had some merits. When Clausius introduced the Greek word "entropy" in 1865[2] to replace his earlier *Aequiva-*

*Department of History, University of Kansas, Lawrence, Kansas 66044.
1. Peter Guthrie Tait, *Sketch of Thermodynamics* (Edinburgh, 1868), p. 29.
2. Rudolf Clausius, "Ueber verschiedene für die Anwendung bequeme Formen der Hauptgleichungen der mechanischen Wärmetheorie," *Annalen der Physik,* *125* (1865), 353–400. Clausius had first thought of calling S the "*Verwandlungsinalt* des Körpers," the transformational content of the body, but settled on entropy, derived from the Greek word for transformation, because it was more universal in character and similar to the word energy. See *Die mechanische Wämetheorie,* 1st ed. (Braunschweig, 1867), *2*, 34, where the above article is reprinted.

lenzwerth, he explained that entropy meant transformation and thus was close to the meaning of the earlier term; namely, a measure of the ability of heat to transform reversibly into work at different temperatures.[3] It could have been, therefore, as Tait said, "an excellent term" to represent availability. In fact, Clausius never did protest Tait's use of the term, not even when Maxwell followed Tait's lead and also identified entropy with available energy in his *Theory of Heat* in 1871.[4] Clausius' protests, when they came,[5] concerned his rights to certain equations and ideas, not to a name.

When Maxwell, alerted by J. Willard Gibbs, realized the errors of his ways in 1873, he accused Tait of leading him astray.

It is only lately under the conduct of Professor Willard Gibbs that I have been led to recant an error that I had imbibed from your $\theta\Delta^{cs}$, namely that the entropy of Clausius is *unavailable energy* while that of T′ is available energy. The entropy of Clausius is neither the one nor the other, it is only Rankine's thermodynamic function and if we compare the vocabulary

Thermodynamic Function	Entropy (Clausius)
Entropy (Tait)	Available Energy

I think we shall prefer the 2nd column. Available energy there is none in a system of uniform temperature and pressure.[6]

3. Rudolf Clausius, "Ueber eine veränderte Form des zweiten Hauptsatzes der mechanischen Wärmetheorie," *Annalen der Physik, 93* (1854), 481–506. The English translation, "On a Modified Form of the Second Fundamental Theorem in the Mechanical Theory of Heat," appeared in the *Philosophical Magazine, 12* (1856), 81–98. Clausius called it the theorem of the equivalence of transformations (see pp. 85, 90, 92).
  4. James Clerk Maxwell, *The Theory of Heat* (London, 1871), p. 186.
  5. Rudolf Clausius, "Zur Geschichte der mechanischen Wärmetheorie," *Annalen der Physik, 145* (1872), 132–146. The main body of the article also appeared as "A Contribution to the History of the Mechanical Theory of Heat," *Philosophical Magazine, 43* (1872), 106–115. The introductions differ significantly, the covering letter to the *Philosophical Magazine* being more tactful than the original, where Clausius observed that, "Es zeigt sich gegenwärtig in England bei mehreren physikalischen Schriftstellern ein stark hervortretenden Streben, die mechanischen Wärmetheorie so viel, wie möglich, für ihre Nation in Anspruch zu nehmen" (p. 132).
  6. J. C. Maxwell to P. G. Tait, December 1873, Maxwell Collection, Cambridge University Library. T′ refers, of course, to Tait, second in command to William Thomson, the other half of T and T′. Maxwell signed the letter $dp/dt;$ this nickname seems to have first appeared in a postcard from Tait dated 1/2/71, where he addressed Maxwell, "Dr. J. C. M. [= dp/dt, (T′ 's Thermodyn^{es} 162)]." (Maxwell Collection, Cambridge University Library.) In Section 162 of Tait's book, we find the equation $dp/dt \cdot 1/M = JC$. (*Sketch of Thermodynamics, op. cit.* [note 1], p. 91.)

Maxwell was right, of course, to point out the error of identifying entropy with a form of energy—a typical case of Tait's grammatical slovenliness[7]—but he was judging Tait's *Thermodynamics* a bit too severely, nonetheless. In the first place, Maxwell should have realized that his recent discovery that entropy was closely related to Rankine's thermodynamic function would not be news to Tait, for Tait had already observed in 1868 in his *Sketch of Thermodynamics* that the "*Aequivalenzwerth* [entropy] of Clausius is nearly identical with the Thermodynamic Function of Rankine."[8] Secondly, Maxwell apparently still did not realize that Clausius had used entropy to analyze the problem of lost work, so that Tait had been right in indicating a relationship between unavailability of energy and entropy in Clausius' thought.

Certain features of British thermodynamics contributed to the communication tangles. While Clausius used the entropy concept for both reversible and irreversible processes, Rankine, who did have an identical function, applied it only to reversible cycles, and Thomson, who did not formulate an analogous function, dealt with the irreversible case. Add the further complication that, perhaps because of his conflict with Tyndall over claims for the German Mayer, Tait over-reacted to the protests of the German Clausius, and we have a tangle of major proportions.

7. Actually, Tait never explicitly stated that entropy was equivalent to unavailable energy; it was simply implied by his rather loose phrasing. The passage reads: "It seems more convenient, however, to treat as the Entropy of a substance the availability of its contained heat, etc., for the production of work, than its unavailability; so that we shall . . . use the excellent term Entropy in the opposite sense to that in which Clausius has employed it,—viz., so that the *Entropy of the Universe tends to zero*, which is Thomson's theory of dissipation, rather than the unmodified nomenclature of Clausius, according to which the *Entropy tends to a maximum*" (*ibid.*, p. 29). Tait should have realized that entropy could not be equivalent both to Rankine's function and to unavailable energy, and he nowhere said that he would use entropy with the opposite meaning, only with the opposite sense, namely, as denoting a tendency toward zero rather than toward a maximum. Tait may, of course, have suffered a lapse in logic rather than in English. When Maxwell reviewed the second edition of Tait's *Thermodynamics*, he observed that on the very first page Tait had introduced four categories, matter, force, position, and motion, to which, according to Tait, every distinct physical conception should be referred, only to add, before finishing the page, that heat did not belong to any of those categories, but to a fifth, energy. Maxwell commented: "This sort of writing, however unlike what we might expect from the conventional man of science, is the very thing to arouse the placid reader, and startle his thinking powers into action." (Maxwell, "Tait's Thermodynamics," *Nature, 17* [1878], 257–259, 278–280, esp. p. 257.)

8. *Ibid.*

In order to clarify the picture, I will first consider the controversy between Clausius and the British, especially between Clausius and Tait. I will then turn to the evolution of the concepts of dissipation and irreversible entropy. I will show that Tait did make some earlier efforts to be genuinely fair to Clausius, that William Thomson committed a glaring *faux pas* with his blanket approval of Tait's treatment of energy dissipation, that Clausius was the first to show the relationship between lost work and irreversible entropy increase, and, finally, that both Clausius and Thomson share credit for the next major step in thermodynamic thought, the explanation of spontaneity and equilibrium as developed by J. Willard Gibbs. I will argue that, despite Gibbs's obvious preference for Clausius' over competing thermodynamic concepts, Gibbs proved to be the catalyst for the first explicit and legitimate merger of entropy and dissipation in Maxwell's final revised version of his classic work on heat.

I. CLAUSIUS AND THE BRITISH

Tait's Thermodynamics

If there were a psychological-historical index to scientific biographies, Tait would no doubt be cross-referenced under Anglophile and Germanophobe. The motivation for each of the three editions of his *Lectures on Some Recent Advances in Physical Science*[9] can be directly traced to one or the other of these prejudices. The first edition in 1876 was practically a *verbatim* record of the lectures he had given to friends in Edinburgh in the spring of 1874,[10] at the very time when his renewed controversy with Tyndall over Mayer's claims was at its height.[11] The second edition followed within months, strengthening his case against Mayer and reviewing his conflict with Clausius.[12] The third edition of 1885 was a tardy response to German reactions to Tait's earlier priority claims for the British.

9. Peter Guthrie Tait, *Lectures on Some Recent Advances in Physical Science* (London, 1876), 2nd ed. (London, 1876), 3rd ed. (London, 1885).
10. *Ibid.*, 1st ed., p. vi.
11. For a recent discussion of the Tait-Tyndall controversy, see my forthcoming article, "The Energy Concept at Mid-Nineteenth Century," to be published as part of the symposium, "Energy and Society," held at the Dallas AAAS Meeting, December 1968.
12. Tait, *Lectures*, 2nd ed., *op. cit.* (note 9), pp. vi–xiv.

In the 1885 preface, Tait criticized du Bois-Reymond, which might have sparked an interesting match between two sardonic wits had their remarks not been so unfortunately distant both in time and space (Tait was replying to an 1878 address!). Apparently to assure objectivity, Tait delegated to his colleague Crum Brown[13] the translation of a passage from du Bois-Reymond's address. This passage, after a number of suggestive contrasts between how Germans and foreigners approach priority questions,[14] paraphrased sections from Tait's *Lectures*.

For if an Irish physicist living in England and a Scottish physicist (who need no such addition to their fame) had Spectrum Analysis in their pocket ten years before Kirchhoff and Bunsen, why did they not make out of it what Bunsen and Kirchhoff did? Why? A Scottish man of science, whose name has been recently much before us, tells us in his *Lectures on Some Recent Advances in Physical Science*. The German investigator knows all that is going on in Science, or at least has some one by him who does. If a German comes on a new idea, he can at once see, or be told, whether another has it or not, and in the latter case he can print the idea, and so secure the priority; the poor Britons, on the other hand, make the most splendid discoveries in the world without ever guessing that they have struck on anything new . . . and let the priority slip them. The wily Germans! who, instead of contenting themselves like other innocent folks with their mother tongue, sneak into foreign languages to spy out the discoveries that are being made.[15]

13. Tait, *Lectures*, 3rd ed., *op. cit.* (note 9), p. viii.

14. *Ibid.* As a sample of du Bois-Reymond's suggestive comparisons, consider the following: "Foreign investigators, in their ignorance of the German language, often discovered for the second time things long known to us. Not unfrequently, even when better informed, they took advantage of the presumed right of independent discovery to cite their German predecessors only by the way or not at all. The Germans, on the other hand, showed a perfect national impartiality which was far more to their credit than their linguistic superiority."

15. *Ibid.*, pp. vii–ix. Du Bois-Reymond was referring to Tait's claim that William Thomson and Stokes had earlier discussed a correspondence between the dark lines in the solar spectrum and the bright spectral lines in the flames of elements, and his paraphrase was a caricature of Tait's explanation, quoted in the *Lectures* from an address he had made to the Royal Society of Edinburgh in 1871, as to why their discovery had remained unknown:

The question of priority just alluded to illustrates in a very curious way a singular and lamentable, though in one sense honourable, characteristic of many of the highest class of British scientific men; i.e., their proneness to consider that what appears evident to them *cannot but* be known to others. . . . Their foreign competitors, on the other hand (especially the Germans), are often profoundly aware of all that has been done, or, at least, have some one at hand who is, and can thus, when a new idea occurs to them,

325

Tait raised the Bible and cried Pharisee! "Is this not conceived very much in the spirit of the well-known passage: —Ich danke dir, Gott, dass Ich [sic] nicht bin wie andere Leute, Räuber, Ungerechte, Ehebrecher . . . ,"[16] a sarcastic ploy that reveals Tait's final hardened posture toward the Germans.[17] It would be unfair to Tait, however, to overlook the fact that the first edition of his *Sketch of Thermodynamics* in 1868 was in great measure a venture at reconciliation, a tempering of his 1864 remarks contra Tyndall and Mayer.

Tait began in 1867 to revise two articles on the theory of heat which had appeared anonymously as book reviews in the *North British Review* in 1864.[18] Written in the midst of his running argument with Tyndall in the *Philosophical Magazine,* Tait represented Mayer as a graven image worshiped only by apostates.

> . . . Mayer gives, as an analogy to the compression of a body and the consequent production of heat, the fall of a stone to the earth or the impact of a number of gravitating masses and the consequent heating of all. This, we need scarcely say, is simple nonsense. . . . [The] paucity of data led Mayer to choose a substance which Joule afterwards showed was capable of giving, even with the erroneous hypothesis, a result not far from the truth; but, even if Mayer had in 1842 possessed accurate data, and therefore been lucky enough to obtain an approximate result instead of a very inexact one, his determination could never have been

---

at once recognize, or have determined for them, its novelty, and so instantly put it in type and secure it. Neither Stokes nor Thomson, in 1850, seems to have had the least idea that he had hit on anything new . . . —the matter appeared so simple and obvious to them—and, but for the fact that Thomson has given it in his public lectures ever since (at first giving it as something well known), they might have thus forfeited all claim to mention in connection with the discovery (Tait, *Lectures,* 2nd ed., *op. cit.* [note 9], pp. 193–194).

16. Tait, *Lectures,* 3rd ed., *op. cit.* (note 9), p. x.
17. There is, for instance, an unpublished poem, dating from 1873, entitled " 'Alice' in Deutschland," where Tait plays on the pun of "Alles" and "Alice," in verses such as

> "Liebchen, Alles ist vergessen,
> Alles, ja, vergehen ist."
> Thinks't thou *this* her scorn will lessen,
> Dreifach Esel, der du bist.
> "Deutschland, Deutschland, über Alles,
> Ueber Alles, in der Welt!"
> Deutschland, writing about Alice,
> Don't know how her name is spelt.

(Tait's "Scrapbook," Microfilm, Edinburgh University Library.)
18. Peter Guthrie Tait (Anonymous), *North British Review, 40* (1864), 21–37, 177–193.

called more than a happy guess founded upon a total neglect of correct reasoning. When we hear, as has lately been our lot, that Mayer is the author of the Dynamical Theory of Heat; and that he deduced in 1842, by a simple calculation, as accurate a value of the dynamical equivalent as Joule arrived at in 1848, after seven years of laborious experiment, we wonder whether language has any meaning to those who thus abuse it.[19]

But 1867 was the year of his tentative reconcilation with Tyndall,[20] and Tait revised the text of his *Sketch of Thermodynamics* to read, "it is difficult to perceive the grounds on which such statements are made," in place of the more inflammatory, "we wonder whether language has any meaning to those who thus abuse it."[21]

Tait sought counsel from Helmholtz and Clausius in Germany, lest, in seeking to do justice to Joule and Thomson, he may have done injustice to others.[22] When Helmholtz replied and spoke eloquently of Mayer's right to recognition as an independent discoverer of the conservation of energy, Tait responded by including Helmholtz' letter in the preface to his book.[23] Tait apparently did not wish

19. *Ibid.*, p. 28.
20. Tyndall wrote in his diary that, at the British Association meeting at Dundee in June 1867, "Thomson met me in the Kinnaird Hall; blocked my passage, smiled and stretched out his hand. I grasped it, expressed in a word my gratification at meeting him and walked on. Shook hands with Tait afterwards at St. Andrews. They were very cordial to me." Quoted in Arthur S. Eve and C. H. Creasey, *Life and Work of John Tyndall* (London, 1945), p. 124. Tait and Tyndall apparently exchanged letters, and Tait began to refer warmly to Tyndall as T". (Cargill G. Knott, *Life and Scientific Work of Peter G. Tait* [Cambridge, 1911], p. 75.)
21. Tait, *Thermodynamics, op. cit.* (note 1), p. 18.
22. Knott, *Life of Tait, op. cit.* (note 20), p. 216. Tait wrote on 2 Feb. 1867 to Helmholtz as follows: "Herewith I send copies of the first two chapters of a little work which I intend soon to publish. Its main object is to serve as a text-book for students till Thomson and I complete our work on Natural Philosophy. . . . My object in sending this to you at present is to ask you and through you Prof. Kirchhoff, whether in attempting to do justice to Joule and Thomson I have done injustice to you or your colleague."
23. Tait, *Thermodynamics, op. cit.* (note 1), pp. v–vii. Tait frankly admitted that he might be writing a one-sided history, but used the analogy of a courtroom to justify such inevitable partiality. "I cannot pretend to absolute accuracy, but I have taken every means of ensuring it, to the best of my ability; though it is possible that circumstances may have led me to regard the question from a somewhat too British point of view. But, even supposing this to be the case, it appears to me that unless contemporary history be written with some little partiality, it will be impossible for the future historian to compile from the works of the present day a complete and unbiased statement. Are not both judge and jury greatly assisted to a correct verdict by the avowedly partial statements of rival pleaders? If not, where is the use of counsel?" (*Ibid.*, p. v.)

to tamper with his original text more than necessary, since it had been set in type and already used for his classes at Edinburgh under the title *Historical Sketch of the Dynamical Theory of Heat*.[24] But feeling perhaps somewhat guilty upon learning from Helmholtz that Mayer's mental illness had been due to his rejection by the scientific world,[25] he changed his characterization of Mayer's reasoning from "simple nonsense" to the more neutral "inadmissible."[26] And in another place in his book, where he discussed the applications of energy laws to the solar system, he tacked on to the phrase, "numerous and beautiful though they have been," the generous line, "especially in the writings of Mayer."[27]

All of these were, however, rather minor changes compared to the revisions that criticisms from Clausius and Rankine called forth. Although all of the references to Clausius in the *North British Review* had been positive, Clausius looked so unfavorably upon the articles that Tait sought additional assurance from Helmholtz. "Is it fair to ask you whether you think with Clausius that my little pamphlet will only do me harm—or with Thomson and Joule (who, of course, are interested parties) as well as Stewart who have reported favourably on it? I wish to avoid strife and to produce a useful little text-book; but, if Clausius is right, I had better burn it at once."[28] Even in Britain all was not well, for, when Tait brought the issue to Maxwell's attention, he referred to "Clausius & others" who were critical of his book, and the context suggests that he had Rankine in mind when he wrote "others." "Are you sufficiently up to the history of thermodynamics to critically examine and put right a little treatise I am about to print . . . ? You would greatly oblige me by doing so, as Clausius & others have cut up very rough about bits referring to them. I don't pretend to know the subject thoroughly and would be glad of your help. The fact seems that both Clausius &

---

24. Knott, *Life of Tait, op. cit.* (note 20), p. 213.
25. Tait wrote in his reply to Helmholtz: "As to Mayer I had no idea that his illness was due to the cold way in which his papers were received; nor, had I known this, would I have written so strongly against his claims to the *establishment* of the Conservation of Energy." (*Ibid.*, p. 216.)
26. Tait, *Thermodynamics, op. cit.* (note 1), p. 18.
27. Tait, *North British Review* articles, *op. cit.* (note 18), p. 192, and Tait, *Thermodynamics, op. cit.* (note 1), p. 86.
28. Quoted in Knott, *Life of Tait, op. cit.* (note 20), p. 217.

Rankine are about as obscure in their writings as anyone can well be."[29]

In any event, Tait broke apart the original text of his *North British Review* articles to expand his few remarks about Rankine and Clausius into major discussions of their ideas.[30] Furthermore, in a completely new section of the book on the mathematical theory of heat, he supplied condensed versions of the key concepts and equations of both Rankine and Clausius.[31] Clausius, the German, fared even better than Tait's Scottish compatriot. For example, he said that "the grand point of Clausius' work is his proof that Carnot's principle of reversibility still holds, though on other grounds than those from which Carnot deduced it. This was a step of the utmost importance to thermodynamics, and sufficient (had he done no more) to entitle him to a foremost place in the history of the subject."[32]

Admirable as these efforts were, such paper and paste insertions could not change the fundamental theme of the book, which was, along with the defense of Joule against Mayer, the promotion of the thermodynamic approach of William Thomson. For example, in the *North British Review* article where Tait listed the significant steps in the development of the new theory of heat, he gave as the seventh step "the adaptation, by Clausius and Rankine, and subsequently, with greater generality and freedom from hypothesis, by Thomson, of Carnot's methods to the true theory; with Joule's experimental verification of Thomson's general results."[33] In the revised version, the passage still leaned towards Thomson, despite the addition of the phrase, "the re-establishment of the second law by Clausius," following the semicolon.[34] Actually, Tait tended always to take back what he gave. Thus, the seventh step had become the eighth because of the additional footprint of Thomson, namely, "Thomson's introduction of an *absolute* Thermodynamic scale of thermometry (1848)."[35] More significant, immediately after praising Clausius, Tait made an insertion commenting on the tentative character of the work of both

29. Tait-Maxwell Correspondence, Cambridge University Library.
30. Tait, *Thermodynamics, op. cit.* (note 1), pp. 24–29.
31. *Ibid.*, pp. 111–114.
32. *Ibid.*, p. 29.
33. Tait, *North British Review* articles, *op. cit.* (note 18), p. 35.
34. Tait, *Thermodynamics, op. cit.* (note 1), pp. 44–45.
35. *Ibid.*, p. 44.

Clausius and Rankine. "The investigations of both these writers fundamentally involve various hypotheses, which may or may not be found by experiment to be approximately true, and which render it difficult to gather from their writings what parts of their conclusions, especially with reference to air and gases, depend merely on the necessary principle of the dynamical theory."[36] The paragraph was not placed in quotes, though it should have been, coming as it did directly from Thomson's 1851 paper on the dynamical theory of heat.[37] Quite obviously, Thomson was king for Tait, or as T. H. Huxley put it, "Tait worships him [Thomson] with the fidelity of a large dog—which noble beast he much resembles in other ways."[38]

Tait's primary purpose in writing his *Sketch of Thermodynamics* seems to have been to secure credit for Thomson for the theory of the dissipation of energy, a theme that sounded numerous times in the *North British Review* articles.[39] At least this became the key issue over which Clausius tangled with Tait for the next ten years. Clausius kept his peace, however, until Maxwell's *Theory of Heat* appeared in 1871.[40]

Clausius' Critique of Maxwell's Theory of Heat

Clausius sent off a letter to the *Philosophical Magazine* with comments on the history of the mechanical theory of heat.[41] He opened by saying that "there has recently appeared a very valuable book, entitled *Theory of Heat*,"[42] a more tactful reference than appeared in his German version. His criticisms, however, were no less trenchant. Overlooking Maxwell's erroneous equating of entropy with

36. *Ibid.,* p. 29.

37. William Thomson, "On the Dynamical Theory of Heat, with Numerical Results Deduced from Mr. Joule's Equivalent of a Thermal Unit, and M. Regnault's Observations on Steam," *Philosophical Magazine, 4* (1852), 8–21, 105–117, 168–176. Reprinted in his *Mathematical and Physical Papers, 1* (Cambridge, 1882); see p. 201.

38. Quoted in Leonard Huxley, *Life and Letters of Sir J. D. Hooker* (London, 1918), pp. 165–166.

39. Tait, *North British Review* articles, *op. cit.* (note 18); see pp. 27–28, 33, 35, 37, 182, 190.

40. Clausius, "Zur Geschichte der mechanischen Wärmetheorie," *op. cit.* (note 5), pp. 132–133.

41. Clausius, "A Contribution to the History of the Mechanical Theory of Heat," *op. cit.* (note 5).

42. *Ibid.,* p. 106.

unavailable energy, Clausius complained that, apart from references to his work in kinetic theory, "my name occurs only once, when it is said that I introduced the word *entropy;* but it is added that the theory of entropy had already been given by W. Thomson."[43] He then turned to specifying sins of omission; even where Tait had counted Clausius in, Maxwell had left him out.

> Besides, W. Thomson in his memoir, when speaking of my demonstration, says:—"The following is the axiom on which Clausius's demonstration is founded:—It is impossible for a self-acting machine, unaided by external agency, to convey heat from one body to another at a higher temperature." The proposition here printed in italics is in Mr. Maxwell's book (p. 153) quoted in exactly the same words in which Thomson has clothed it; but here, instead of the introductory words "The following is the axiom on which Clausius's demonstration is founded," we have: —"Carnot expresses this law as follows." Thus, while in the remainder of the quotation Thomson's words are used, my name is replaced by that of Carnot, without a word of explanation for the alteration. This is so mysterious to me, that I cannot help supposing there must be a misprint. However, of course I must leave it to Maxwell to clear up the matter.[44]

Tait had been fairer, introducing the same axiom by saying, "Clausius proved the proposition in 1850, by a process strictly analogous to that of Carnot . . . but based on the additional axiom. . . ."[45] It is small wonder then that, in his German preface, Clausius referred to Maxwell's book, not to Tait's, as "viel rücksichtloses gegen die Deutschen."[46]

Clausius specified two more places where Maxwell should have acknowledged him.[47] First, with reference to the discovery of the negative specific heat of saturated steam, Maxwell wrote, "It appears from the experiments of M. Regnault . . . that heat leaves the saturated steam as its temperature rises, so that its specific heat is *nega-*

43. *Ibid.* For an illuminating discussion of the interaction between Clausius and Maxwell, and Maxwell and Gibbs, see Martin J. Klein's article, "Gibbs on Clausius," *Historical Studies in the Physical Sciences, 1* (1969), 127–149.
44. *Ibid.,* p. 111.
45. Tait, *Thermodynamics, op. cit.* (note 1), p. 31.
46. Clausius, "Zur Geschichte der mechanischen Wärmetheorie," *op. cit.* (note 5), p. 133.
47. Clausius, "A Contribution to the History of the Mechanical Theory of Heat," *op. cit.* (note 5), p. 111.

tive."[48] Second, with regard to the first attempts to calculate the departure of the behavior of saturated steam from the behavior predicted by the ideal gas laws, Maxwell said, "In the meantime Rankine has made use of the formula in order to calculate the density of saturated steam."[49] Clausius concluded that Maxwell had intentionally suppressed his name.[50]

Maxwell was prompt to respond, not with a protest, but with a second edition in 1872. All three of the criticized passages, plus a bonus fourth, were revised so that they now read in Clausius' favor. Where the first edition said, "Carnot expresses this law as follows . . . ," the second said, "Clausius, who first stated the principle of Carnot in a manner consistent with the true theory of heat, expresses this law as follows."[51] To the phrase, ". . . its specific heat is *negative*," Maxwell added, "a result pointed out by Clausius and Rankine."[52] The bonus was the addition of "this was first shown by Clausius and Rankine" to an earlier discussion of the behavior of saturated steam.[53] Finally, Maxwell introduced Clausius' name in the sentence regarding the calculation of the density of saturated steam.[54] No amendments were made, however, to the erroneous portrayal of entropy; they came only with the fourth edition in 1876.[55]

Although Maxwell made these corrections from a just respect for Clausius' achievements, he enjoyed jesting with Tait about the assorted names that Clausius imposed on the theory of heat—disgregation, ergal, virial, and ergon.[56] At the time the critique from Bonn appeared, he wrote to Tait:

48. Maxwell, *Theory of Heat, op. cit.* (note 4), p. 169.
49. *Ibid.,* p. 173.
50. Clausius, "A Contribution to the History of the Mechanical Theory of Heat," *op. cit.* (note 5), p. 111.
51. James Clerk Maxwell, *Theory of Heat,* 2nd ed. (London, 1872), p. 153.
52. *Ibid.,* p. 169.
53. *Ibid.,* p. 134.
54. *Ibid.,* p. 173.
55. James Clerk Maxwell, *Theory of Heat,* 4th ed. (London, 1876). Maxwell introduced an entirely new section on entropy (pp. 162–165) and completely rewrote his discussion of available energy (pp. 187–193).
56. Clausius coined the term ergal for the force function in mechanics and proposed ergon to represent work which, along with the new name "entropy," suggested a delightful trinity—energy, entropy, and ergon. For a discussion of disgregation and virial, see my article, "Atomism and Thermodynamics," *Isis, 58* (1967), 293–303, especially pp. 301–303. Martin Klein has recently questioned certain aspects of my interpretation of disgregation in his article, "Gibbs on Clausius," *op. cit.* (note 43).

As for C. though I imbibed my $\theta\Delta^{cs}$ from other sources, I know that he is a prime source and have in my work for Longman been unconsciously acted on by the motive not to speak about what I don't know. In my spare moments I mean to take such draughts of Clausiustical Ergon as to place me in that state of disgregation in which one becomes conscious of the general sum of Entropy. Meanwhile till

> Ergal & Virial from their thrones be cast
> and end their strife with suicidal yell.[57]

Maxwell thus aired his minor grievances in private with humor. Tait, however, could not contain his pique and confronted Clausius openly and with rancor.

Tait vs. Clausius

Tait, perhaps because he had taken some pains to do justice to Clausius, turned the wrath of his wounds upon him and sought to deny him the priority he had earlier granted in his *Sketch of Thermodynamics;*[58] namely, that of being the first to reconcile Carnot's theorem and the new mechanical theory of heat. He published an open letter to Clausius in the *Philosophical Magazine* in 1872, where he said:

As regards the question to whom is due the credit of first correctly adapting Carnot's magnificently original methods to the true theory of heat, it is only necessary to compare the *Axiom* of Professor Clausius' first

57. Tait-Maxwell Correspondence, Cambridge University Library. Postcard dated 12 Feb. 1872. In an undated postcard, Maxwell followed the same pattern, a confession of ignorance plus a playful jest. "Observe how my inexpressible ignorance of certain modes of thought has caused Clausius to disagree with me (in a digestive sense) as that I failed in my attempts to boif [sic] him down, and he does not occupy the place in my book on heat to which his other virtues entitle him. If he can get himself assimilated now, I shall appear in a state of disgregation. Ergal lusting against Virial, and Virial against Ergal." This card is also quoted in part by Elizabeth Garber in her interesting article, "James Clerk Maxwell and Thermodynamics," *American Journal of Physics,* 37 (1969), 146–155, p. 150, though without the play on Clausius' words. We seem to disagree on one deciphering of Maxwell's hand; where she reads "boil," I see "boif." There are a number of more important points on which our perspectives differ. For example, I do not believe that it was Maxwell's preference for geometrical methods that made Clausius difficult to "boif," as Mrs. Garber suggests (*ibid.,* p. 150). Maxwell's use of geometry rather than the calculus may be because he was attempting to write a textbook for students. See Maxwell, *Theory of Heat,* 2nd ed., *op. cit.* (note 51), p. vi.

58. Tait, *Sketch of Thermodynamics, op. cit.* (note 1), Section 49.

paper (the only one which has a chance of priority over Thomson) with the behaviour of a thermoelectric circuit in which the hot junction is at a temperature higher than the neutral point, and where therefore heat *does, of itself, pass from a colder to a hotter body*. A thermo-electric battery, worked with ice and boiling water, is capable of raising to incandescence a fine wire, giving another excellent instance of the fallacy of the so-called axiom.[59]

Clausius proved able to blunt the first and break the second argument.[60]

The question which Tait raised regarding the neutral point phenomena of a thermocouple could not be decisively answered, however, because there was still no consensus regarding the exact nature of the behavior of heat in a thermoelectric circuit. The neutral point could be construed as a contradiction of Clausius' axiom only if one adopted the simple view that only the Peltier heats at the two junctions were involved; for if that were the case, then beyond the neutral point where the thermodynamic current reverses, the Peltier heats would also be reversed and heat would now be absorbed at the cold junction and released at the hot. Thus, heat would apparently pass from a lower to a higher temperature. Clausius' tack was to argue that additional exchanges of heat must occur.

According to Clausius, the junction of two different metals generates an electromotive force because of the contact there of different molecular states.[61] Believing that different molecular states exist also within one and the same metal at various temperatures, he supposed that Peltier-type heats should also arise at the places of contact of these different molecular states within each metal of the thermocouple. Thus, the neutral point phenomena were probably not simply limited to heats at the hot and cold junctions. Clausius concluded that, in the absence of any decisive understanding of the complex behavior of a thermoelectric circuit, Tait's first objection

59. Peter Guthrie Tait, "Reply to Professor Clausius," *Philosophical Magazine, 42* (1872), 338.

60. Rudolf Clausius, "On the Objections Raised by Mr. Tait against my Treatment of the Mechanical Theory of Heat," *Philosophical Magazine, 43* (1872), 443–446. The article also appeared in German, "Ueber die von Hrn. Tait erhobenen Einwände gegen seine Behandlung der mechanischen Wärmetheorie," *Annalen der Physik, 146* (1872), 308–313.

61.Rudolf Clausius, "Ueber die Anwendung der mechanischen Wärmetheorie auf die thermo-elektrischen Erscheinungen," *Annalen der Physik, 90* (1853), 513–544.

should be assigned to limbo.[62] Regarding Tait's second objection, Clausius' answer was far more clear-cut. The fact that a thermoelectric wire may reach incandescence, even though operating only between the temperatures of ice and steam, is no different from the case where the work produced by a steam engine generates frictional temperatures far beyond that of the original steam.[63] Furthermore, Clausius continued, the proposed phenomenon would not even fall under his axiom, since it does not involve heat moving by itself counter to a temperature gradient. A compensating movement of heat between the hot and cold junctions accompanies the process.[64]

Refuted in his attempt to undercut Clausius' priority by raising supposed contradictions to the axiom, Tait scored a debating point by showing that Clausius nowhere had stated his axiom in 1850 in the form that had proved crucial to his reply; namely, that heat by itself ("von selbst") could not go counter to a temperature gradient.

The following are, as far as I can see, the words to which Professor Clausius refers as implicitly containing his axiom, which is nowhere explicitly stated in his first paper. They do not contain the phrase "von selbst" to which he assigns so important and extensive a meaning. "Durch Wiederholung dieser beiden abwechselnden Processes könnte man also, ohne irgend eine andere Veränderung, beliebig viel Wärme aus einem *kalten* Körper in einem *warmen* schaffen, und das widerspricht dem sonstigen Verhalten der Wärme, indem sie überall das Bestreben zeigt, vorkommende Temperatur—differenzen auszugleichen und also aus den wärmeren Körpern in die kälteren überzugehen."[65]

Clausius had blundered in trying to make retroactive a later form of his axiom,[66] though the idea had been implicit in his work in 1850.

62. Clausius, "On the Objections Raised by Mr. Tait," *op. cit.* (note 60), p. 445.
63. *Ibid.*, pp. 443–444.
64. *Ibid.*, p. 444.
65. Peter Guthrie Tait, "On the History of the Second Law of Thermodynamics, in reply to Professor Clausius," *Philosophical Magazine, 43* (1872), 516–518, esp. p. 517.
66. In his first protest, Clausius had said, "This axiom, in its briefest form is:—'Heat cannot of itself pass from a cooler into a hotter body' " ("A Contribution to the History of the Mechanical Theory of Heat," *op. cit.* [note 5], p. 108). In replying to Tait's objections, he implied that this was the actual form of the axiom in 1850. "As already mentioned in my previous article, the axiom employed by me for the demonstration of Carnot's theorem modified [was] that heat cannot pass by itself from a cooler into a hotter body." ("On the Objections Raised by Mr. Tait

Tait pressed this point in his continuing efforts to confer priority on Thomson, writing in 1876 in his *Lectures on Some Recent Advances in Science:* "What I have just given you is, in a much amplified form, the gist of some of Sir W. Thomson's remarks of 1851 on this point. Clausius, in the preceding year, had endeavoured to supply this defect in Carnot's work by an appeal to the general behaviour of heat, i.e., its always striving to pass from a warmer body to a cold one. I have elsewhere given reasons which seem to show this proof inadmissible."[67]

Such rumblings in Edinburgh on the dating of Clausius' axiom brought no response from Bonn. However, Clausius had his seismograph tuned to all that came from Tait's pen. For example, when Tait made a minor addition of one sentence to the second edition of his *Lectures* in 1876,[68] Clausius did not let it pass.[69] Tait said that the well-known behavior of those intelligent creatures known as Maxwell's demons, reversing the usual flow of heat, "is absolutely fatal to Clausius' reasoning." While admitting chance and local departures from uniformity, Clausius would not permit dramatic reversals. After all, he concluded, "mein Satz sich nicht darauf bezieht, was die Wärme mit Hülfe von Dämonen thun kann, sondern darauf, was sie für sich allein thun kann."[70]

His anger never abating, Tait struck back in 1877 with a second edition of his *Sketch of Thermodynamics,* stripping acknowledgments to Clausius from the text as far as possible. For example, Section 56 had originally read: "He [Thomson] seems to have been at first unwilling to encounter the new problems suggested by the true theory. In consequence of this he was anticipated by Clausius in the publication of the second law; but having, in ignorance of what Clausius had done, obtained its demonstration for himself, he advanced

---

..., " *op. cit.* [note 60], p. 443.) Clausius actually had not delineated his axiom in any formal way until 1854, when he wrote, "Heat can never pass from a colder to a warmer body without some other change, connected therewith, occurring at the same time." ("On a Modified Form of the Second Fundamental Theorem . . . ," *op. cit.* [note 3], p. 86.)

67. Tait, *Lectures, op. cit.* (note 9), p. 118.
68. *Ibid.,* 2nd ed., p. 119.
69. Rudolf Clausius, "Ueber eine von Hrn. Tait in der mechanischen Wärme-theorie angewandte Schlussweise," *Annalen der Physik, 2* (1877), 130–133.
70. *Ibid.,* p. 133.

with tremendous strides. . . ."[71] The revised version read more smoothly, without mention of Clausius: "He seems to have been at first unwilling to encounter the new problems suggested by the true theory, but having in 1851 obtained a satisfactory basis for the Second Law, he advanced with tremendous strides. . . ."[72] Tait stooped to such picayune points as reversing the order of names from "Clausius and Rankine" to "Rankine and Clausius."[73] Section 49, which concerned Clausius' priority, disappeared completely. The deletions might have known no bounds if Tait had not been constrained by the fact, openly admitted, that Rankine had actually written most of his passages about Clausius.[74]

Tait joined Maxwell in dropping all former references to entropy as an appropriate word for available energy. Contrary, however, to Maxwell, who had revised his discussion of available energy, Tait left his presentation of the subject untouched, thus ignoring Clausius' foremost complaint in 1872 that Tait had credited Thomson with an equation that rightfully was his.[75] At that time, Tait had

71. Tait, *Thermodynamics, op. cit.* (note 1), pp. 32–33.

72. Peter Guthrie Tait, *Sketch of Thermodynamics*, 2nd ed. (Edinburgh, 1877), p. 40. Where Step 8 in Tait's listing of the developments in the theory of heat had earlier read "the re-establishment of the second law by Clausius," it now said, "the establishment of the second law by Thomson." Compare pp. 44–45 in the first edition with p. 54 in the second.

73. *Ibid.*

74. *Ibid.*, pp. xv–xvi. "The result of my inquiries is not likely to be satisfactory to Professor Clausius, for I have now, after careful revision of the whole documentary evidence, found it necessary to cancel certain additions which I had made to the paragraphs written for me by Rankine. These paragraphs, I now see, were correct as they originally stood. But, thinking them too severe on Professor Clausius' claims, I rashly added some mitigating passages, which I have now been obliged to retract as unsupported by evidence." Clausius later needled Tait by saying that the very idea that a man of science would presume to write a book on a subject about which others would have to do the writing for him "hat mich etwas in Erstaunen gestezt." (Rudolf Clausius, *Die mechanische Wärmetheorie, 2,* 2nd ed. [Braunschweig, 1879].) Clausius had appended to this treatise on electricity a special section entitled "Tendenz des Buches *Sketch of Thermodynamics* von Tait," pp. 324–330, in which he also discussed his personal correspondence with Tyndall and Mayer and an exchange of letters with Tait. In 1867 Clausius had received a note accompanying Tait's articles, saying: "Would you kindly look over the little pamphlet which accompanies this, and which is not yet published, so as to tell me whether in trying to give Joule and Thomson the credit they deserve, and which some of their countrymen [Tyndall] appear indisposed to grant them, I have inadvertently done injustice to you. If such be the case, I shall be delighted to make the necessary corrections before publishing, as my sole object is to be impartial" (*ibid.,* p. 328).

75. Clausius, "A Contribution to the History of the Mechanical Theory of Heat," *op. cit.* (note 5), pp. 113–114.

pronounced confidently: "In common with all the scientific friends I have consulted, I am unable to perceive that Professor Clausius has 'refuted' any one of my former remarks, or that he is likely to be able to refute any of the others—though he says it can be easily done. Let Professor Clausius attempt the refutation. . . ."[76] when the refutation came, Tait sought refuge with Thomson.

II. DISSIPATION AND LOST WORK: THE IRREVERSIBLE PROCESS

A Word of Introduction

Before tackling the complex problem of the equations governing the dissipation of energy in irreversible processes, some preliminary explanation about entropy and its relation to lost work will be helpful. Entropy exhibits two forms of behavior, one for reversible and the other for irreversible changes. If a process occurs reversibly, i.e., proceeds under conditions where the driving forces—the pressure and temperature differentials—are so minute that minor changes may reverse the direction of the process, then all entropy changes appear in pairs, equal but opposite in sign, and the net entropy change is zero. If the process is irreversible, the differences in pressure and temperature are significant, and minor changes cannot reverse the direction of the process; the net change in entropy is then not zero but invariably an increase.

In the irreversible process, an amount of potential work is always lost; that work is equal to the product of the lowest temperature involved in that process and the entropy increase. For example, consider the simplest irreversible process: the conduction of heat q from a higher absolute temperature t to a lower one t_0. The entropy increases at the lower temperature by q/t_0 and decreases at the higher temperature by $-q/t$, with a net increase of $q[1/t_0 - 1/t]$. The irreversible conduction of heat obviously incurs losses in potential work, since the heat could have performed work if transmitted via an engine. The theoretical maximum work for a reversible engine operating under these conditions is $Jq[(t - t_0)/t]$; this is identical with the product of the lowest temperature t_0 and the entropy increase in the

76. Peter Guthrie Tait, "Reply to Professor Clausius," *Philosophical Magazine,* *44* (1872), 240.

irreversible process, $Jq[1/t_0 - 1/t]$, where J is a conversion factor to mechanical units.

There is another characteristic of entropy in irreversible processes that is critically important for the ensuing discussion. The entropy change for such processes cannot be determined in terms of the amounts of heat and the temperatures actually occurring. In order to calculate that change, one must imagine a reversible sequence of processes that would bring about the same overall transformation of pressure, temperature, and volume coordinates. Entropy is defined as q/t only for reversible exchanges (I was guilty in the previous example of failing, for the sake of simplicity, to raise that technicality); when Clausius challenged Tait's claim that Thomson had originated the equation for lost work, he pointed out that Tait had failed in his derivation to observe this canon of thermodynamics.

Tait's Quandary

Tait's derivation of what he called "Thomson's expression for the amount of heat dissipated during the cycle"[77] is too succinct to paraphrase usefully.

178. The real dynamical value of a quantity, dq, of heat is $\int dq$, whatever be the temperature of the body which contains it. But the *practical* value is only (§§ 54, 175)

$$J[(t - t_0)/t]dq$$

where t is the temperature of the hot body, and t_0 the lowest available temperature. This value may be written in the form

$$Jdq - Jt_0 dq/t .$$

Hence, in any cyclical process whatever, if q_1 be the whole heat taken in, and q_0 that given out, the practical value is

$$J(q_1 - q_0) - Jt_0 \int dq/t .$$

Now, if the cycle be reversible, the *practical* value is

$$J(q_1 - q_0)$$

by the first law; so that, in this particular case (at least unless $t_0 = 0$),

$$\int dq/t = 0 .$$

77. Tait, *Thermodynamics, op. cit.* (note 1), p. 122.

But in general this integral has a finite positive value, because in non-reversible cycles the practical value of the heat is always less than

$$J(q_1 - q_0) \, .$$

Hence the amount of heat lost needlessly, i.e., otherwise than to the refrigerator, or in producing work, is

$$t_0 \int dq/t \, .$$

This is Thomson's expression for the amount of heat *dissipated* during the cycle. It is, of course, an immediate consequence of his important formula for the work of a perfect engine (§ 175).[78]

So smooth a derivation would seem beyond the pale of doubt. Clausius, however, detected two critical flaws.[79]

The expression for the work obtained in a cycle was obviously wrong by simple inspection for, as Clausius observed, the net result of any cycle, reversible or irreversible, should always be an amount of work equal to the excess of heat accepted to that rejected, namely, $J(q_1 - q_0)$.[80] Tait's expression, therefore, violated the first law of thermodynamics for all irreversible processes and thus was erroneous.

Clausius delighted in pointing out the further error that, contrary to Tait's claim, the integral $\int dq/t$ would not be positive but negative, leading to the absurd conclusion that more energy would supposedly become available in irreversible cycles.[81] Thus, Tait's equation would contradict the second law of thermodynamics as well as the first. Stumbling towards defeat, Tait published an urgent letter to Thomson, asking him to break his silence, for "Clausius has challenged your claim to the well-known expression for the amount of heat dissipated in a non-reversible cycle."[82] Accompanying the letter, however, was not Section 178 in its original form, but a patched version that represented Tait's unsuccessful effort to escape from Clausius' checkmate. The amended portion of the text now read:

Now the realized value is

$$J(q_1 - q_0)$$

78. *Ibid.*, pp. 121–122.
79. Clausius, *Die mechanische Wärmetheorie*, 2nd ed., *op. cit.* (note 74), pp. 321–322.
80. *Ibid.*, p. 322.
81. *Ibid.*, p. 322.
82. Peter Guthrie Tait, "On the Dissipation of Energy," *Philosophical Magazine*, 7 (1879), 344–346, esp. p. 344.

by the first law; and if the cycle be reversible, this must be equal to the extreme practical value. Hence, in this particular case,

$$\int dq/t = 0 .$$

But in general this integral has a finite negative value, because in non-reversible cycles the realized value of the heat is always less than

$$J(q_1 - q_0) - Jt_0 \int dq/t$$

which is the extreme practical value.

Hence the amount of heat lost needlessly, i.e., rejected in excess of what is necessarily rejected to the refrigerator for producing work, is

$$-t_0 \int dq/t .$$

This is Thomson's expression for the amount of heat dissipated during the cycle.[83]

Thus, Tait made three moves. First, he introduced a new term, the realized value, to represent $J(q_1 - q_0)$, securing the first law for all cases. Second, he admitted his error; the integral $\int dq/t$ should generally be negative. Third, he coined another term, the extreme practical value, to represent the expression $J(q_1 - q_0) - Jt_0 \int dq/t$, so that he could argue that the extreme or reversible value exceeds the realized value by the second term.

When Tait confessed that he had originally been "somewhat slipshod" in this section and offered his revision as penance, Thomson gave his blessing.

The passage quoted, with amendments, by Professor Tait from his *Thermodynamics*, seems to me perfectly clear and accurate. Taken in connection with the sections which preceded it in the original, its meaning was unmistakable; and a careful reader could have found little or no difficulty in making for himself the necessary corrections with which Professor Tait now presents it. It is certainly not confined to reversible cycles; but, on the contrary, it gives an explicit expression for the amount of energy dissipated, or as I put it, "absolutely and irrecoverably wasted" in operations of an irreversible character.[84]

Thomson had, unfortunately, "absolutely and irrecoverably," made a public thermodynamic blunder, for Clausius' key criticism was still

83. *Ibid.*, p. 345.
84. William Thomson, "Note by Sir W. Thomson on the Preceding Letter," *Philosophical Magazine*, 7 (1879), 346–348, esp. p. 346.

unanswerable, since Tait never amended that part of Section 178 where he had committed his fundamental mistake. His mistake was to use the integrated form of a differential equation, which was valid only for reversible processes, as though it were applicable to irreversible ones. Quite clearly this was the key issue, for Tait twice stated the question explicitly in his letter to Thomson, first at the beginning of his revision,[85] and again more strongly in the conclusion: "From Professor Clausius's comments it appears, as I have already said, that he considers the method I have adopted from you to be one which cannot be applied except to *reversible* cycles, and which, it is absurd to employ in any argument connected with the dissipation of energy."[86] Thomson simply overrode the contention, since he believed that Tait had obtained an expression which first appeared in his paper on dissipation:[87] "I think Professor Tait quite right in referring also to that paper for the formula $t_0 \int dq/t$. The whole matter is contained in the formula

$$we^{-1/J} {}_T^S \int \mu dt$$

which is given explicitly in that paper."[88] The whole matter was not contained in that formula, however, for its similarity to Tait's formula was purely deceptive.

Thomson's Theory of Dissipation in an Engine

The challenge and response to Clausius by Thomson and Tait in the pages of the *Philosophical Magazine* in 1879 was not their first discussion of the claims from Bonn. There is a fragment of an interesting letter[89] from Thomson dealing with these claims. The fragment begins abruptly with an integral, in the midst of a passage quoted from Thomson's 1852 paper on dissipation. If the earlier sentences are provided, the passage reads as follows:

85. Tait, "On the Dissipation of Energy," *op. cit.* (note 82), p. 344. ". . . Professor Clausius asserts that the method I employ (and which I certainly obtained from your paper in 1852) is inapplicable to any but reversible cycles. This, I think, is equivalent to denying altogether your claims in the matter."

86. *Ibid.*, p. 346.

87. William Thomson, "On a Universal Tendency in Nature to the Dissipation of Mechanical Energy," *Philosophical Magazine, 4* (1852), 256–260.

88. Thomson, "Note . . . on the Preceding Letter," *op. cit.* (note 84), p. 347.

89. Edinburgh University Library. The letter is quoted in full by Knott in *Life of Tait, op. cit.* (note 20), pp. 223–224.

Let S denote the temperature of the steam . . . ; T the temperature of the condenser; μ the value of Carnot's function, for any temperature t; and R the value of [the fragment begins here]

$$e^{-1/J\,{}_T^S\!\int \mu dt}$$

"Then $(1 - R)w$ expresses the greatest amount of mechanical effect that can be economized in the circumstances from a quantity w/J of heat produced by the expenditure of a quantity w of work in friction, whether of the steam in the pipes and entrance ports, or of any solids or fluids in motion in any part of the engine; and the remainder, Rw is absolutely and irrecoverably wasted, unless some use is made of the heat discharged from the condenser." The whole thing is included in this illustration. . . . I don't believe Clausius yet to this day understands as much of the *fact* of dissipation of energy as is stated in that first paper in which the theory is propounded and the name given, and it does not appear that he has ever made any acknowledgment whatever of T[homson] in the matter. This *must* be because he does not understand it; *not* because he would consciously appropriate what is not his own.[90]

A little amateur sleuthing reveals not only that tradition has been correct to view the undated fragment as addressed to Tait, but also that the date of its appearance must lie sometime prior to the second edition of Tait's *Thermodynamics*. Tait, in the Preface to that edition, obviously paraphrased Thomson, even parroting his "The whole matter is contained in. . . ."

Professor Clausius ought to have seen at once that the problems proposed and solved in that article of Thomson's *must* have involved in their solution the expression in question—even if formulae had not been given. Thomson distinctly states in the article the conditions under which energy is dissipated, in connection with the economic working of a steam engine, and gives the expression

$$Rw \quad \text{or} \quad we^{-1/J\,{}_T^S\!\int \mu dt}$$

for the portion of the heat w, which is "absolutely and irrecoverably wasted." The whole matter is contained in this.[91]

But it wasn't.

90. *Ibid.*, p. 223. Thomson was not always so charitable and spoke at one point of "the bone over which Clausius snarls" (*ibid.*, p. 224).
91. Tait, *Thermodynamics*, 2nd ed., *op. cit.* (note 72), pp. xvi–xvii.

When Thomson considered, in his paper on dissipation, the pressure losses incurred by steam in flowing through pipes and valves, he believed that he could consider such dissipation of mechanical energy as equivalent to additions of heat and then treat that heat just as he would any external addition. Thus, if mechanical energy w were dissipated, an amount of reversible work $(1 - R)w$ would be obtainable, where

$$R = e^{-1/J} {}_T^S \int \mu dt \ .$$

Thomson's nomenclature may need some comment here. The expression $(1 - R)$ is simply the proportion of heat converted into work by a reversible Carnot engine, or $(t - t_0)/t$ in Tait's symbols. Thomson, prior to his definitive experiments with Joule, was not willing to assume that Carnot's function μ is inversely proportional to the ideal gas temperature scale,[92] and he chose to evaluate μ from Regnault's data for steam, using Centigrade temperatures T and S for condenser and boiler. Substituting J/t for μ and absolute temperatures t_0 and t for the limits of the integral reduces the expression to the usual form, t_0/t.[93] Testing Thomson's supposition against

92. See Thomson, "On the Dynamical Theory of Heat . . . ," *op. cit.* (note 37); *Papers, 1,* pp. 196, 215. "I cannot see that any hypothesis, such as that adopted by Clausius fundamentally in his investigations on this subject, and leading, as he shows to determination of the densities of saturated steam at different temperatures, which indicate enormous deviations from the gaseous laws of variation with temperature and pressure, is more probable, or is probably nearer the truth, than that the density of saturated steam does follow these laws as it is usually assumed to do. In the present state of science it would be perhaps wrong to say that either hypothesis is more probable than the other [or that the rigorous truth of either hypothesis is probable at all]" (fn., p. 196). The bracketed material was a later annotation (see p. vi).

93. Thomson believed that Clausius failed to understand this correspondence. "As for the very letters of the formula, T in the same article says, 'If the system of thermometry adopted be such that

$$\mu = J/t + a \ .'$$

Accepting Clausius' statement that 'neither the expression $(t_0 \int dq/t)$ nor anything of like meaning can be found in the article referred to by T',' the only conclusion is that he is ignorant of the fact that

$$e^{-1/J} {}_T^S \int \mu dt = T + a/S + a \ ,$$

and so had his eyes closed to the fact that Rw/J means the same as

$$(T + a)dq/(S + a)$$

or $t_0 \ dq/t$ according to the notation of T'." (Quoted in Knott, *Life of Tait, op. cit.* [note 20], p. 223.)

a hypothetical case of dissipation will show that although one obtains the correct value for the heat rejected to the condenser, the work supposed to be derived from the originally dissipated mechanical energy is grossly in error.

To see this, consider a Carnot cycle with an ideal gas operating between absolute temperatures t and t_0. If v_1 and v_2 represent the initial and final volumes of the isothermal expansion at t, the heat added at t is

$$kt \ln v_2/v_1 \, ,$$

where k is some constant depending on the amount of gas and the units chosen. The work obtainable from the reversible cycle would then be

$$Jkt \ln v_2/v_1[(t - t_0)/t] = J(t - t_0)k \ln v_2/v_1 \, .$$

Suppose now that this reversible cycle is replaced by an irreversible one that is identical to the first except that the gas undergoes a Gay-Lussac expansion from v_1 to v_2. No work will be done and no heat will be absorbed, and, for an ideal gas, the final temperature will still be t. Since the cycles are otherwise identical, the amount of heat rejected to the condenser will be the same as before, namely,

$$kt_0 \ln v_2/v_1 \, ,$$

but instead of obtaining work from the cycle, an amount of work equivalent to the rejected heat will be required to complete it. Now according to Thomson, an amount of heat Rw should be dissipated to the condenser, i.e.,

$$(t_0/t)kt \ln v_2/v_1 = kt_0 \ln v_2/v_1 \, ,$$

which is correct. But he would also consider that a proportion of the dissipated mechanical energy $(1 - R)w$ is available for work during the reversible remainder of the cycle, namely,

$$J(t - t_0/t)kt \ln v_2/v_1 = J(t - t_0)k \ln v_2/v_1 \, .$$

In other words, if Thomson's analysis were true, throttling should not dissipate any energy, since this work is precisely that obtained from a reversible cycle.

Everything was not, therefore, contained in Thomson's formula. And $t_0 \int dq/t$ was not Thomson's expression as Tait had claimed;

it was Clausius'. When Tait admitted in his letter to Thomson that he had erred in assigning a positive value to the integral, he was somewhat at a loss to discover why he had done it. "I cannot altogether complain of Prof. Clausius's comments because I cannot account for my having called the above integral (in the way in which I have employed it) a *positive* quantity, except by supposing that in the revision of the first proof of my book I had thoughtlessly changed the word 'negative' to 'positive'."[94] The reason for Tait's error is not hard to find; it turns on the likelihood that Tait, who frankly admitted he had "bagged" from Rankine and Thomson,[95] had poached from Clausius as well.

Clausius on Lost Work

In his 1854 paper, where he first introduced the idea that $\int dq/t$ should be zero for a reversible cycle, Clausius mentioned that the integral could only be greater than zero for an irreversible cycle. The argument was quite simple. He had defined the transformation value q/t such that a positive change in it represented a transfer of heat from a higher to a lower temperature. If the transformation value were negative for a cycle, therefore, heat would be moving against a temperature gradient, contrary to his statement of the second law.[96] He went no further in his analysis of irreversibility until he examined the steam engine in 1855 and developed the controversial $t_0 \int dq/t$.[97] In his first paper, Clausius had defined his symbol for heat q in terms of the reservoirs which acted as sources and sinks for the heat ex-

94. Tait, "On the Dissipation of Energy," *op. cit.* (note 82), p. 344.
95. "I have finished Chap. III of my Sketch of Thermodynamics, and as you are not available I have sent it to Maxwell to look over. You will be rather surprised when you see the quiet way in which I have bagged from you and Rankine. . . ." A postcard from Tait to Thomson dated 13/1/68, quoted by Knott, *Life of Tait, op. cit.* (note 20), p. 219.
96. Clausius, "On a Modified Form of the Second Fundamental Theorem . . . ," *op. cit.* (note 3), p. 96. I am using Tait's symbols here to avoid confusion; Clausius used Q and T, with t reserved for ordinary temperature readings.
97. Rudolf Clausius, "Ueber die Anwendung der mechanischen Wärmetheorie auf die Dampfmaschine," *Annalen der Physik, 97* (1856), 441–476, 513–559; "On the Application of the Mechanical Theory of Heat to the Steam Engine," *Philosophical Magazine, 12* (1856), 241–265, 338–354, 426–443.

changed with the medium performing the cycle. Thus q was negative when a source gave heat to the medium, since the source was then losing heat, and in terms of this convention

$$\smallint dq/t = N > 0 \, .$$

In his second paper, Clausius changed his perspective to now define q in terms of the medium; i.e., q was negative when the medium, not the reservoir, lost heat. The above integral, consequently, was reversed in sign. Since he had earlier treated N as a positive number, he chose to continue that convention,[98]

$$\smallint dq/t = -N \, .$$

When N appeared in Clausius' equation for work, therefore, it represented a positive value, but not $\smallint dq/t$. Tait apparently overlooked this subtle change in convention and inadvertently assigned a positive value to the integral, revealing thereby the probable source of his ideas.

Clausius began his treatment of the steam engine by indicating that he would not consider the irreversibility introduced by the large temperature differentials between boiler steam and hot flue gases.[99] Thus he restricted himself to the throttling type of irreversibility; i.e., to adiabatic dissipations of pressure differentials, the problem which Thomson had considered. Clausius' approach was not to calculate the dissipated heat but to estimate the expected work. Suppose, he said, that all of the heat q_0 rejected by the system in the cycle occurs at a single temperature t_0 while the addition of heat q_1 to the system may proceed at a variety of higher temperatures t. Then the sum of transformation values is

$$\int_0^{q_1} dq/t + q_0/t_0 = -N \, ,$$

so that

$$q_0 = -t_0 \int_0^{q_1} dq/t - t_0 N \, .$$

98. *Ibid.*, p. 245.
99. *Ibid.*, p. 246. See the footnote on this page.

The first law, Clausius clearly stated, requires that the work w be equal to the net heat,[100]

$$w = J(q_1 + q_0) ,$$

(where the signs of course are positive for additions and negative for rejections of heat), so that the work for a cycle becomes

$$w = J\left(q_1 - t_0\int_0^{q_1} dq/t - t_0 N\right) .$$

Since $N = 0$ for a reversible cycle, Clausius concluded that $- Jt_0N$ represented the loss of work due to irreversibility.[101]

Testing Clausius' equation for work by the same irreversible cycle used to examine the validity of Thomson's, one finds that it yields the correct value,

$$w = - Jt_0N = Jt_0q_0/t_0 = Jq_0 = - Jkt_0 \ln v_2/v_1 ,$$

where the minus sign signifies that the work must be supplied from outside the system. Clausius triumphed where Thomson had failed. For even Thomson's correct estimate of the dissipated heat rejected to the condenser had been based on faulty analysis. His mistake was in failing to mark the critical distinction between internally dissipated mechanical energy and external additions of heat.

Clausius and Thomson on Irreversibility

As the quotations from Tait's *Thermodynamics* suggest, Thomson had hesitated initially before adopting the mechanical theory of heat. His conversion probably dates from his publication of the letter he sent to Joule from Paris on 15 October 1850.[102] In this letter he related, somewhat excitedly, that he could now resolve an apparent dilemma in the mechanical theory; the apparent dilemma was as follows. Rankine had recently proposed that, contrary to the assumptions of the caloric theory, a portion of saturated steam should

100. Clausius' own expression was $W = 1/A (Q_1 + Q_0)$, where A of course was the heat equivalent of work.

101. *Ibid.*, pp. 248–249.

102. William Thomson, "On a Remarkable Property of Steam Connected with the Theory of the Steam-Engine," *Philosophical Magazine, 37* (1850), 386–389. Reprinted in *Papers, 1,* 170–173.

condense upon expanding adiabatically. The resulting wet steam should scald, according to experience. The dilemma arose from the well-known fact that boiler steam, when throttled through a safety valve, does not scald. Thomson found a resolution of this in Joule's thesis that heat is generated by friction in liquids. He wrote to Joule that

> [Rankine's] conclusion can, I think, be reconciled with the known facts only by means of your discovery, that heat is evolved by the friction of fluids in motion. For it is well known that the hand may be held with impunity in a current of steam issuing from the safety valve of a high pressure boiler. . . . But, according to Mr. Rankine's proposition, steam allowed to expand from saturation will, *if no heat be supplied to* it, remain saturated, except a small portion which becomes liquified. Either then Mr. Rankine's conclusion is opposed to the facts, or *some heat must be acquired by the steam as it issues from the boiler.* . . . There is no possible way in which the heat can be acquired except by the friction [fluid friction] of the steam as it rushes through the orifice. Hence I am justified in saying that your discovery alone can reconcile Mr. Rankine's discovery with known facts.[103]

The same satisfying argument which overcame Thomson's resistance to the mechanical theory also led him into some faulty thermodynamic logic.[104] Clausius never committed the error; in fact, he assured the victory of the mechanical theory by making a stalwart defense against the admission of internally created heat. He did this not in a debate with Thomson, but with a fellow German.

J. Bauschinger maintained that the integral $\int dq/t$ should also be zero for an irreversible steam engine cycle, and that Clausius had overlooked this because he had ignored the heat generated within the steam itself. "The steam which remains in the boiler, as it expands to take the place of that which has streamed forth, performs work which is imparted to the exiting steam as kinetic energy [lebendiger Potenz]. When the latter comes to rest in the cylinder, this work is transformed into heat. Clausius has failed to include that heat in his calculation of N."[105] Bauschinger reinvestigated the

103. *Ibid.,* p. 171.
104. Thus Thomson equated the internal dissipation of mechanical energy with an external addition of heat.
105. J. Bauschinger, "Ueber das Integral $\int dQ/T$," *Zeitschrift für Mathematik und Physik, 11* (1866), 152–162, esp. p. 155.

irreversible steam engine cycle that Clausius had analyzed, and, by including the internally generated heat, he showed that $\int dq/t = 0$. Clausius did not bother to take issue point by point with Bauschinger's analysis, since he was able to answer[106] all the points with the simple illustration he had earlier appended to a reprint of his steam engine article.[107] Here Clausius had considered a Gay-Lussac expansion of a perfect gas from volume v_1 to volume v_2. Since his rule for the summation of transformation values required the consideration of a cycle, he calculated N by imagining that the gas was restored to its original condition via a reversible isothermal compression. Since no heat had been exchanged with the surroundings during the free expansion,

$$N = -\int dq/t = -1/J \int_{v_2}^{v_1} pdv/t = k \ln v_2/v_1 .$$

Clausius argued that since the reversible compression could not increase the transformation value (thus, implicitly including the surroundings in the analysis), N represented the uncompensated transformation value for the original irreversible expansion.

Clausius granted Bauschinger's contention that if the heat generated within the gas by the irreversible expansion were also included in the calculation of N, the integral $\int dq/t$ would be zero. He declined this approach, however, because it would deprive the irreversible case of any further theoretical significance.[108] He therefore emphasized and reiterated his point that the calculation of N should be restricted to externally exchanged heat: "It is not necessary to take cognizance of processes which occur within the changing substance and thereby consume and create quantities of heat during a cycle; one need only consider, in the formation of the integral $\int dq/t$ for calculation, those quantities of heat which the substance during the cycle receives from or rejects to the surroundings."[109] Thus, Clausius not only established the relation between irreversible entropy in-

106. Rudolf Clausius, "Ueber umkehrbare und nicht umkehrbare Vorgänge in ihrer Beziehung auf die Wärmetheorie," *Zeitschrift für Mathematik und Physik, 11* (1866), 445–462.

107. Rudolf Clausius, *Die mechanische Wärmetheorie, 1,* 1st ed. (Braunschweig, 1865).

108. Clausius, "Ueber umkehrbare und nicht umkehrbare Vorgänge . . . ," *op. cit.* (note 106), pp. 456–460.

109. *Ibid.,* p. 459.

crease and lost work, but he also secured the theoretical foundation for all further elaborations and applications of the concepts of entropy and irreversibility.

No further elaborations came from Clausius' pen other than his 1865 declaration that the entropy of the universe tends to a maximum. That declaration was not very helpful in theoretical analyses, and in any case Thomson had already voiced a comparably general statement in 1852, when he had talked about the trend toward total dissipation of mechanical energy and uniform dispersion of heat in the universe.[110] The decisive elaboration came with Gibbs's solution of the puzzle of thermodynamic equilibrium, and it was Thomson who helped prepare the way for this by his discussion of a simpler system than the universe.

Entropy and Dissipation United in Maxwell and Gibbs

Gibbs would, perhaps, admit no debt to Thomson, for every evidence suggests that he sided with Clausius in the thermodynamic controversy. He criticized the British, especially Maxwell, in 1873 for introducing confusion over the meaning of entropy,[111] and he prefaced his famous 1875 paper with Clausius' phrasing of the two laws of thermodynamics.[112] Although Gibbs never once mentioned Thomson in his work, he was indebted, I believe, to Thomson's concept of dissipation of energy via the good offices of Maxwell and his *Theory of Heat*. Maxwell, in turn, was indebted to Gibbs in the revision of his treatment of available and unavailable energy in his *Theory of Heat*, thereby uniting the two traditions of entropy and dissipation.

Thomson, in contrast to Clausius, did not limit his consideration of dissipation to the throttling losses of pressure in steam engines; he also considered the amount of work that might be culled from

110. Thomson, "On a Universal Tendency in Nature to the Dissipation of Mechanical Energy," *op. cit.* (note 87), p. 260.

111. J. Willard Gibbs, "A Method of Geometrical Representation of the Thermodynamic Properties of Substances by Means of Surfaces," *Transactions of the Connecticut Academy, 2* (1873), 382–404. Reprinted in *The Scientific Papers of J. Willard Gibbs, 1* (New York, 1961), 33–54. See footnote, p. 52.

112. J. Willard Gibbs, "On the Equilibrium of Heterogeneous Substances," *Transactions of the Connecticut Academy, 3* (1875–1878), 108–248, 343–524. Reprinted in Gibbs, *Scientific Papers, 1,* 55–349. See p. 55.

a system of bodies at nonuniform temperatures.[113] Maxwell grouped both sources of dissipation, frictional and thermal, under the single rubric of available energy, and he raised the question of the maximum mechanical work that might be obtained from a system at constant volume with no access to additional sources of heat.[114] He attributed the approach that grouped both forms of dissipation to Thomson, and it was this approach that Gibbs found so useful in his initial breakthrough into the theory of equilibrium.

Gibbs reveals his debt to Maxwell's treatment of available energy in assuming more in his derivation than his statement of conditions would legitimately imply. He imagined a small system coming to equilibrium with a comparatively large fluid mass at absolute temperature T and pressure P. The small system and large fluid mass are separated by a special envelope; in order to assure both a final equilibrium at T and P as well as a constant temperature and pressure throughout the surrounding medium, he attributed to the envelope seemingly contradictory properties; e.g., a complete responsiveness to all pressure and temperature differentials, but a responsiveness that could proceed only at an extremely slow pace. The membrane was to have no volume and no heat capacity.[115] Starting from these conditions alone, Gibbs proposed the following derivation of the basic equation of the theory of equilibrium:

> If we regard, as we may, the medium as a very large body, so that imparting heat to it or compressing it within moderate limits will have no appreciable effect upon its pressure and temperature, and write V, H, and E, for its volume, entropy, and energy, equation (1) [the first law] becomes
>
> $$dE = TdH - PdV$$
>
> which we may integrate regarding P and T as constants, obtaining
>
> $$E'' - E' = TH'' - TH' - PV'' + PV',$$
>
> where E', E'', etc., refer to the initial and final states of the medium. Again as the sum of the energies of the body and the surrounding

113. Thomson, "On a Universal Tendency in Nature to the Dissipation of Mechanical Energy," *op. cit.* (note 87), pp. 258–260.

114. Maxwell, *Theory of Heat*, 2nd ed., *op. cit.* (note 51), pp. 185–188.

115. Gibbs, "A Method of Geometrical Representation . . . ," *op. cit.* (note 111), p. 39.

medium may become less, but cannot become greater (this arises from the nature of the envelope supposed) [sic], we have

$$\epsilon'' + E'' \leq \epsilon' + E' .$$

Again as the sum of the entropies may increase but cannot diminish

$$\eta'' + H'' \geq \eta' + H' .$$

Lastly it is evident that

$$v'' + V'' = v' + V' .$$

These four equations may be rearranged with slight changes . . . [and] by addition we have[116]

$$\epsilon'' - T\eta'' + Pv'' \leqq \epsilon' - T\eta' + Pv' .$$

This derivation actually relies upon more than the stated conditions. The envelope has no bearing on the possible changes in the total energy of the combined small system and large fluid mass. Gibbs has said nothing about how this larger, total system is related to the universe, and that relation is critically important for any assertions about the behavior of the total energy, total entropy, and total volume. Since Gibbs does not state these related conditions, we must infer them from his equations. Since the energy can only remain constant or decrease, the total system can perform mechanical work but receive none, and it can lose heat but gain none. Since the entropy can only increase, the possibility of losing heat is excluded. The only total system for which Gibbs's proof makes sense is the system proposed by Maxwell for analyzing available work, a system isolated with respect to changes of volume and exchanges of heat, but not isolated with respect to mechanical energy. The system can perform mechanical work on the universe; for example, it can lift weights external to itself. Without that provision, there is no sense in saying the total energy can decrease, and that provision was basic to Maxwell's treatment of available energy. It is clear that Gibbs brought together the two traditions of entropy and dissipation.[117] The unification was, however, only implicit, since Gibbs talked only of entropy, not of availability. The first explicit for-

116. *Ibid.*, p. 40.
117. A further tribute to Thomson appears in Gibbs's term for the equilibrium surface, "the surface of dissipated energy." (*Ibid.*, pp. 49–50.)

mulation was Maxwell's, in his revised discussion of available energy in 1876.

Maxwell imagined a completely reversible arrangement for extracting all available energy from an isolated system at nonuniform temperature, and he asked how much of that available energy would be lost if an amount of heat H escaped the reversible engines and went by direct conduction from absolute temperature θ_1 to absolute temperature θ_2. Assuming that H was small enough not to alter the final uniform temperature θ of the system, he argued that the loss in available energy would be equal to the product of the entropy increase due to the conduction of H, namely, $H\,(1/\theta_2 - 1/\theta_1)$, and the temperature θ.[118] Thus Clausius' conception of irreversible entropy increase and his equation for lost work entered Maxwell's discussion of available energy and dissipation.

Maxwell, wisely perhaps, opened no new wounds by attempting to attribute the equation to any one in particular, choosing rather to give credit to both theories and their authors: "Processes of this kind, by which while the total energy remains the same, the available energy is diminished, are instances of what Sir W. Thomson has called the Dissipation of Energy. The doctrine of the dissipation of energy is closely connected with that of the growth of entropy, but is by no means identical with it."[119] To which we might add in closing, "Amen—So be it," for the kingdom of science was enriched thereby.

ACKNOWLEDGMENTS: The research for this paper has been supported by National Science Foundation Grant No. GS-1923, and this support is gratefully acknowledged. I wish to thank Russ McCormmach and Martin Klein for offering perceptive comments on the first draft of this paper and the former for wielding a helpful editorial pen on the final draft. I also wish to express my appreciation to Miss Phyllis Downie, Reference Librarian, and Mr. Charles Finlayson, Keeper of the Manuscripts at Edinburgh University Library, for their kind assistance in my long-distance search for manuscript materials from the plains of Kansas.

118. Maxwell, *Theory of Heat,* 4th ed., *op. cit.* (note 55), pp. 191–192.
119. *Ibid.,* p. 192.